Principles of Modeling Uncertainties in Spatial Data and Spatial Analyses

T0225321

Principles of Modeling Uncertainties in Spatial Data and Spatial Analyses

Wenzhong Shi

CRC Press
Taylor & Francis Group
Boca Raton   London   New York

CRC Press is an imprint of the
Taylor & Francis Group, an **informa** business

CRC Press
Taylor & Francis Group
6000 Broken Sound Parkway NW, Suite 300
Boca Raton, FL 33487-2742

First issued in paperback 2020

© 2010 by Taylor and Francis Group, LLC
CRC Press is an imprint of Taylor & Francis Group, an Informa business

No claim to original U.S. Government works

ISBN-13: 978-0-367-57724-7 (pbk)
ISBN-13: 978-1-4200-5927-4 (hbk)

Library of Congress Cataloging-in-Publication Data

Shi, Wenzhong.
    Principles of modeling uncertainties in spatial data and spatial analysis / Wenzhong Shi.
        p. cm.
    Includes bibliographical references and index.
    ISBN 978-1-4200-5927-4 (alk. paper)
    1.ˉ Geographic information systems--Data processing. 2.ˉ Geographic information systems--Mathematical models.ˉ I. Title.

G70.212.S545 2008
910.285--dc22   2008005717

Visit the Taylor & Francis Web site at
http://www.taylorandfrancis.com

and the CRC Press Web site at
http://www.crcpress.com

Contents

SECTION II   *Modeling Uncertainties in Spatial Data*

SECTION III   Modeling Uncertainties in Spatial Model

SECTION IV *Modeling Uncertainties in Spatial Analyses*

SECTION V Quality Control of Spatial Data

SECTION VII   Epilogue

Foreword

Michael F. Goodchild

On February 3, 1998 a military aircraft of the U.S. Marine Corps struck and severed the cables supporting a gondola in Cavalese, Italy, leading to the deaths of 20 people. The plane had wing and tail damage but was able to return to its base. It subsequently became clear that the cable-car was not shown on the maps being used by the pilot, and in the subsequent trial the pilot claimed also that the aircraft's height-measuring system had malfunctioned. Errors, omissions, and uncertainties in geographic information do not often result in international incidents of such significance, but it is an inescapable fact that any knowledge of the Earth's surface is subject to uncertainty of some kind, whether it be in the positions of features, their existence, or their description. It is impossible to know the exact location of anything on the Earth's surface, since our methods of measurement are always subject to error; uncertainties creep into our maps, databases, and written records through a wide range of additional mechanisms. Many of the classification schemes used to map aspects of the Earth's surface, from soils to land use, are inherently uncertain, with the result that two people mapping the same area will never produce identical maps. Despite centuries of progress in mapping technology, the creation of geographic information remains as much an art as a science.

From this perspective it is important for users of geographic information to have some awareness of the uncertainties that are likely to influence their decisions—to know what the database does *not* tell them about the real world, and about the reliability of what it *does* tell them. As geographic information technologies have developed over the past few decades it has become clear that one ignores issues of uncertainty at one's peril. For example, regulations in many countries now make it difficult to construct in areas classified as wetland. Decisions are made daily about the uses to which private land can be put. But such decisions are clearly open to legal challenge if they can be shown to have been based on maps that are inherently uncertain.

Research on the description and modeling of uncertainty in geographic information began in earnest in the late 1980s, and has accelerated over the past two decades. Today, a large literature describes successful efforts to address the issue, and tools are increasingly available to allow the effects of uncertainty to be propagated, so that uncertainties can be associated with the results of analysis as well as with the inputs.

John Shi has been one of the leaders in this research area. He has made very significant contributions, particularly in the modeling of uncertainties in geographic features of complex geometry, and has also made a very valuable contribution as the organizer of a series of conferences on spatial data quality, the International Symposia on Spatial Data Quality, that have occurred every 2 years since 1999. The conferences provide a forum for a very broad-based discussion of recent research on uncertainty that is not limited to any single paradigm or theoretical framework, because experience over the past two decades has shown that this problem of uncertainty is so pervasive, and so multidimensional, that no single approach can possibly address it.

The contents of this book provide an excellent introduction to this multidimensional problem. Early research tended to focus on error and accuracy, on the grounds that the creation of geographic information was similar to any problem of scientific measurement, and could be addressed through the application of the theory of errors. Although this is conceptually simple, in reality geographic information tends to have some very awkward properties that complicate the approach enormously. Maps are not collections of independent measurements, but instead represent the culmination of a long and complex process that induces very strong correlations in errors. All positions will be subject to error, but nearby positions will have more similar errors than distant positions—in other words, relative errors of position over short distances tend to be much less than absolute errors. To handle this, models of uncertainty need to incorporate strong spatial autocorrelations and to require comparatively advanced mathematics.

By the mid-1990s, however, it was clear that some aspects of the problem of uncertainty derived in part from the inherent vagueness of definitions, and that these could be handled much more effectively using the theoretical constructs of fuzzy sets. For example, we may not know exactly what is meant by wetland, but nevertheless it may make sense to be able to say that this area is more like wetland than that area. The approach was immediately attractive to many users of geographic information, who found it more intuitive and accessible than the statistical approach.

Both statistical and fuzzy frameworks are covered in this book, which provides a comprehensive overview of the current state of the field. At the same time, it strongly reflects John Shi's own approaches, and the very significant contributions he has made. It should be indispensable reading for anyone interested or actively engaged in this research area, and desirable reading for anyone using geographic information to solve real problems. We have made much progress in the past two decades, and today few users of geographic information systems are willing to assume that their outputs are exact and correct *just because they came from a computer.* Additionally, we are still some distance from the goal of placing a plus or minus on every output, and of incorporating uncertainty into every decision made with geographic information. But this book may help us get closer to that goal.

Preface

This book presents four major theoretical breakthroughs in uncertainty modeling, which have resulted from the writer's investigation of uncertainty modeling in spatial data and spatial analysis. They are (a) advances in spatial object representation, from determinedness to *uncertainty*-based representation of geographic objects in geographic information science (GISci); (b) from uncertainty modeling for *static* spatial data to *dynamic* spatial analyses; (c) from uncertainty modeling for spatial *data* to spatial *models*; and (d) from error *description* of spatial data to spatial data quality *control*.

GISci, compared to classical sciences such as mathematics or physics, is a newly emerged science. Its theoretical foundations have been recently formulated and are in a vital position of readiness for further development. This book intends to contribute to the progress of one fundamental area of GISci—data quality and uncertainty modeling for spatial data and spatial analyses.

GISci covers the following major fundamental areas: (a) space and time, (b) data quality and uncertainty, and (c) spatial analysis. Of these three areas, data quality and uncertainty, area (b) is considered both independently and also in relation to areas (a) and (c).

There are two types of geographic entities and phenomena in the real world: the precise and the uncertain. Cognition and representation of space and time, area (a), is considered not only in relation to precise spatial objects, but also in relation to uncertain spatial objects. Uncertainty in spatial representation is unavoidable, as it is not yet possible to represent the infinite natural world in a finite way. All measurements contain a certain degree of error, no matter how high the accuracy of modern measurement technologies. Thus, it is to be expected that, given the infancy of GISci, the currently used determined approaches must have the potential for further development to encompass uncertain-based approaches in cognition and representation of space and time.

Uncertainty modeling and quality control are two key issues in area (b). Uncertainty modeling depicts the error in spatial data, and uncovers the nature of uncertainty distribution. Quality control aims to control the overall quality of spatial data, with the possibility of reducing the spatial data error to a desired level.

Spatial analysis in GISci area (c) is, likewise, not uncertainty free. Uncertainty in source data and limitations of spatial analysis models may propagate, or any original uncertainties may become amplified. The quality of a decision, based on a spatial analysis, is affected by the quality of the original data, the quality of the spatial analysis model, and the degree of uncertainty that is propagated or generated in the spatial analysis.

A framework for handling uncertainties in spatial data and spatial analyses is outlined in this book. Covered are uncertainties in the real and natural worlds, uncertainties in the cognition of natural objects, modeling errors in spatial object measurement, modeling uncertainty in spatial models, uncertainty propagation in spatial analyses, quality control for spatial data, and finally a presentation of uncertainty information. Theories and methods for handling these uncertainties are given and, as such, form principles for modeling uncertainties in spatial data and spatial analyses.

In line with the logic of uncertainty generation and handling outlined above, this book is organized in seven sections: (I) overview, (II) modeling uncertainties in spatial data, (III) modeling uncertainties in spatial model, (IV) modeling uncertainties in spatial analyses, (V) quality control of spatial data, (VI) presentation of data quality information, and (VII) epilogue. Within each section are several chapters. The outline of the contents of the seven parts, listed above, is as follows.

Section I (Chapters 1–3) provides an overview of the principles of modeling uncertainty of spatial data and spatial analysis, with the principles of uncertainty in general together with the concepts of uncertainty in spatial data and analyses introduced in Chapter 1. Various uncertainty sources are summarized in Chapter 2. They include uncertainty inherited in the natural and real worlds, uncertainty due to human cognition, uncertainty introduced in data capture processes, and uncertainty arising from spatial data processing and spatial analyses. Mathematical foundations of uncertainty modeling in spatial data and analyses are summarized in Chapter 3. They include probability theory, statistics, evidence theory, fuzzy set and fuzzy topology, and rough set theory, as well as information theory and entropy.

The developed methods for handling uncertainties in spatial data are introduced in Section II (Chapters 4 to 6). Uncertainties in spatial data can be classified into the following five types: (a) positional uncertainty, (b) attribute uncertainty, (c) temporal uncertainty, (d) incompleteness, and (e) logic inconsistency. Since temporal information can be regarded as a form of attribute information, the methods for handling attribute uncertainty are potentially applicable to handling temporal uncertainty. Positional uncertainty modeling is the focus of Chapter 4 while attribute uncertainty modeling is introduced in Chapter 5. Uncertainty modeling for integrated positional and attribute uncertainty, an issue which is critical in multisource data integration, is introduced in Chapter 6.

Error models for positional uncertainty have been comprehensively studied and are presented systematically, as uncertainty models for points, line segments, polylines, curves, polygons, and area objects in Chapter 4. Line feature errors are modeled through (a) the confidence region, (b) error distribution, and (c) error of points on the lines with correlated and independent cases. Of these error models, the confidence region model of a line segment forms the corner stone of positional error modeling of objects in object-based GIS.

As previously indicated, geographic phenomena can fall into two classes: the determine-based and the uncertain-based. Previous solutions have mainly provided determine-based representation of space and time in Euclidean space, where determined points, lines, areas, and volumes are described. Uncertainty modeling of spatial data presented in Chapter 4 forms a basis for uncertainty-based representation of the geographic entities covering points, line segments, polylines, curve lines, and polygons.

The properties of attribute uncertainty and the methods to model the attribute uncertainties in GIS and remote sensing are introduced in Chapter 5. These methods include the rate of defect model, the use of which is straightforward for practical applications; the probability vector models, which can indicate uncertainty spatial distribution; and the commonly used error matrix method.

The method for integrating positional and attribute uncertainties is introduced in Chapter 6. It should be noted that modeling the integrated error is a critical issue in multisource data integration. The theoretical model "S-band" and the two solutions of the model are given: (a) probability theory-based solution and certainty factor-based solution.

Modeling uncertainties in spatial models is introduced in Section III (Chapters 7 and 8). Modeling uncertain topological relationships is covered in Chapter 7. The concepts for topological relationships between objects in GIS and modeling the topological relationships are introduced. In addition, fuzzy topology is adapted to model uncertain topological relationships between spatial objects in GIS. Geographic objects can, in fact, be represented by either spatial data or spatial models. Two typical examples of a spatial model are the digital elevation model (DEM) and curve functions in GIS. Natural terrains in the natural world can be represented by digital elevation models. In this case, the model can be either regular, such as a regular tessellation like a square, or irregular, such as a triangulated irregular network (TIN). A curve entity in the natural world—for instance, a curved portion of a road— can be represented by either a regular curve, such as circular curve, or an irregular curve, approximated by the third-order spline curve function.

Since error models for a curve are introduced in Chapter 4 (Section II), Chapter 8 concentrates mainly on handling model error in DEM. Two advances have been made along the lines of DEM model accuracy estimation: (a) a formula to estimate the average model accuracy of a TIN, and (b) accuracy estimation of the bicubic interpolation model.

Modeling uncertainties in spatial analyses is the focus of Section IV (Chapters 9 to 11). Here, spatial analyses mainly refer to GIS spatial analyses, such as overlay analysis, buffer analysis, and line simplification analysis. Each analysis is a transformation, based on one or more original spatial data set(s). Uncertainty inherited from the original data set(s) are further propagated or even amplified through such a spatial analysis. In many cases, new uncertainties are also generated. In addition to modeling uncertainty in static spatial data, a logical further step is to model uncertainty in spatial analyses.

Covered in Chapter 9 is modeling uncertainties in overlay spatial analyses where analytical and simulation approaches for determining uncertainty propagated through an overlay analysis are given. Modeling uncertainty in buffer spatial analysis is covered in Chapter 10. A solution is provided for quantifying uncertainty propagation through a buffer spatial analysis, where four error indicators are proposed for quantifying the uncertainties. In Chapter 11, the newly proposed uncertainty modeling method for line simplification spatial analysis is given.

Focus in the first chapters is on descriptions of uncertainties, including descriptions of uncertainty in spatial data, descriptions of uncertainty in spatial analysis, and descriptions of uncertainty in spatial models. To ensure an understanding of handling uncertainty, description is a necessary first step; a further step is to control or even reduce the uncertainties in the spatial data, analyses, or models, if possible.

Therefore, Section V introduces another theoretical breakthrough in handling uncertainties in spatial data. This is from uncertainty description to spatial data quality control. In this regard, the quality control for spatial data is introduced. An explanation of quality control for object-based spatial data is given in Chapter 12. The aim is to control the overall geometric quality of vector spatial data by the least squares adjustment method with an example for vector cadastral data. Quality control for field-based spatial data—the aim being to geometrically rectify high resolution satellite imageries by point- and line-based transformation models—is presented in Chapter 13. Quality control for digital elevation models, designed to improve DEM accuracy with the newly proposed hybrid interpolation model and bidirectional model, is described in Chapter 14.

The methods for presenting data quality information are described in Section VI (Chapters 15 to 17). Visualization techniques for uncertainty information are provided in Chapter 15. Examples given are the error ellipse, arrow-based approach, grayscale map-based, color map, symbol-based approach, and the three-dimensional visualization approach. An object-oriented metadata system for managing metadata on uncertainty information is introduced in Chapter 16. A Web service-based data quality information solution for disseminating uncertainty information on the Internet is introduced in Chapter 17.

Finally, in Section VII, (Chapter 18), a summary and comments on principles of modeling uncertainties in spatial data and spatial analyses are given. Suggestions for further development in the field of modeling uncertainties in spatial data and analyses are outlined.

Wenzhong Shi
Hong Kong

Acknowledgments

I would like to thank my family, Wendy, Julia, and Daniel, for their support in writing this book. I would like also extend my thanks to my colleagues, Tracy, Yan, Kim, Elaine, Jacqueline, Eryong, Xiaohna, Yumin, and many others for their assistance.

The work described in this book was supported by the Young Professor Scheme of The Hong Kong Polytechnic University; the Distinguished Young Scholar Fund (B) from the National Natural Science Foundation, China (No. 40629001), the CRC Program of Hong Kong Research Grants Council (No. 3_ZB40), and the Chang Jiang Scholars Program, Ministry of Education, China.

Wenzhong Shi
Hong Kong

Section I

Overview

1 Introduction

Section I of this book presents an overview of the contents of this book, covering the basic concepts of uncertainty in Chapter 1, the sources of uncertainties in Chapter 2, and mathematical foundations in Chapter 3. This introductory first chapter is designed to the provide the basic concepts of uncertainty that underpin this book's aim to further contribute to the already established concepts and methods related to the modeling of uncertainty in spatial data and analyses. The concepts described include uncertainties, errors, dimensions of spatial data, data quality components, and elements of spatial data. These concept descriptions form the foundation upon which the later chapters of the book are built and may be useful reference points.

1.1 UNCERTAINTY

Mandelbrot (1967) raised the question, "How long is the coast line of Britain?" and in the same vein I would like to raise the question, "How long is the coastline of China?" It was reported as 9,000 km in the 1920s and then as 11,000 km in the 1950s. Even later, in the 1960s, it was measured as 14,000 km, based on an aerial photogrammetric survey, with a map scale of 1:100,000, and 18,000 km, based on aerial photogrammetric survey with a map scale of 1:50,000. Can it be possible that the length of the coastline of China has doubled in size over the last 50 years? The element of uncertainty is obviously present in the measured value of the coastline length. This uncertainty may be caused by such factors as uncertainty inherited from insufficient knowledge of the coastal regions of the natural world, uncertainty in our cognition or interpretation of the coastline, error in the measurement technology and measurement operational procedure, or even uncertainty in the digital representation of the coastline in different scales in the computer environment.

1.1.1 THE CONCEPT OF UNCERTAINTY

"Uncertainty" can refer to vagueness, ambiguity, or anything that is undetermined. As indicated above, by the broadness of the express "the natural world," uncertainty is evident in many academic fields such as physics, statistics, economics, metrology, psychology, and philosophy. However, the connotation of the term *uncertainty* can differ from one field to another. Uncertainty, as used in geographic information science (GISci), refers to the connotation mainly used in the fields of metrology and statistics.

In GISci, the uncertainty of a measurement can be described by a range of values that possibly include the true value of the measured object. The measurement can be, for instance, the width of a building, the length of a road, or the area of a

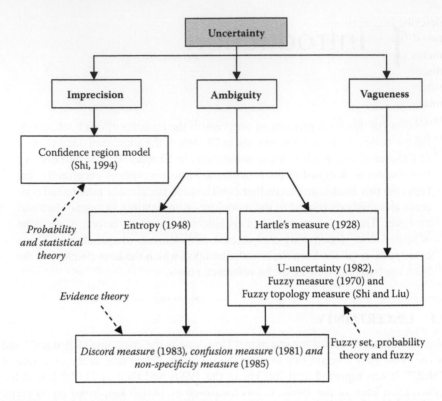

FIGURE 1.1 Uncertainties measured based on varies mathematical theories. (Adapted and further extended based on Klir, G. J. and Folger T. A., 1988. *Fuzzy Sets, Uncertainty, and Information.* Prentice Hall, Englewood Cliffs, New Jersey.)

land parcel. For example, the National Bureau of Surveying and Mapping, China announced in October 2005 that the latest measured height of Mount Everest (or Jo-mo Glang-ma, in Chinese), was 8844.43 meters with an accuracy of ±0.21 meter. This means that the true value of the mountain is possibly a value within the range [8844.22–8844.64]. The uncertainty of the mountain height measurement is indicated or given by the range.

Uncertainty can be in terms of either imprecision, ambiguity, or vagueness. These interpretations of uncertainty and their corresponding mathematical theories for handling the uncertainties are summarized in Figure 1.1.

Imprecision refers to the level of variation associated with a set of measurements or the lack of quality precision. Ambiguity is associated with either one or many relationships, or with a form of lack of clarity, which implies one or more meanings. For example, ambiguity can cause difficulty in deciding to which class, in a classification of satellite images, an object belongs. Vagueness refers to a lack of clarity in meaning and is normally associated with the difficulty of making a sharp or precise distinction in relation to an object in the real world.

Different aspects of uncertainty are addressed by correspondingly different mathematical theories. For instance, probability and statistical theory are employed to describe imprecision—one type of uncertainty. Fuzzy set theory is used to assess two different uncertainty categories. It has been demonstrated that fuzzy measurements can be used to describe vagueness, whereas ambiguity can be quantified by discordance measures, confusion measures, and nonspecificity measures. A measurement of information was introduced by Hartley in 1928 (known as Hartley's measure) to describe ambiguity in a crisp set. Shannon's entropy measures uncertainty in information theory, based on the probability theory. Fuzzy topology theory is used to model uncertain topologic relationships between spatial objects (Liu and Shi, 2006).

1.1.2 The Concept of Errors

In many cases, error is used as a synonym for uncertainty. However, it should be realized that the concept of error is, in fact, narrower than uncertainty. Unlike error, uncertainty is a neutral term. Uncertainty may be caused by mistakes, and it may also be caused by incomplete information. In this book, our focus is on uncertainty/error modeling for spatial data. Error in the measurements of spatial data can be quantified by statistical deviation, the instability of the observer, limitations of measurement instruments, and unfavorable observation conditions—the major sources of error in the spatial data capture process. Errors of measurement can normally be classified as: (1) random error, (2) systematic error, and (3) gross error.

1.1.2.1 Random Error

When a set of measurements is obtained under the same conditions, and the measurements contain irregular errors in terms of magnitude and sign, these irregular errors are known as random measurement errors. Random errors are due to occasional factors affecting the measurement instrument, such as a change in the measurement environment. It is difficult to estimate the level of effect on the measured value of each occasional factor because of the irregular error pattern. Although the random errors affect the measured value by chance, the error pattern has certain statistical characteristics if the measurement number is large.

1.1.2.2 Systematic Error

After performing a series of measurements under the same conditions, it may be noticed that the magnitude and type of error in the measurements follow a regular pattern. This type of error is called *systematic error*. Systematic error may be due to the performance error of measurement equipment. Systematic error can be potentially eliminated by improving measurement equipment performance or by removal of the error (if observed) in the later data processing process.

The impact of systematic error on measurement results is normally larger than that of random error. Thus, as far as possible, there is a pressing need for the complete elimination of systematic error. Common methods of reducing systematic errors include: (a) calibrating the measurement instrument, (b) adding a correction number to the measurements in the data processing process, and (c) using a more appropriate measurement method to avoid or reduce the chance of generating systematic error.

1.1.2.3 Gross Errors

Gross errors are mistakes made during measurement or data processing processes, such as those caused by the observer misidentifying the target to be measured, or the introduction of human error in the computation process. As gross errors are generally larger than random errors and may significantly affect the reliability of measured results, precautions should be taken to reduce their generation. For example, in traverse surveying, if the backward direction is identified incorrectly due to an observer's mistake, the closing error of traverse will greatly exceed the acceptable limit; this is gross error. Therefore, if such errors are to be avoided, a well-designed measurement scheme is necessary. One solution is to introduce redundant measurements in the measuring process.

1.2   UNIVERSALITY OF UNCERTAINTY

Uncertainty exists widely in the natural world, whereas certainty is conditional and relative. For example, in the micro world, an electron within a molecule follows a random path around the atom. The examples in the macro world are the shared uncertain boundaries between urban and rural areas and the uncertain pattern of climate distribution. In fact, uncertainty is universal and, as indicated above, exists widely in many fields, such as mathematics, physics, and economics. Brief examples of uncertainty in several fields are given below.

One hundred years ago Albert Einstein, who was awarded the 1921 Nobel Prize in Physics, criticized the nature of randomness in quantum physics by stating that "God does not play dice."

However, due to lack of full knowledge of the universe, in recognition of uncertainty in quantum theory, Werner Karl Heisenberg, the founder of quantum mechanics and who was awarded the Nobel Prize in Physics in 1932, stated that the position and speed of a particle could not be measured exactly. This led to the introduction of his famous uncertainty principle in quantum theory. Later, Stephen Hawking pointed out, "Not only does God play dice with the Universe—he sometimes casts them where they can't be seen." In March 1927, Heisenberg published a paper in the *Zeitschrift für Physik* entitled "On the Perception Content of Quantum Theoretical Kinematics and Mechanics," which was translated into English by Wheeler and Zurek in 1983. That paper presented his formulation of the uncertainty principle in quantum mechanics. This uncertainty principle was controversial at the time, because it did not follow the orthodox laws of physics. Nowadays the uncertainty principle is highly regarded by physicists.

In mathematics, Gödel's first incompleteness theorem (1931) states,

In any consistent formalization of mathematics that is sufficiently strong to axiomatize the natural numbers, one can construct a true statement that can be neither proved nor disproved within that system itself.

His second theorem states, No consistent system can be used to prove its own consistency.

These theorems arise from the problem of uncertainty in the axiomatic system. The incompleteness theory has made a deep impact on the development of mathematics, computing science, and even philosophy.

Ilya Prigogine, a Belgian scientist, received the 1977 Nobel Prize in Chemistry for his dissipative structure research and for his contributions to the understanding of nonequilibrium thermodynamics and reinterpretation of the second law of thermodynamics. He asserted the universality of uncertainty (which is an inherent cosmic expression deeply embedded within the core of reality) and indicated that certainty in classical physics is conditional. Prigogine believed that we live in an ascertainable probability world where life and object evolved along a time dimension; making certainty itself illusory. He further pointed out that all existing planes from cosmography to molecular biology generate an evolution pattern of uncertainty. Prigogine also stated that science can provide only possibility rather than certainty. Einstein considered time to be illusory and Prigogine considered certainty to be illusory.

Classical mechanics is a certainty-based description of the natural world, indicating that the natural world is ordered and regular. This order, however, is based on certainty governed by certain conditions. For instance, the *Pioneer 10* spacecraft (launched on March 2, 1972) as it was traveling through the solar system was affected by an anomalous gravitational force, and slowed down more quickly than anticipated. That is, it did not behave in accordance with Newton's law of gravity. NASA planetary scientists and physicists explained that limitations of Newton's law of gravity made it invalid in the scale of the cosmos.

In the field of economic research, classical economics, sometimes called Newtonian economics, was dominant. The introduction of the uncertainty principle had a knock-on effect in the development of the science of uncertainty in physics, based on Prigogine's theory of dissipative structure. In "the information age," it is acknowledged that the essence of information is uncertainty, and information is measured by entropy. This post-Newtonian economics, or Prigogine's economics, is developed based on the uncertainty principle by using Prigogine's theory of dissipative structure. One research interest of Daniel Kahneman (the recipient of the 2002 Nobel Prize in Economics) is the influence of utility in making choices. His research indicates that experienced utility could be measured. This result affirms research achievements of the post-Newton, or Prigogine's economics.

Nobel laureate Kenneth J. Arrow and other information economists used Shannon's entropy to define information, hence formally indicating that the uncertainty principle has an impact on the area of economic research.

Examples of uncertainty in different fields include: the constitution of Gödel's incompleteness theorem in mathematics, Heisenberg's uncertainty principles in physics, "the end of certainty" presented by Prigogine, and the value theory with uncertainty in economics. Thus, it can be seen that, uncertainty exists widely in the fields of natural and human sciences and is indeed, a ubiquitous phenomenon in the natural world and part of human life.

1.3   UNCERTAINTY IN SPATIAL DATA

To return to the main purpose of this book: the modeling of uncertainty in spatial data and analysis in the GISci domain, it is necessary to reiterate that the real world, modeled by spatial data, is a complex, diversified, and nonlinear system. This system contains a diversity of uncertainty, which results from four major factors: (a) inherent uncertainty in the real world, (b) the limitation of human knowledge in cognition of the real world, (c) the limitation of measurement technologies for obtaining spatial data, and (d) the potential of generating and propagating uncertainty in the spatial data processing and analysis, based on captured spatial data.

This book specifically addresses the latter two uncertainty factors: uncertainty due to limitations in measurement technologies and uncertainties generated and propagated in spatial analysis. The limitations in the measurement technologies lead to error-prone measurement. Accuracy of a measured result can, therefore, only be within certain level. However, a deeper understanding of natural phenomenon can be deduced from further analyses of captured data, namely, spatial analyses in GISci. In GISci, three fundamental disciplinary issues are of importance: (a) space and time in geography, (b) uncertainty and spatial data quality, and (c) spatial analysis. The second fundamental discipline issue is addressed in this book.

It is estimated that phenomena up to and over 70% in the real world are position-related, and are able to be described by spatial data. Spatial data can depict the location, shape, and size of a spatial entity, and relationships between spatial entities. The location of the spatial entity is usually described in two- or three-dimensional space. The spatial data can be further grouped into two categories: geometric and relational. The former type labels a spatial object geometrically as a point, a line, or a polygon. The latter describes relationships between spatial entities in terms of adjacency, containment, contiguity, and connectivity. In descriptions of the natural world, such as a certain phenomenon, process, property, characteristics, or status, spatial data can be in the form of text, numbers, figures, formulae, models, tables, or images.

During the past two decades, uncertainty modeling and the quality of spatial data have been studied by many researchers and organizations; for example, quality elements of spatial data are addressed by SDTS (1994), Goodchild and Gopal (1989), Guptill and Morrison (1995), Shi et al. (2002), and ISO 19113 (2005). The dimensions of spatial data and data quality elements, based on the published achievements of these researchers, together with further developments by the author of this book, are given in the following sections.

1.3.1   Dimensions of Spatial Data

An entity in the real world is commonly described by spatial data in various dimensions. Before uncertainties in spatial data are described, the dimensions of spatial data must first be defined. "Spatial data" is used in this instance as a synonym for geographic data or geometric data for describing position-related phenomenon. A

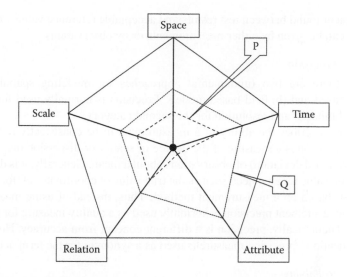

FIGURE 1.2 The objects P and Q are determined in the five-dimension. (From Shi, W. Z. 2009. In GIS Manual, (eds.) M. Madden. ASPRS, 2009, Springer.)

spatial object in GISci can be uniquely determined by a set of spatial data in the following five dimensions (Shi, 2008):

- Space
- Time
- Scale
- Attribute
- Relationships

As illustrated in Figure 1.2, the two objects P and Q are uniquely determined by their "coordinates" within each of these five dimensions. With an understanding of the dimensions of spatial objects indicated above, the quality of spatial data can be given accordingly.

1.3.2 SPATIAL DATA QUALITY

1.3.2.1 Quality Components

Six components—namely, accuracy, precision, reliability, resolution, logical consistency, and completeness—are identified as the specific quality components of spatial data. These components are briefly described as follows.

1.3.2.1.1 Accuracy

Accuracy is normally referred to as the degree to which a measurement is free from bias. It is the extent to which an estimated value approaches the true value or the reference value. ISO 3534-1 (1993) indicates that accuracy is defined as the closeness of the agreement found between test results and acceptable reference values, where the test result can be given by either measurements or by observations.

1.3.2.1.2   Precision

In GISci, there are two fundamental approaches for modeling spatial objects: (a) object-based and (b) field-based models. Precision is normally used for the first category of models, and resolution for the second category.

Precision describes the ability of a measurement to be consistently repeated. In statistics, precision is a measure of the degree of scatter or dispersion (usually measured by standard deviation) of observations about a mean. Generally, it is difficult to obtain a true value for an object, such as the true value of coordinates of, for example, a lamppost, based on a measurement method. Thus, instead of using measurement accuracy, measurement precision is normally used as a quality measure for that measurement. Theoretically, precision is a different concept from accuracy. However, in many applications, precision measure is used as a synonym for the term *accuracy*.

1.3.2.1.3   Reliability

In statistics, reliability refers to the consistency of a set of measurements. In GISci, therefore, reliability refers to the consistency of a set of measurements of spatial objects. Reliability is different from the logical consistency.

1.3.2.1.4   Resolution

Resolution refers to the fineness of detail that can be distinguished in an image. In the field of remote sensing, for example, resolution describes the fineness of a remotely sensed image. Indicators for describing resolution can be pixels per square inch, dots per square inch, or lines per millimeter.

1.3.2.1.5   Logical Consistency

Logical consistency refers to the degree of adherence to logical data structure rules, attributes and relationships. In such a case, data structure can be conceptual, logical or physical (ISO 19113 2005).

1.3.2.1.6   Completeness

The degree of completeness indicates the degree to which the entity objects of a data set cover all the entity instances of the abstract universe. Completeness is used to define the relationship between the objects represented in a data set and the abstract universe of all objects (Morrison, 1988). The definition of completeness has been further extended by Brassel et al. (1995) to give a more comprehensive definition: *completeness* describes whether the entity objects, within a data set, represent all entity instances of the abstract universe.

1.3.2.2   Elements of Spatial Data Quality

Based on the dimensions of spatial data (space, time, scale, attribute and relationships, and the quality components of spatial data: accuracy, precision, reliability, resolution, logical consistency, and completeness), a matrix can be formed with the elements of spatial data quality. According to this matrix (see Table 1.1), a total of

TABLE 1.1

The Matrix of Quality Elements of Spatial Data

	Space	Time	Attribute	Scale	Relationships
Accuracy	Positional accuracy	Temporal accuracy	Attribute accuracy	—	—
Precision	—	—	—	—	—
Reliability	—	—	—	—	—
Resolution	Spatial resolution	Temporal resolution	Thematic resolution	—	—
Logical Consistency	Locational consistency	Temporal consistency	Domain consistency	—	Topologic consistency
Completeness	—	—	—	Completeness	—

30 spatial data quality elements is available. Within these 30 spatial data quality elements, only some of these elements can be clearly defined at this stage. These specific elements are described below. Excluded are those elements to be further studied by GISci scientists, such as relationship accuracy and scale reliability.

1.3.2.2.1 Positional Accuracy

Positional accuracy refers to the accuracy of the location of the spatial objects in GISci. This concept is also widely used in the description of measurement quality in measurement science. Positional accuracy of an object can be affected by three categories of error: random error, systematic error, and gross error as described in detail in Section 1.1.2 of this chapter. In GISci, different types of spatial objects may have different measurement accuracy; for instance, a building corner and a center line of a street are normally surveyed with higher positional accuracy, while another type of feature, such as the boundary of a woodland or soil parcel, may have lower positional measurement accuracy. Two error sources exist in the latter types of measurement: (a) uncertainty in the boundary recognition from the real world, and (b) measurement error. Here, the uncertainty in the boundary cognition (a) is more significant than the error in the measuring process (b).

Positional accuracy can be further classified as (a) absolute accuracy, which refers to closeness of reported coordinate values to the reference values or the values accepted as true and (b) relative accuracy, which refers to the closeness of position of one measurement to that of other measurements for the same point or feature. Positional accuracy is also classified as internal and external accuracy, as well as horizontal and vertical accuracy.

1.3.2.2.2 Attribute Accuracy

Attribute accuracy is determined by the closeness of the attribute values to their true value or reference value. Two types of attribute data exist: quantitative data (e.g., the pH value of the soil within a land parcel); and qualitative data (e.g., a land use type of "forest" assigned to a land parcel). Correspondingly, qualitative accuracy can be distinguished from quantitative accuracy for attribute data in GISci.

1.3.2.2.3   Temporal Accuracy

Temporal accuracy refers to the accuracy of the temporal attributes and temporal relationships of features (ISO 19113, 2005).

1.3.2.2.4   Logical Consistency

According to ISO 19113, logical consistency refers to the degree of adherence to logical rules of data structure, attribution, and relationships. In such a case, data structure can be conceptual, logical, or physical. Logical consistency can be further classified as conceptual consistency (adherence to rules of the conceptual schema); domain consistency (adherence of values to the value domains); format consistency (degree to which data is stored in accordance with the physical structure of the data set); and topological consistency (correctness of the explicitly encoded topological characteristics of a data set).

1.3.2.2.5   Completeness

Completeness refers to the presence or absence of features, the attributes and relationships of spatial data in comparison with that which is in the data model or that which exists in the natural world. Completeness can be indicated by the following two error indicators: (a) errors of omission (data absent from a data set), and (b) errors of commission (excess data present in a data set).

1.4   SUMMARY

Uncertainty is regarded as a generic phenomenon in the natural world, and is illustrated above by examples of uncertainty in several disciplines. This chapter has introduced basic concepts related to uncertainty modeling in spatial data and spatial analysis, covering uncertainty, errors, dimensions of spatial data, quality components and elements of spatial data. Spatial data are specified by its five dimensions: space, time, scale, attribute, and relation. Error in spatial data can be classified as random error, systematic error, and gross error. Quality components of spatial data are defined, including accuracy, precision, reliability, resolution, logical consistency, and completeness in these five dimensions. The elements of spatial data quality are thus formed and can be used, practically, in the quality assessment of spatial data in various application fields.

REFERENCES

Brassel, K., Bucher, F., Stephan, E.-M., and Vckovski, A., 1995. Completeness. In *Elements of Spatial Data Quality,* ed. Guptill, S. C., and J. L. Morrison, Elsevier Science, Oxford, 81–108.

Dubois, D. and Prade, H., 1985. Theorie des possibilities. Masson, Paris (English translation, Plenum Press, New York, 1987).

Gödel, K., 1931. Uber formal unentscheidbare Satze der Principia Mathematica and verwandter Systeme (English: On Formally Undecidable Propositions in Principia Mathematica and Related Systems). *Monatshefte fur Mathematik und Physik* 38: 173–198.

Goodchild, M. and Gopal, S., 1989. *Accuracy of Spatial Databases*. London: Taylor & Francis.

Guptill, S. C. and Morrison J. L., 1995. *Elements of Spatial Data Quality*. Oxford: Elsevier.

Hartley, R. V. L., 1928. Transmission of information. *Bell Systems Tech. J.* 7: 535–563.

ISO 19113. 2005. Geographic Information—Quality Principles, BS EN ISO 19113:2005.

ISO 3534-1. 1993. Statistics—Vocabulary and symbols—Part 1: Probability and general statistical terms.

Klir, G. J. and Folger, T. A. 1988. Fuzzy Sets, Uncertainty, and Information. Prentice Hall, Englewood Cliffs, New Jersey.

Liu, K. F. and Shi, W. Z., 2006. Computing the fuzzy topological relations of spatial objects based on induced fuzzy topology. *Int. J. Geog. Inform. Sci.* 20(8): 857–883.

Mandelbrot, B., 1967. How long is the coast of Britain? Statistical self-similarity and fractional dimension. *Science,* New Series, 156(3775): 636–638.

Morrison, J., 1988. The proposed standard for digital cartographic data. *Amer. Cartog.* 15: 129–135.

SDTS, National Institute of Standards and Technology. 1994. Federal Information Processing Standard Publication 173. (Spatial Data Transfer Standard Part 1. Version 1.1) U.S. Department of Commerce.

Shannon, C. E., 1948. The mathematical theory of communication. *Bell System Tech. J.* 27: 379–423, 623–656.

Shi, W. Z., 2009. Spatial data quality and uncertainty. In *GIS Manual* (eds.) M. Madden. ASPRS, 1352 pages.

Shi, W. Z., Fisher, P. F., and Goodchild, M. F. (ed.), 2002. *Spatial Data Quality.* London and New York: Taylor & Francis. 313 pages.

2 Sources of Uncertainty in Spatial Data and Spatial Analysis

This chapter provides a description and analysis of the various sources of uncertainty in spatial data and spatial analyses. A basic concept of uncertainty has been introduced in Chapter 1. A clear determination of the uncertainty source is necessary in order for the nature of that uncertainty to be more precisely understood and subsequently analyzed. This understanding forms the theoretical foundation for the modeling of uncertainties in spatial data described in Section II, spatial analyses in Section IV, and quality control in Section V of this book.

2.1 INTRODUCTION

Uncertainty ubiquity is due to the complexity and uncertainty of nature, limited human knowledge and cognition of the natural world, limited measurement technology levels, and approximated methods used in spatial data processing and analysis. The ultimate spatial data and analysis results are normally obtained through a series of processes, including spatial data capture, storage, update, transition, querying, and analysis. Spatial data uncertainties are introduced, propagated, accumulated, or even amplified during these processes. Uncertainties in captured spatial data are related to many factors, such as the quality of the data capture instruments, the environment state during the data capture, the particular focus of observers, the map projection methods, and the coordinate systems used.

One of the fundamental reasons for the generation of uncertainty in spatial data is a result of the nature of the differences between the complex and continuous nature of the real world and the simplified and discrete representation of that world in a computer environment, namely, the Geographic Information System (GIS).

The sources of spatial data uncertainty are classified as follows: (a) the inherent uncertainty in nature, (b) uncertainty due to limitations of human cognition of the natural world, (c) object measurement error, and (d) uncertainties arising through spatial analyses and spatial data processing. A spatial data set may be affected by more than one of these four categories. The uncertainty sources and their impacts on spatial data and spatial analysis are illustrated in Figure 2.1 and described in detail below.

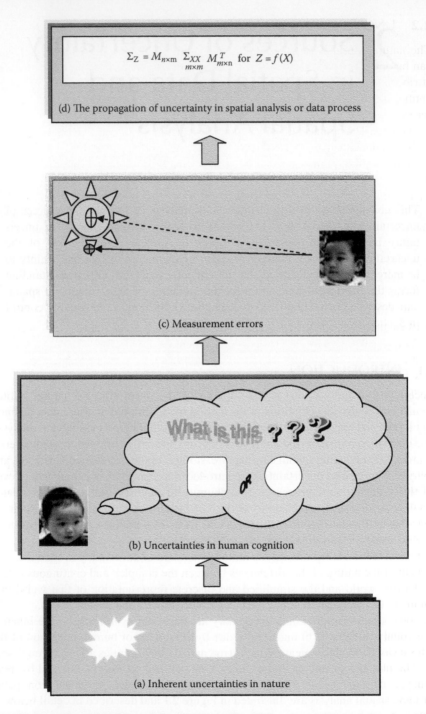

FIGURE 2.1 **(See color figure following page 38.)** Main sources of uncertainty in spatial data and analyses (Adapted from Shi, W. Z. 2005. *Principle of Modeling Uncertainties in Spatial Data and Analysis. Science* Press, Beijing [in Chinese]).

2.2 UNCERTAINTIES INHERENT IN THE NATURAL WORLD

The natural world is complex, huge, and nonlinear. Objects in the natural world can be described as spatial entities in GIS in the dimensions of space, time, scale, attributes, and relationships (a detailed description is given in Chapter 1). Spatial entities described in these dimensions possess uncertainty. In the space dimension, for example, spatial entities may not have determined, distinct, or easily identified boundaries. The boundaries are unlikely to be crisp, but are more of a transition zone with a gradual change in class types, such as forest and grassland.

In the attribute dimension, the spectrum characteristics of a satellite image are determined by the corresponding substance component in the real world. Spectrum characteristics, appearing in a remotely sensed image, are used to classify the image covering different substances. In theory, one type of substance should possess a unique spectrum line. Therefore, a specific spectrum in a satellite image should correspond to a unique class on the ground. In reality, however, the same substance may exhibit a different spectrum. Hence, the same spectrum in a satellite image may correspond to a different classification in the natural or real world. This leads to misclassification in satellite image analysis. The fundamental reason behind this misclassification is the existence of substance spectrum uncertainty; the spectrum performance of the same substance is not unique.

In the scale dimension, feature measurement results in the real world are uncertain and change in accordance with the scale in which they have been measured. The measurement of identical objects in the real world, using the same measurement technology but at a different scale level produces very different measurement results. In the example given in Chapter 1, the length of the coastline of China was measured to be 14,000 km in the scale of 1:100,000, but measured to be 18,000 km in the scale of 1:50,000.

Uncertainty, as well as existing in spatial data dimensions in GISci, also exists in various other disciplines. For example, in ecology, the interface between the media of different gasses and water is dynamic and protean, and the boundary between them is uncertain. Vegetation distributions of different botanical classes in different climates change gradually within a vague boundary. In the natural world, the spatial entities associated with the ecological environment may not have a crisp boundary, possibly because of uncertainty in the spatial distribution of the ecological environment.

Uncertainty also exists in regional studies, for example, the boundary between an urban region and the neighboring rural region or a city and its adjoining area or regional boundary such as found in the Pearl River Delta region of China. There is a gradual transition from one zone to another without a distinct boundary evident. This causes boundary vagueness and uncertainty.

The ubiquity of uncertainty in biology is seen in species evolution and degeneracy, gene variance, and population tides. According to Charles Darwin's theory of evolution, biological evolution is caused by "descent," "variation," and "natural selection." Descent gives rise to the existence of a species and forms a basis of biological evolution, and at the same time guarantees certain and specific common characteristics. However, any two examples of that species may have tiny variations.

These variations are due to random genetic mutation and form the basis of discrimination, subsequently providing a condition for biological evolution. Without random genetic mutations, no species could evolve. Natural selection determines the evolution of a particular species, by enabling "descent' and "variation." This, over long periods of time, can result in the evolution of a more complex higher-level species or, conversely, extinction. However, such biological evolution is full of uncertainty and affected by many random factors, and therefore does not completely develop according to a predefined pattern.

2.3 UNCERTAINTY RESULTING FROM HUMAN COGNITION LIMITATIONS

Geospatial cognition is an essential first step in spatial data capture in GISci. As indicated above, the real world is complex, and the level of human cognition of this world is still at a preliminary stage, both regarding knowledge and the methodologies employed to gain further understanding. Hence, probably only a small portion of the real world so far is recognized and understood.

Currently, only relevant spatial features are selected, measured, and represented owing to the current limited information-handling ability of GIS, as well as the complexity of spatial entities in the real world. The selected spatial features are measured by available measurement technologies. The measurement results, which are in fact an abstraction and approximation of the spatial features in the real world, are then represented in a spatial database in GIS. These approximations are in accordance with the restrictions of current human cognition ability. They are not a full and complete description. Therefore, representations in a spatial database are less detailed and are approximations of the whole object set in the real world. In such cases, the uncertainty of the spatial entities, represented in the spatial database in GIS, can be due either to classification uncertainty or spatial sampling limitations in the cognition of the spatial objects. For instance, the surety that a shrub class object, acquired from a classification of remotely sensed images, is indeed a shrub area, is not, in fact, 100% sure. This is either because of insufficient evidence from the satellite image or because of limitations of the applied image classification method. Such uncertainties are a result of cognition inadequacies.

2.4 UNCERTAINTY RESULTING FROM MEASUREMENT

2.4.1 SPATIAL DATA

As described in Section 1.3.1 of Chapter 1, spatial data are used to describe spatial objects in the dimensions of space, time, scale, attribute, and relationships. The object in a space dimension can be uniquely determined by the coordinates of the object in a specified coordinate system. The attribute of an object can be specified by the class of the object or an attribute value; the object in a time dimension can be depicted by its valid period, that is that defined by the starting time and ending time of the object. The relationships between spatial objects can be described based on topology or fuzzy topology theory.

Objects in the real world can be modeled by the following three fundamental models in GISci: (a) object-based, (b) field-based, and (c) mathematical function-based. In the object-based model, points, lines, and polygons are used as the basic elements for describing either simple or complex spatial objects. In a field-based model, the basic elements are shells, such as a triangular or square shell. A remotely sensed image, with its basic element of a pixel—a square, is an example of a field-based model representation of the real world. In a mathematical function-based model, spatial objects are described by a series of mathematical functions. For example, the real-world terrain can be described based on the level of complexity of the terrain or by linear polynomial functions, nonlinear functions, or hybrid interpolation functions (Shi and Tian, 2006).

2.4.2 SPATIAL DATA MEASUREMENT METHODS

To represent the real world in GIS by spatial data, that data must first be captured by instruments. Historically, the invention of the telescope motivated the development of astronomy; the microscope led to advances in biology and medicine; the aeronautical engine contributed to the first industrial revolution; the electrometer led to the second industrial revolution mainly in Germany during the late 19th to early 20th century. Today, advances in computing technology are leading to an information revolution. Scientific inventions orchestrate a leap in human cognition of the real world, which, in turn, produce new corresponding uncertainty problems.

The evolution of spatial data capture technologies can be seen in ground surveys, aerial photogrammetry, and high-resolution satellite mapping. The accuracy level of the captured spatial data is increasing. However, this increase is restricted by the limitations of measurement technologies.

Spatial data, the lifeblood of GIS, is generated from data conversion; interpolation methods; map digitization (including vector digitization and rater scanning); photogrammetry (including analogy, analytical, and digital photogrammetry); field surveying method (including utilizing total station, distance measuring equipment, and global positioning system); laser scanning (including terrestrial and airborne laser scanning); remote sensing technologies (including using multispectral images, high spatial resolution images, and radar images), and others.

GIS spatial data capture methods are typically divided into the following two categories (Shi, 1994): (a) direct method and (b) indirect method. In the case where the direct method is applied, such as a total station instrument, the point coordinate error is the original measurement error propagation result, such as the distance and angle errors generated when measurement is conducted. Usually, no further error is introduced. However, if a point in GIS is captured by an indirect method such as map digitization, point coordinate errors will include both measurement errors and further processing errors. The classification of the spatial data capture method is illustrated in Figure 2.2.

The GIS spatial data sets captured by the above methods can be grouped into the following four categories: (a) single point data, such as data captured from a total station instrument; (b) two-dimensional vector data, such as data captured from map digitization; (c) two-dimensional raster data, such as satellite images; and

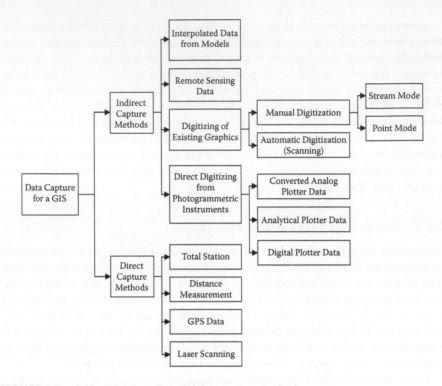

FIGURE 2.2 A classification of spatial data capture methods.

(d) three-dimensional point cloud data, such as three-dimensional points from terrestrial or airborne laser scanning.

2.4.3 UNCERTAINTIES IN SPATIAL DATA CAPTURE

GIS data capture methods, including map digitization, photogrammetry, laser scanning, total station, GPS, and remote sensing, have been summarized in the above sections. In the following sections, error sources and the accuracy levels reached by these data capture methods are further discussed and analyzed. The characteristics of each method in respect to the potential obtainable accuracy level and spatial data type captured are individual and unique. Hence, it is not surprising that these characteristics must undergo careful consideration in the selection of a suitable data capture method for a specific application. The cost factor also has to be considered.

A spatial data capture method normally includes the following components: survey equipment, a data capture scheme, a post-data processing method, and manually operated equipment. However, spatial data capture technology and the related accuracy is restricted by the current status level of metrology technology; the performance of the data capture schemes and data processing methods can vary from one to another. Likewise, the reliability of human operators may also be different from one person to another. Hence, the quality of the captured spatial data may not necessarily be assured.

A description of the spatial data capture methods in common use and their associated spatial data accuracy, plus the main error sources of these methods are summarized and analyzed as follows. Included are map digitization, photogrammetry, laser scanning, total station, GPS, and remote sensing.

2.4.3.1 Map Digitization

Map digitization is a common method for capturing digital spatial data, from maps, for GIS. Two map digitization methods are used: (a) manual digitization for paper maps and (b) on-screen digitization for scanned raster maps. Two modes can be set for a digitization process: the point mode and the stream mode.

As indicated above, the accuracy of map digitization is related to many factors. Such factors vary from the quality of the raw paper map used (the latter includes inherent source errors and errors due to paper deformation caused by conditions such as a humid climate) to the density and complexity of spatial features on the map; the skill and operation method of the digitization operator; instrument quality (such as that of the digitization instrument or scanner); the data processing software functions regarding the handling of errors; and also the accuracy of control points used for coordinate transformation.

2.4.3.2 Aerial Photogrammetry

Aerial photogrammetry is a mapping technology based on stereo aerial photographs. The technology has evolved by means of the following three consecutive stages: analog photogrammetry, analytical photogrammetry, and digital photogrammetry, the latter being the mainstream photogrammetry technology of today. Aerial photogrammetry is appropriate for mapping large areas. The input for photogrammetry mapping includes: (a) stereo aerial photographs with a certain level of overlap and (b) group control points of the area to be mapped. The outcomes of digital aerial photogrammetry are: (a) digital topographic maps, (b) digital orthophotos, and (c) digital elevation models. The fundamental theoretical basis of aerial photogrammetry is the collinear equation. The major workflow of this technology covers the following: aerial triangulation, relative orientation, absolute orientation, stereo mapping, and the production of DEM and digital orthophotos.

The accuracy of the photogrammetry product is affected by the following factors: accuracy and scale of the aerial photographs, accuracy of ground control points, reliability of the algorithms used for the topographic mapping, including image matching, image registration, automatic aerial triangulation, and internal and external orientation. Although digital photogrammetry involves many automatic processes, human operators are still involved in practical photogrammetry mapping production. Therefore, the human operator is also one of the factors affecting the accuracy of the photogrammetric products. Photogrammetry has, at present, reached the geometric accuracy of decimeter level, a state in which horizontal accuracy is normally higher than that of vertical accuracy.

2.4.3.3 The Total Station

The total station is currently the most widely used instrument for ground surveying. The station has the ability to measure both angle and distance. Three-dimensional coordinates are then computed, based on these two original measurements. Meanwhile, captured spatial data is recorded in a storage card, internal storage, or an electronic notebook. This storage function avoids the inconvenience of manual recording in field surveying, which is a potential source of gross error introduction. The latest total stations may even have the ability to automatically search the targets. This greatly reduces the labor intensity of an observer and provides a solution to enable safe surveying in a dangerous area. Such total stations are usually in the form of intelligent robots.

A total station measures target angles and distances by integrated optical and electronic technologies. The accuracy of a total station's observations is thus affected by optical and electronic element factors. Such factors that affect the station's measurement accuracy include instrument function, operator's skill, measurement schemes, and instrumental and weather conditions during measurement. Among these factors, weather conditions have the greatest impact on the overall accuracy of the station's observations; for instance, variations in humidity and temperature alter the index of refraction. Frequency, air pressure, and temperature influence the accuracy of distance measurement. Weather conditions, such as rain or snow, affect the efficiency of the electronic distance measurement.

At present, a high-performance total station has an angle measuring accuracy of 0.5″ and a distance measuring accuracy of at least 1 mm ± 1 ppm.

2.4.3.4   Satellite Positioning Technology

The Global Positioning System (GPS), is a highly accurate positioning technology based on pseudo range measurements between the ground receiver and the tracked satellites.

Factors affecting GPS accuracy may include the atmosphere (where the ionosphere and troposphere cause GPS signal delay); the multipath effect (a result of signals from a satellite reaching the antenna over more than one path due to different surface reflections); the number of GPS satellites available at the current location; the geometric distribution of the satellites; GPS and receiver clock errors; GPS orbital errors; environmental disturbance such as a high voltage power transformer station or vegetation with an elevation angle greater than 15 degrees from the horizontal; and human factors, such as the selective availability (SA) policy at the beginning of the operation, which can introduce intentional, slowly changing random errors of up to 100 meters into the publicly available navigation signals. The SA policy has now been disabled.

GPS positioning accuracy is closely related to the positioning mode and measurement type used. GPS positioning accuracy is around 10 to 20 meters, with a standalone GPS positioning mode using pseudo range measurement. A differential GPS (DGPS) mode utilizes a reference station to reduce errors in GPS measurements and the positioning accuracy can reach meter level. The GPS Real-Time Kinematic (RTK) is a special GPS positioning technique using a carrier phase. The RTK can be used in real time with positioning accuracy of centimeter level. However, the carrier phase measurement can reach millimeter level, thus GPS positioning accuracy, while the carrier phase can reach centimeter level or higher.

2.4.3.5 Laser Scanner

Laser scanning is a new technology for capturing three-dimensional spatial data. The principle of capturing ground surface data with a laser scanner involves the estimation of the distance from a laser source to a target point by computing the echo time of the laser beams emitted. The beams are then converted into the three-dimensional coordinates x, y, z of the target point. During this laser transmission process, the reflective intensity (I) of the target is also detected and recorded. Therefore, the result of laser scanner usage is a four-dimensional coordinate (x, y, z, I) for each target point. Laser scanning technology generally falls into two categories: terrestrial laser scanning and airborne laser scanning. Airborne laser scanning is integrated with two other technologies: an inertia measurement unit (IMU) that records the aircraft orientations and a GPS unit that records the aircraft position during scanning.

The accuracy of airborne laser scanning data is affected by several factors: accuracy and reliability of GPS and IMU, the power and accuracy of the laser emitter, and methods applied for processing the laser-scanned data originally obtained. In many cases, the point set obtained from a laser scanner cannot be used directly. It needs further processing to enable the extraction of the required information such as buildings or terrains. The performance of the object extraction algorithms greatly affects the quality of the obtained information.

The point cloud scanned by a terrestrial laser scanner can reach a geometric accuracy ranging from millimeters to centimeters. The airborne laser scanner, in general, provides the point accuracy of a decimeter, depending on flight height and quality of the laser scanner technology used.

2.4.3.6 Remote Sensing Technology

Remote sensing normally refers to an imaging technology that captures images of the Earth or other planet surfaces by using the satellite sensors. Signals from such satellite images contain both spatial and spectral information. The satellite images can be classified as panchromatic, multispectral, microwave, radar, and others. Thematic information can be classified or extracted from these captured images either manually or automatically.

Remote sensing technology originally broke through the limitation of restricted cognition of the Earth as point-based and for small areas, and at a predefined time spot or period. It can continuously survey the real world at almost any given time. In addition, infrared or microwave remote sensing can benefit human cognition of the natural world.

It is known that the accuracy of spatial data captured from remote sensing is subject to the spatial resolutions of the satellite images. The atmosphere conditions and the instability of remote sensors and platforms directly cause errors in remotely sensed images. Such errors are sometimes difficult to control. Although it is possible to eliminate some errors by an appropriate spectrum and geometric correction method, it is not possible to eliminate all. Many factors affect the quality of remote sensing images. Such quality relies mainly on raw image resolutions (including spatial, spectrum, and temporal resolutions). With the development of remote sensing technology, the spatial resolution of a civilian remotely sensed image is continuously improved: from 60 m for Landsat, 30 m for TM, 10 m for SPOT, 5 m for SPOT-5, 1 m for IKONOS, 0.67 m for QuickBird, and up to 0.50 m for WorldView-1 image. The spectrum resolution has been developed from the multispectrum of several bands to the hyperspectrum of over 100 bands.

More satellite image products can be produced based on the original satellite images. However, the quality of the satellite image products, generated from the original remote sensing images, are further affected by other factors such as the accuracy of ground control points, the reliability of classification methods, the quality of mathematical models for geometric correction, and terrain conditions.

Additionally, uncertainty in captured data is subject to the currency of the data. For example, if an out-of-date satellite image (i.e., year 2000) for a land resource inventory in the year 2007 is used, the out-of-date satellite image employed may cause uncertainty in the results presented in the land resource inventory.

2.5   UNCERTAINTIES IN SPATIAL ANALYSES AND DATA PROCESSING

2.5.1   UNCERTAINTIES IN SPATIAL ANALYSES

Spatial analyses refer to the techniques for analyzing, simulating, forecasting, and controlling geographical events and spatial processes based on spatial data and their distribution. New spatial information can be generated for GIS applications from the analysis of the raw spatial data. Spatial analyses can be classified as map-based spatial analyses, such as the GIS buffer and overlay analysis; analyses of geo-processes, such as site selection or land price modeling; and spatial, information-based analyses. As stated at the beginning of this chapter, each of these analyses methods may introduce, propagate, or even amplify the original spatial data uncertainties.

Errors generated from the spatial analyses operations will significantly affect the quality of the resulting data set(s). This leads to the research development in error propagation in spatial analyses, and such research aims to develop methods to quantify the errors propagated through spatial analysis or spatial operation, based on spatial data set(s).

Error propagation in spatial analysis is defined as a process whereby error is propagated from the original data set(s) to the resulting data set(s) generated by a spatial operation. The concept of error propagation in spatial analyses is illustrated in Figure 2.3.

Error propagation in a spatial analysis is dependent upon (a) errors in source data set(s), (b) the spatial analysis operation used, and (c) the presentation method for the data resulting from the spatial analysis operation.

The two approaches for modeling error propagation in spatial analyses are the analytical approach (such as that based on error propagation law in statistics) and simulation approach (such as Monte Carlo simulation) for either raster-based or vector-based spatial analysis environments.

Details concerning spatial analysis uncertainty modeling are introduced in Chapters 9, 10, and 11 of Section IV of this book. Covered are overlay, buffer, and line simplification uncertainty modeling.

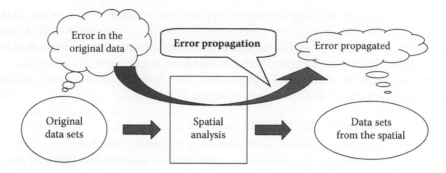

FIGURE 2.3 Error propagation through a spatial analysis. (Adapted from Shi, W. Z., 2007, In Karen Kemp (ed.): *Encyclopedia of Geographic Information Science*, Sage Publications, Thousand Oaks.)

2.5.2 Uncertainty in Spatial Data Processing

2.5.2.1 Uncertainty Introduced from Computation and Equipment

Phenomena in the natural world, such as the terrain on the Earth's surface, are normally continuous. However, in a computer environment, such phenomena are presented in a discontinuous binary form. This approximation to the natural world is the fundamental source of uncertainty causing the difference between what is presented in GIS and the reality apparent in nature. Other factors related to computation and equipment may also cause uncertainties; they include physical and logical modeling in GIS, data encoding, data processing, and computation and output methods.

In GIS, the aim of modeling is to present abstractions of aspects of the natural world through selection, simplification, and symbolization. Therefore, it can be expected that the modeling process itself also causes uncertainty. For instance, if the size and shape of a bus station are not the requirements of the end users of a GIS, the bus station will be represented as a zero-dimensional point rather than a two-dimensional area object, despite the fact that it exists as a two-dimensional area in the natural world.

Error can result in data processing as a consequence of the mishandling of equipment. For example, if a number is larger than the largest number storable in a 32-bit computer, and this number is to be stored and processed in the computer used, an error will occur. One solution is to store that number by a double-precision variable. Uncertainty is also associated with the "rounding off" process. For example, 0.385 rounded off to two decimal places is 0.39. The difference between 0.385 and 0.39 is a rounding error. Rounding errors cause uncertainty in the manipulation of a mass of measured spatial data.

Surveying equipment can only provide measurements of limited precision. Even if the precision of the measured quantity is sufficient, the output result may not be suitably accurate if the output devices (such as the computer monitor or the printer) can only reach a certain accuracy or resolution level. Hence, the selection of equipment that can accurately store, manipulate, and output a mass of exact values is necessary to ensure the presentation of reliable and exact measured results.

In some applications, measurement equipment with a high level of accuracy and reliability is used to depress any uncertainty evident in the result. However, measurements from such equipment can still give an approximation of the true value. It is unlikely that a measured quantity that is equal to the true value can be obtained. Therefore, it can be true to say that no infinite, accurately measured quantity exists. Each measurement is subject to a certain degree of uncertainty.

2.5.2.2   Uncertainty Due to Heterogeneous Data Fusion

When heterogeneous data are fused and a new data set formed, uncertainty in the fused data has to be addressed. To construct a comprehensive database, the GIS user may find it necessary to integrate various data sets from multiple sources and with different accuracy and formats. An ordinary spatial database can perform a regional analysis but may not be as successful in a global analysis. If an application requires a global analysis, the fusion of various regional data has to be conducted, and inconsistency between the data sets may appear. Therefore, the heterogeneous data fusion adds a further example to the previously mentioned sources of uncertainties. Interoperability is an important issue for spatial data fusion, as is data conversion from one GIS data format to another. The multisources of spatial data should be interoperable before they can be well integrated. However, data format conversion may also introduce errors into the converted GIS data. The fundamental reason for this uncertainty is that the basic data model for one GIS may be different from that of another GIS.

2.6   SUMMARY

The sources of uncertainty in spatial data and spatial analyses have been summarized in this chapter. These sources are divided into inherent uncertainty in the natural world, uncertainty generated during human cognition, uncertainty introduced in data capture, and uncertainty arising from spatial analysis and data processing. Errors arising from spatial data capture are very critical and have been analyzed in detail, where the data capture methods cover map digitization, photogrammetry, laser scanning, total station, GPS, and remote sensing.

REFERENCES

Shi, W. Z., 1994. *Modeling Positional and Thematic Uncertainties in Integration of Remote Sensing and Geographic Information Systems*. Enschede: ITC Publication, No. 22, The Netherlands. 147 pages.

Shi, W. Z., 2005. *Principle of Modeling Uncertainties in Spatial Data and Analysis*. Science Press, Beijing. 408 pages (in Chinese).

Shi, W. Z. and Tian, Y., 2006. A hybrid interpolation method for the refinement of regular grid digital elevation model. *Int. J. Geog. Inform. Sci.* 20(1): 53–67.

Shi, W. Z., 2007. Error Propagation. In Karen Kemp (ed.): *Encyclopedia of Geographic Information Science*, Sage Publications, Thousand Oaks. 584 pages.

3 Mathematical Foundations

Mathematical theories form the basis of spatial uncertainty modeling. According to the nature and category of the uncertainties in spatial data and spatial analyses, corresponding mathematical theories have been adopted or applied. To enable a better understanding of the development of uncertainty modeling given in later chapters, a brief summary of the related mathematical foundations are provided in this chapter.

3.1 INTRODUCTION

Uncertainties are classified into three categories: imprecision, ambiguity, and vagueness. *Imprecision* means the level of variation associated with measurement. *Ambiguity* refers to the ambiguous boundary of the region or object of interest, while *vagueness* indicates category determination difficulty, that is, the difficulty in determining to which category the objects of interest belong. An example of vagueness is found in the lack of sharpness in the definition of a boundary.

As indicated, various uncertainties are modeled by using different mathematical theories. Probability theory was the first theory employed for handling imprecision uncertainty. The fuzzy set and fuzzy measure theories are used to assess two different categories of uncertainty. Fuzzy set theories provide a natural framework for handling vagueness uncertainty. Fuzzy measure is used to describe vagueness, while ambiguity is treated with discord, confusion, and nonspecificity measures. A measure of information, known as Hartley's measure, was introduced by Hartley in 1928, to describe ambiguity in a crisp set. Shannon introduced a general uncertainty measure, known as Shannon's entropy, based on probability theory. (Mathematical theories to assess uncertainties in spatial data are summarized in Figure 1.1, Chapter 1.)

The mathematical theories used to model uncertainties in spatial data and spatial analyses are described in this chapter as follows: probability theory, statistics, information theory and entropy, rough set, evidence theory, fuzzy set, and fuzzy topology.

3.2 PROBABILITY THEORY

Probability theory was introduced to study random phenomena. Each random phenomenon is mathematically modeled, with fundamental rules of events. The probability of an event is studied in each model. This theory can be applied to study random error in spatial data. The elementary theorems in the probability theory, together with theory details are found in the work of Jaynes (2003).

3.2.1 Probability of a Random Event

A random event or, simply, an event may or may not occur in cases where uncertainty exists. An event is a collection of outcomes of an experiment. Events consisting of a single outcome are known as simple or elementary events or samples. The collection of all possible outcomes for an experiment is defined as *sample space*. Sure and impossible events are denoted by Ω and $\emptyset$, respectively.

3.2.1.1 Operations of Events

Let A and B be two events. Relationships between these two events can be defined as follows:

1. *Inclusion:* If event B occurs, event A must occur. This is indicated by $A \supset B$.
2. *Equivalence:* If $A \supset B$ and also $B \supset A$, the two events are equivalent. This is indicated by $A = B$.
3. *Product:* If both events A and B occur simultaneously, this is called product of A and B, denoted by $A \cap B$ (or AB).
4. *Sum:* If either event A or event B occurs, this is called a sum of A and B, denoted by $A \cup B$ (or $A + B$).
5. *Difference:* If event A occurs and event B does not occur, this is called a difference between A and B, denoted by $A|B$ (or $A - B$).
6. *Exclusion:* Events A and B are said to be mutually exclusive if they do not occur simultaneously. This case is indicated by writing $AB = \emptyset$.
7. *Complement:* If events A and B are mutually exclusive ($AB = \emptyset$) and an outcome of an experiment is either A or B ($A + B = \Omega$), then B is called the complement of A and is denoted by $B = A^C$.
8. *Exhaustion:* If at lease one of the events $A_1, A_2, ..., A_n$ must occur when the experiment is performed ($A_1 \cup A_2 \cup \cdots \cup A_n = \Omega$), then the set of events, $\{A_1, A_2, \cdots, A_n\}$, is called exhaustion.

3.2.1.2 Definition of Probability

By repeating a random experiment n times, the probability of event A occurring can be estimated. The ratio m/n is called the relative frequency of event A in these n times of the experiment where m represents the number of occurrence of A in the experiment. When n is sufficiently large, by central limit theory the ratio m/n approaches to a constant p. That is,

$$\lim_{n \to \infty} \frac{m}{n} = p. \tag{3.1}$$

The constant p is the probability of the event A, denoted by $\Pr(A) = p$.

The value of p provides an objective description for whether event A occurs, where $0 \le \Pr(A) \le 1$. The probability of the sample space is $\Pr(\Omega) = 1$, and the probability of any impossible event is $\Pr(\Omega) = 0$. Note that the ratio m/n is subject to the result of repeated experiments, and this result does not change the value of p. Therefore, the probability (m/n) is an approximation of the true value of the probability of event A. The larger the value of n, the higher the precision of the approximation.

3.2.1.3 Basic Axioms of Probability

Let A and B be two events. The following axioms hold:

1. $0 \le \Pr(A) \le 1$ and $0 \le \Pr(B) \le 1$.
2. $\Pr(\Omega) = 1$ and $\Pr(\emptyset) = 0$.
3. $\Pr(A \cup B) = \Pr(A) + \Pr(B) - \Pr(A \cap B)$.
4. If $A \supset B$, $\Pr(A) \ge \Pr(B)$ and $\Pr(A) - \Pr(B) = \Pr(A|B)$.
5. For any event A, $\Pr(A^C) = 1 - \Pr(A)$.
6. If events $A_1, A_2, \ldots, A_n$ are mutually exclusive and exhaustive events, the probability of their union is equal to 1, $\Pr(A_1 \cup A_2 \cup \cdots \cup A_n) = \Pr(A_1) + \Pr(A_2) + \ldots + \Pr(A_n) = 1$.

3.2.1.4 General Formula of Probability

3.2.1.4.1 Conditional Probability

A random event is an event with a probability of occurrence determined by some probability distribution. Let A and B be two random events where $\Pr(B) \ne 0$. A conditional probability is the probability of event A given that event B has occurred. It is denoted by $\Pr(A|B)$. Its mathematical expression is given as

$$\Pr(A|B) = \Pr(AB)/\Pr(B).$$

It follows that

$$\Pr(AB) = \Pr(A)\Pr(B|A) = \Pr(B)\Pr(A|B). \tag{3.2}$$

For n random events $A_1, A_2, \ldots, A_n$, we have

$$\begin{aligned} &\Pr(A_1 A_2 \ldots A_n) \\ &= \Pr(A_1)\Pr(A_2|A_1)\Pr(A_3|A_1 A_2)\ldots\Pr(A_n|A_1 A_2 \ldots A_{n-1})\Pr(A_1 A_2 \ldots A_{n-1}). \end{aligned} \tag{3.3}$$

3.2.1.4.2 Independence

When the occurrence of event A does not change the probability of event B, $\Pr(A|B) = \Pr(A)$ where $\Pr(B) \ne 0$, the events A and B are said to be independent. The equivalent condition of independence is

$$\Pr(A \cap B) = \Pr(A)P(B). \tag{3.4}$$

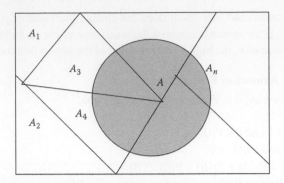

FIGURE 3.1   A sketch of the law of total probability.

3.2.1.4.3   The Law of Total Probability

Let a sample space Ω be partitioned into n mutually exclusive and exhaustive events $A_1, A_2, ..., A_n$ (refer to Figure 3.1). That is, for $i, j = 1, 2, ..., n$ and $i \neq j$,

$$\begin{cases} A_i \cap A_j = \phi \\ A_1 \cup A_2 \cup \cdots \cup A_n = \Omega \end{cases}, \tag{3.5}$$

where the probability of each event A_i is larger than 0.

For any event A consisting of n subevents, we have

$$\Pr(A) = \sum_{i=1}^{n} \Pr(A_i) \Pr(A|A_i). \tag{3.6}$$

This result is called the *law of total probability*.

3.2.1.4.4   Bayes' Formula

Suppose that events $B_1, B_2, ...$ satisfy the following conditions:

$$\begin{cases} B_i \cap B_j = 0, \quad i \neq j \\ \Pr(B_i) > 0 \\ \Pr(B_1 \cup B_2 \cup \cdots) = 1 \end{cases}, \tag{3.7}$$

where $i, j = 1, 2, ...$

For any event A with nonzero probability, based on the law of total probability, we have

$$\Pr(B_i/A) = \frac{\Pr(B_i)\Pr(A/B_i)}{\sum\limits_{i=1}^{\infty}\Pr(B_i)\Pr(A|B_i)}. \tag{3.8}$$

This is the Bayes' theorem.

3.2.1.4.5   Normal Approximation to Bernoulli

Suppose that the probability of an event A is p in one Bernoulli trial, and that that trial is an experiment, the outcome of which is random and can have either of two possible outcomes, "success" and "failure." Suppose also the number of occurrences of event A to be k, when a series of n independent Bernoulli trails are preformed. For $k = 1, 2, \ldots, n$, the probability that "event A occurs k times" is

$$P_{n,k} = \binom{n}{k} p^k (1-p)^{n-k}. \tag{3.9}$$

When both k and n are large, the probability can be approximated by the normal distribution. It follows

$$P_{n,k} \approx \frac{1}{\sigma\sqrt{2\pi}} e^{-\frac{x^2}{2}}, \tag{3.10}$$

where

$$x = \frac{k - np}{\sigma}$$

and

$$\sigma = \sqrt{np(1-p)}.$$

3.2.1.4.6   Poisson Approximation

Poisson distribution can provide an approximation to binomial distribution when n is large and p is small. Then,

$$P_{n,k} \approx \frac{\lambda^k}{k!} e^{-\lambda}, \tag{3.11}$$

where $\lambda = np$.

3.2.2   RANDOM VARIABLE AND DISTRIBUTION FUNCTION

3.2.2.1   Random Variable and Its Probability Distribution Function

An outcome of a random experiment with a sample space can be represented by a variable. Although this variable may be of a different value in each experiment, that variable follows the laws of probability theory. Such a variable is called a *random variable*. It quantifies the random experiment.

Suppose that we have a random experiment with sample space Ω. A random variable X is a real-valued function defined on the sample space of the experiment. The probability $\Pr(X = x)$ of the random variable X equal to a constant x is called the *probability density function* of X abbreviated p.d.f. Moreover, the probability $\Pr(X \leq x)$ of the random variable X not greater than the constant x is called the cumulative distribution function (sometimes called the *distribution function*) of X and is abbreviated as c.d.f. It is denoted by $F(x)$: for $-\infty < x < \infty$,

$$F(x) = \Pr(X \leq x).$$

(3.12)

3.2.2.2   Properties of the Distribution Function

The mathematical definition of a distribution function is a function with the following properties:

1. $F(x) = \sum_{y \leq x} f(y)$ in a discrete space or $F(x) = \int_{-\infty}^{x} f(y)dy$ in a continuous space.
2. For $x_1 < x_2$, we have $F(x_1) \leq F(x_2)$. That is, the distribution function $F(x)$ is a monotonic function.
3. For any real number x and $h > 0$, $\lim_{h \to 0} F(x + h) = F(x)$. This means that $F(x)$ is a right-continuous function.
4. For $a,b \in (-\infty,+\infty)$, $\Pr(a < X \leq b) = F(b) - F(a)$.
5. For $h > 0$, $\Pr(X = a) = F(a) - \lim_{h \to 0} F(a - h)$.

3.2.2.3   The Distribution Function of a Function of the Random Variable

Firstly, discrete and continuous random variables are defined. The possible values for a discrete random variable can be listed as a sequence. In particular, any random variable with a finite number of possible values is discrete. However, the possible values for a continuous random variable form an interval (or intervals) with a positive length.

Let Y denote a random variable in the form of a function of the random variable X with space Ω, $Y = u(X)$. In the discrete case, the distribution function of X is

$$F(x) = \Pr(X \leq x) = \sum_{y \leq x} f(y).$$

(3.13)

In the continuous case, the distribution function is

$$F(x) = \Pr(X \leq x) = \int_{-\infty}^{x} f(y)dy.$$

(3.14)

Suppose that the function $y = u(x)$ is a continuous function of X with an inverse function $x = w(y)$. Thus, the distribution function of Y is

$$G(y) = \Pr(Y \le y) = \Pr(u(X) \le y) = \Pr(X \le w(y)) = \sum_{z \le w(y)} f(z) \qquad (3.15)$$

for the discrete case or

$$G(y) = \Pr(Y \le y) = \Pr(u(X) \le y) = \Pr(X \le w(y)) = \int_{-\infty}^{w(y)} f(x) dx \qquad (3.16)$$

for the continuous case.

3.2.3 Statistical Properties of the Random Variable

3.2.3.1 Expectation and Variance

Let X be a random variable having a p.d.f. $f(x)$. The expectation of this random variable is denoted by $E(X)$. The expectation of the random variable $(X - E(X))^2$ is called the *variance* of X, which is written as $Var(X)$. The variance means the *dispersion* or "spread" of the distribution; small variance means that the distribution is concentrated near the population mean (μ), whereas a large variance indicates a wide spread. The square root of $Var(X)$ is called the standard deviation of X denoted by the symbol σ; that is

$$\sigma = \sqrt{Var(X)}.$$

The expectation and the variance are, in the discrete case,

$$\begin{cases} E(X) = \sum_x xf(x) \\ Var(X) = \sum_x (x - E(X))^2 f(x) \end{cases} \qquad (3.17)$$

or, in the continuous case,

$$\begin{cases} E(X) = \int_{-\infty}^{\infty} xf(x) dx \\ Var(X) = \int_{-\infty}^{\infty} (x - E(X))^2 f(x) dx \end{cases} \qquad (3.18)$$

3.2.3.2   Properties of the Expectation and the Variance

The expectation and the variance have the following properties:

1. $Var(X) = E(X)^2 - [E(X)]^2$.
2. For any constant a, $E(a) = a$ and $Var(a) = 0$.
3. For any constant c, $E(cX) = cE(X)$ and $Var(cX) = c^2 Var(X)$.
4. $E(XY) = E(X) E(Y) + E[(x - E(X)) (y - E(Y))]$.
5. Let $x_1, x_2, \ldots, x_n$ be n random variables, the expectation and the variance of their sum are

$$E\left(x_1 + x_2 + \cdots + x_n\right) = E\left(x_1\right) + E\left(x_2\right) + \cdots + E\left(x_n\right)$$

and

$$Var\left(x_1 + x_2 + \cdots + x_n\right) = \sum_{i,j=1}^{n} E\left[\left(x_i - E\left(x_i\right)\right)\left(x_j - E\left(x_j\right)\right)\right].$$

6. Let $x_1, x_2, \ldots, x_n$ be n independent random variables, the expectation of their product is $E(x_1 x_2 \cdots x_n) = E(x_1)E(x_2)\cdots E(x_n)$.
7. Let $x_1, x_2, \ldots, x_n$ be n independent random variables. Suppose that the mean and the variance of each of these random variables are $E(x_k) = \mu$ and $Var(x_k) = \sigma^2$ for all $k = 1, 2, \ldots, n$. The expectation and the variance of the random variable

$$y = \frac{1}{n}\sum_{k=1}^{n} x_k$$

are $E(y) = \mu$ and

$$Var\left(y\right) = \frac{\sigma^2}{n}.$$

3.2.3.3   Chebyshev's Inequality

An upper bound for a certain probability can be estimated based on Chebyshev's theory of inequality. Let a random variable X have the mean $E(X)$ and finite variance $Var(X)$, then, for any positive real number ε,

$$Pr\left(\left|X - E\left(X\right)\right| \geq \varepsilon\right) \leq \frac{Var\left(X\right)}{\varepsilon^2}. \tag{3.19}$$

The right side of the equation is an upper bound for the probability $Pr(|X - E(X)| \geq \varepsilon)$, which is not necessarily close to the true value of the probability.

3.2.3.4   Median and Mode

When a value m satisfies both $\Pr(X \le m) \ge 1/2$ and $\Pr(X \ge m) \ge 1/2$, this value is called the *median* of a random variable X. It has the following properties:

$$\begin{cases} \Pr\left(X \le m\right) \ge \Pr\left(X > m\right) \\ \Pr\left(X \ge m\right) \ge \Pr\left(X < m\right) \end{cases}. \tag{3.20}$$

A value of x that maximizes the probability $\Pr(X = x)$ is called the *mode of the distribution* of the random variable X.

3.2.3.5   Covariance and Correlation Coefficient

Suppose the expectation and the variance of random variables X_1 and X_2 exist, then the covariance, denoted by $\sigma_{X(1),X(2)}$ or $Cov(X_1,X_2)$, of these two variables is given as

$$\sigma_{X(1),X(2)} = E\left[\left(X_1 - E\left(X_1\right)\right)\left(X_2 - E\left(X_2\right)\right)\right]. \tag{3.21}$$

The correlation coefficient of X_1 and X_2 is defined by the covariance of X_1 and X_2 dividing by the standard deviations of X_1 and X_2

$$\rho_{X(1),X(2)} = \frac{E\left[\left(X_1 - E\left(X_1\right)\right)\left(X_2 - E\left(X_2\right)\right)\right]}{\sqrt{Var\left(X_1\right)}\sqrt{Var\left(X_2\right)}}, \tag{3.22}$$

where $\rho \in [-1,1]$.

3.2.4   LAWS OF LARGE NUMBERS AND CENTRAL LIMIT THEORY

3.2.4.1   Laws of Large Numbers

In probability theory, various theorems are known as the laws of large numbers. These theorems are usually classified into two major categories: weak laws and strong laws. Both laws concern bounds on sample size, accuracy of approximation, and degree of confidence. The weak laws deal with limits of probabilities involving the sample mean, while the strong laws deal with probabilities involving limits of the sample mean.

Bernoulli's law states that a large number of events for a random experiment will have population statistics. Suppose that the relative frequency of an event A equals m/n in n experimental trails. This relative frequency approaches to the probability p of the event A in the sense that for any positive real number ε,

$$\lim_{n \to \infty} \Pr\left(\left|\frac{m}{n} - p\right| < \varepsilon\right) = 1. \tag{3.23}$$

This is an example of a weak law.

Let X_1, X_2, ..., X_k be k independent random variables. For $i = 1, 2, ..., k$, consider case 1 that their expectation and variance are given as $E(x_i) = \mu$ and $Var(X_i) = \sigma^2$ and case 2 that the variables have identical distribution with a finite mean $E(X_i) = \mu$. Both cases show that

$$\frac{1}{k}\sum_{i=1}^{k} X_i$$

converges to the expectation $E(X_i) = \mu$ based on probability. That is, for any positive real number ε, we have

$$\lim_{k \to \infty}\left[\Pr\left(\left|\frac{1}{k}\sum_{i=1}^{k} X_i - \mu\right| < \varepsilon\right)\right] = 1. \tag{3.24}$$

This result is one form of the weak law.

3.2.4.2   The Central Limit Theorem

If X_1, X_2, ..., X_k are k independent and identically distributed random variables with finite mean μ and positive variance σ^2, the random variable

$$\frac{(1/k)\sum_{i=1}^{k} X_i - \mu}{\sigma/\sqrt{k}}$$

approximates having the normal distribution with mean zero and variance 1, $N(0,1)$. This is called the *central limit theorem*.

Using the central limit theorem for population, we have

$$\lim_{k \to \infty}\Pr\left[\frac{\frac{1}{k}\sum_{i=1}^{k} X_i - \mu}{\frac{1}{\sqrt{k}}\sqrt{\frac{1}{k}\sum_{j=1}^{k}\left(X_j - \frac{1}{k}\sum_{i=1}^{k} X_i\right)^2}} \leq x\right] = \int_{-\infty}^{x}\frac{1}{\sqrt{2\pi}}e^{-\frac{v^2}{2}}dv. \tag{3.25}$$

and for sample, we have

$$\lim_{k \to \infty} \Pr\left[\frac{\frac{1}{k}\sum_{i=1}^{k}X_i - \mu}{\frac{1}{\sqrt{k}}\sqrt{\frac{1}{k-1}\sum_{j=1}^{k}\left(X_j - \frac{1}{k}\sum_{i=1}^{k}X_i\right)^2}} \leq x \right] = \int_{-\infty}^{x} \frac{1}{\sqrt{2\pi}} e^{-\frac{v^2}{2}} dv \qquad (3.26)$$

3.3 STATISTICS

3.3.1 INTRODUCTION OF STATISTICS

The natural world is full of uncertainties, with randomness being a type of uncertainty that can be dealt with by statistics. Statistical inference draws sensible and accurate conclusions from uncertainties by analyzing data collected from the natural world. Probability theory is the backbone of sophisticated statistical inference, enabling data analysis by a model-based approach. Given the distribution of a random variable, probability theories can be applied to derive different distribution forms (such as distribution function, distribution law, and density function) and to study quantitative properties of the random variable (such as the mathematical expectation, variance, and correlation coefficient). Statistical theory provides methods for estimating the distribution and the quantitative properties of a random variable based on experiment outcomes.

Probability theories form the mathematical basis for statistics. The statistics are to collect, organize, and interpret numerical data from sampling effectively. Studies of the impacts of the randomness of data are also enabled. Based on finite data, statistics also enable the inference of the statistical properties of data to be found and the outcome of a problem of interest to be forecast. Statistical inference is an inference about a population from a random sample by means of estimation and hypothesis testing. Estimation emphasizes the assignment of a numerical value to a population parameter, based on sample data. Hypothesis testing is used to decide whether assumptions of the parameters, or the form of a population, are accepted or rejected.

Statistics provide a solution to predict the population behavior from samples. This classical theory has been widely applied

3.3.2 DISTRIBUTION FUNCTION OF THE RANDOM VARIABLE

3.3.2.1 Commonly Used Discrete Distribution

3.3.2.1.1 Binomial Distribution

A Bernoulli experiment is a random experiment, the outcome of which can be classified in one or two mutually exclusive and exhaustive ways. When a Bernoulli experiment is performed, independently several times, the occurrence number X of an event A in these Bernoulli trials is a random variable.

Suppose that there are n Bernoulli trials. If the probability of the occurrence of the event A, is let to be a constant p, the possible values of X are 0, 1, 2, ..., n. The probability of $X = d$ is given as

$$\Pr\left(X = d\right) = C_n^d p^d q^{n-d},\tag{3.27}$$

where $q = 1 - p$ and $0 \leq p \leq 1$. The right side of this equation is the result of the binomial expansion of $(p + q)^n$. Therefore, the random variable X is called a *binomial distribution* and denoted by $b(n, p)$:

$$X \sim b\left(n, p\right)\tag{3.28}$$

with $b(d; n, p) = C_n^d p^d q^{n-d}$. If the values of n and p are fixed, $b(d; n, p)$ is a function of d and gives a plot of $n + 1$ discrete points over $d = 0, 1, 2, ..., n$.

The expectation and the variance of the binomial distribution are

$$\begin{cases} E\left(X\right) = \mu_x = np \\ Var\left(X\right) = \sigma_x = np\left(1 - p\right) \end{cases}.\tag{3.29}$$

When $n = 1$, the random variable X follows a Bernoulli distribution. Its expectation and variance are

$$\begin{cases} E\left(X\right) = \mu_x = p \\ Var\left(X\right) = \sigma_x = p\left(1 - p\right) \end{cases}.\tag{3.30}$$

If Y_1, Y_2, ..., Y_n are independent variables having the Bernoulli distribution with the fixed probability of success p, the sum of these variables will have a binomial distribution. That is,

$$x = y_1 + y_2 + \ldots\ldots + y_n \sim b\left(n, p\right).\tag{3.31}$$

Some distributions can approximate the binomial distribution. When $p \leq 0.10$ and $n \geq 30$, for example, the binomial distribution can be approximated by the Poisson distribution:

$$b\left(d; n, p\right) = C_n^d p^d \left(1 - p\right)^{n-d} \approx \frac{\lambda^d}{d!} e^{-\lambda},\tag{3.32}$$

where $\lambda = np$. In addition, when n is sufficiently large, the following normal approximation is given

$$b(d;n,p) \approx \Phi\left(\frac{d-np}{\sqrt{np(1-p)}}\right) - \Phi\left(\frac{d-1-np}{\sqrt{np(1-np)}}\right), \tag{3.33}$$

where

$$\Phi(x) = \int_{-\infty}^{x} \frac{1}{\sqrt{2\pi}} \exp\left(-\frac{x^2}{2}\right) dx$$

is the distribution function of the standard normal distribution.

3.3.2.1.2 Poisson Distribution

When distinct events occur in such a way that the number of events occurring in a given period of time depend on the length of that time period, this number of events has a Poisson distribution with mean λt, where t is the length of the time period and λ is a positive constant.

A random variable X that has the following form of $f(x)$ is said to have a Poisson distribution with mean λ:

$$f(x) = \Pr(X = x) = \frac{\lambda^x}{x!} e^{-\lambda}, \tag{3.34}$$

where $x = 0, 1, 2, \ldots$ That is denoted by $X \sim P(\lambda)$. The expectation and the variance of this variable are then

$$\begin{cases} E(X) = \mu_X \\ Var(X) = \sigma_X = \lambda \end{cases}. \tag{3.35}$$

3.3.2.1.3 Hypergeometric Distribution

Suppose that a population consists of two types of objects: good or defective ones. Let N be the total number of objects and D be the number of defective objects in the population. A number n of the objects is chosen randomly from the population without replacement. The number of defective objects in this sampling is a random variable, say, X. The probability that there are d defective objects in the n samples is

$$\Pr(X = d) = \frac{C_D^d C_{N-D}^{n-d}}{C_N^n}. \tag{3.36}$$

If we say that X is $h(n, D, N)$, we mean that X has the hypergeometric distribution p.d.f.

$$f(x) = \Pr(X = x) = \frac{C_D^x C_{N-D}^{n-x}}{C_N^n},$$ (3.37)

where

$$x = 0, 1, 2, \ldots, D, \quad C_N^n = \frac{N!}{n!(N-n)!}.$$

If N and one of x, n and D are relatively large, it may become difficult to compute the probability. An alternative way to estimate the probability is based on approximations to the hypergeometric distribution. For $n/N \leq 0.10$, the binomial approximation is

$$h(x; n, D, N) \approx b(x; n, p) = C_n^x p^x (1-p)^{n-x},$$ (3.38)

where $p = D/N$.

For $n/N \leq 0.10$ and $D/N \leq 0.10$, the Poisson approximation to the hypergeometric distribution equals

$$h(x; n, D, N) \approx P(x, \lambda) = \frac{\lambda^x}{x!} e^{-\lambda},$$ (3.39)

where $\lambda = nD/N$.

3.3.2.2 Commonly Used Continuous Distribution

Several commonly used continuous distributions are listed in Table 3.1.

3.3.3 ESTIMATION OF A POPULATION PARAMETER

3.3.3.1 Population and Sample

The entire outcome of interest collection is known as the *population*. A sample is a subset of the population. Portions made up of selections of population outcomes x_1, x_2, ..., x_n is called the *sample*. The number of outcomes in the sample makes up the sample size. Let n be the sample size.

3.3.3.2 Statistic of the Population and the Sample

Let X be a random variable from a population distribution and x_1, x_2, ..., x_n denote a random sample of size n taken from the population. Several statistics of the sample are listed in Table 3.2. These statistics include mean, variance, standard deviation, beta coefficient, and coefficient of skewness.

TABLE 3.1
Commonly Used Continuous Distributions of One-Dimensional Random Variables

Distribution	Probability Density Function $f(x)$	Expectation $E(X)$	Variance $Var(X)$
Uniform distribution $u(a, b)$	$p_u(x) = \begin{cases} \dfrac{1}{b-a}, & a \le x \le b \\ 0, & x < a \text{ or } x > b \end{cases}$ where $-\infty < a < b < \infty$	$\dfrac{a+b}{2}$	$\dfrac{(b-a)^2}{12}$
Standard normal distribution $N(0,1)$	$p_N(x) = \dfrac{1}{\sqrt{2\pi}} e^{-\frac{x^2}{2}}$ where $-\infty < x < \infty$	0	1
Normal distribution $N(\mu, \sigma^2)$	$p_N(x) = \dfrac{1}{\sqrt{2\pi}\sigma} e^{-\frac{(x-\mu)^2}{2a^2}}$ where $-\infty < x < \infty$, $-\infty < x < \infty$, and $\sigma > 0$	μ	σ^2
Rayleigh distribution $R(\mu)$	$p_{R(r)} = \begin{cases} \dfrac{x}{\mu^2} e^{-\frac{x^2}{2\mu^2}}, & x \ge 0 \\ 0, & x < 0 \end{cases}$ where $\mu > 0$	$\sqrt{\dfrac{\pi}{2}}\mu$	$\dfrac{4-\pi}{2}\mu^2$
Exponential distribution $e(\mu,\lambda)$	$p_e(x) = \begin{cases} \lambda e^{-\lambda(x-\mu)}, & x \ge \mu \\ 0, & x < \mu \end{cases}$	$\mu + \dfrac{1}{\lambda}$	$\dfrac{1}{\lambda^2}$
Beta distribution $\beta(p,q)$	$p_\beta(x) = \begin{cases} \dfrac{x^{p-1}(1-x)^{q-1}}{B(p,q)}, & 0 < x < 1 \\ 0, & x \le 0 \text{ or } x \ge 1 \end{cases}$ where $p > 0, q > 0$	$\dfrac{p}{p+q}$	$\dfrac{pq}{(p+q)^2(p+q+1)}$
Gamma distribution $\Gamma(\alpha,\beta)$	$p_\Gamma(x) = \begin{cases} \dfrac{\beta^\alpha}{\Gamma(\alpha)}(x-c)^{\alpha-1} e^{-\beta(x-c)}, & x > c \\ 0, & x \le c \end{cases}$ where $\alpha > 0, \beta > 0$	$\dfrac{\alpha}{\beta} + c$	$\dfrac{\alpha}{\beta^2}$

(continued on next page)

TABLE 3.1 (continued)
Commonly Used Continuous Distributions of One-Dimensional Random Variables

Distribution	Probability Density Function $f(x)$	Expectation $E(X)$	Variance $Var(X)$
Logarithmic normal distribution $Ln(\mu, \sigma^2)$	$p_{Ln}(x) = \begin{cases} \dfrac{1}{\sqrt{2\pi}\,\sigma x}e^{-\frac{(\ln x - \mu)^2}{2\sigma^2}}, & x > 0 \\ 0, & x \leq 0 \end{cases}$ where $-\infty < \mu < \infty,\ \sigma > 0$	$e^{\mu + \frac{\sigma^2}{2}}$	$e^{2(\mu+\sigma^2)} - e^{2\mu+\sigma^2}$
Chi-square distribution $\chi^2(n)$	$p_{\chi^2}(x)$	n	$2n$
T-distribution $F(m,n)$	$p_t(x)$	$0\ (n > 1)$	$\dfrac{n}{n-2}$ $(n > 2)$
F-distribution $F(m,n)$	$p_F(x)$	$\dfrac{n}{n-2}$ $(n > 2)$	$\dfrac{2n^2(m+n-2)}{m(n-2)^2(n-4)}$ $(n > 4)$
Weibull distribution $W(m,a)$	$p_W(x)$ where $m > 0,\ \alpha > 0$	$\alpha^{\frac{1}{m}}\Gamma\left(1+\dfrac{1}{m}\right)$	$a^{\frac{2}{m}}\left[F\left(1+\dfrac{2}{m}\right)-\Gamma^2\left(1+\dfrac{1}{m}\right)\right]$
Cauchy distribution $C(\mu,a)$	$\dfrac{\alpha}{\pi}\dfrac{1}{\alpha^2 + (x-\mu)^2}$ where $\alpha > 0$	Nonexistence	Nonexistence

3.3.3.3 Population Parameter Estimation

Let X be a random variable from a population distribution and $x_1, x_2, \ldots, x_n$ denote a random sample of size n taken from the population. The parameters of that population can be estimated from the statistics of the sample by using the method of moments and the method of maximum likelihood.

3.3.3.3.1 Point Estimation with the Method of Moments

The method of moments assigns the value of the rth sample moment to the rth population moment. Let a random variable X have a distribution function with k unknown parameters $(\theta_1, \theta_2, \ldots, \theta_k)$. For a sample X, the distribution function is expressed in the form of $F(x; \theta_1, \theta_2, \ldots, \theta_k)$. Suppose that the rth moment of the distribution exists and is a function of the unknown parameters,

TABLE 3.2

Statistic of a Population and Its Sample

Characteristic	Random Sample	Population
Mean	$\bar{X} = \dfrac{1}{n}\sum_{k=1}^{n} X_k$	$\mu = E(X)$
Variance	$S^2 = \dfrac{1}{n-1}\sum_{k=1}^{n}\left(X_k - \bar{X}\right)^2$	$\sigma^2 = Var(X)$
Standard deviation	$S = \sqrt{\dfrac{1}{n-1}\sum_{k=1}^{n}\left(X_k - \bar{X}\right)^2}$	$\sigma = \sqrt{Var(X)}$
Beta coefficient	$C_v = \dfrac{S}{\bar{X}}$	$C_v = \dfrac{\sqrt{Var(X)}}{E(X)} = \dfrac{\alpha}{\mu}$
Coefficient of skewness	$C_s = \dfrac{n}{(n-1)(n-2)S^3}\sum_{k=1}^{n}(X_k - \bar{X})^3$	$C_s = \dfrac{\mu^3}{\left(\sqrt{Var(X)}\right)^3}$

$$v_r = v_r\left(\theta_1, \theta_2, ..., \theta_k\right) = \int_{-\infty}^{\infty} x^r dF\left(x, \theta_1, \theta_2, ..., \theta_k\right), \tag{3.40}$$

where $r = 1, 2, ..., k$. Let $X_1, X_2, ..., X_n$ be a random sample of size n from the population. The sum

$$\hat{v}_r = \frac{1}{n}\sum_{i=1}^{n} X_i^r \tag{3.41}$$

is the rth moment of the sample.

The method of moments equates $\hat{v}_r$ to v_r from $r = 1$ to the value of r such that there are enough equations to provide unique solutions for $\theta_1, \theta_2, ..., \theta_k$, say, $h_i(\hat{v}_1, \hat{v}_2, \cdots, \hat{v}_k)$ where $i = 1, 2, ..., k$, respectively. That is, for $i = 1, 2, ..., k$, the estimator of θ_i is

$$\tilde{\theta}_i = h_i\left(\hat{v}_1, \hat{v}_2, \cdots, \hat{v}_k\right). \tag{3.42}$$

3.3.3.3.2   Point Estimation with the Method of Maximum Likelihood

Suppose that a population distribution is continuous and its p.d.f. is expressed as $p(x, \theta_1, \theta_2, ..., \theta_k)$ where $\theta_1, \theta_2, ..., \theta_k$ are unknown parameters. Parameter estimation can be estimated as follows.

Let $x_1, x_2, ..., x_n$ be a random sample from a distribution having p.d.f. $p(x, \theta_1, \theta_2, ..., \theta_k)$. When the function

$$L = \prod_{i=1}^{n} p\left(x_i, \theta_1, \theta_2, ..., \theta_k\right) \tag{3.43}$$

has a maximum at $\theta_1 = \hat{\theta}_1, \theta_2 = \hat{\theta}_2, ..., \theta_k = \hat{\theta}_k$, the value $\hat{\theta}_i$ is an estimator of θ_i for $i = 1, 2, ..., k$. The function L and its logarithm function ($\ln L$) are maximized at the same value of $\theta_1, \theta_2, ..., \theta_k$. Thus, the unknown parameters can be estimated by setting the first derivative of $\ln L$ with respect to the parameters equal to zero and solving the resulting equations for $\theta_1, \theta_2, ..., \theta_k$.

The logarithm function of L is

$$\ln L\left(\theta_1, \theta_2, \cdots, \theta_k; x_1, x_2, \cdots, x_n\right) = \ln \prod_{i=1}^{n} L\left(x_i \theta_1, \theta_2, ..., \theta_k\right)$$

$$= \sum_{i=1}^{n} \ln L\left(x_i, \theta_1, \theta_2, ..., \theta_k\right). \tag{3.44}$$

Then, we set

$$\begin{cases} \dfrac{\partial \ln L}{\partial \theta_1} = 0 \\[2mm] \dfrac{\partial \ln L}{\partial \theta_2} = 0 \\[2mm] ... \\[2mm] \dfrac{\partial \ln L}{\partial \theta_k} = 0 \end{cases}. \tag{3.45}$$

By solving this system of equations, the maximum likelihood estimators $\hat{\theta}_1, \hat{\theta}_2, ..., \hat{\theta}_k$ of $\theta_1, \theta_2, ..., \theta_k$ can be obtained.

If the population distribution is discrete, the p.d.f. of the random variable will be expressed as $\Pr(X = x)$. The parameter to this p.d.f. can also be estimated based on the maximum likelihood.

3.3.3.3.3 Evaluation for the Point Estimator of Parameter

Many criteria exist for evaluating the estimator of a parameter. They include unbiasedness, efficiency, and consistency.

If an estimator $\hat{\theta}$ satisfies

$$E\left(\hat{\theta}\right)=\theta, \qquad (3.46)$$

it is then, called an *unbiased estimator* of θ.

Suppose that both $\hat{\theta}$ and $\hat{\theta}'$ are unbiased estimators of θ and the variance of $\hat{\theta}$ is not greater than that of $\hat{\theta}'$. $\hat{\theta}$ is then more efficient as an estimator than $\hat{\theta}'$. When $Var(\hat{\theta})$ is the minimal among unbiased estimators, $\hat{\theta}$ is called the most efficient unbiased estimator of θ.

For any positive real number ε, an estimator $\hat{\theta}_n$ in a sample of size n is consistent if

$$\lim_{n \to \infty} \Pr\left(\left|\hat{\theta}_n - \theta\right| > \varepsilon\right) = 0. \qquad (3.47)$$

The Chebyshev inequality implies that for any $r \geq 0$, an estimator θ_n has the following property:

$$\lim_{n \to \infty} E\left|\hat{\theta}_n - \theta\right|^r = 0 \qquad (3.48)$$

is a consistent estimator of θ.

3.3.3.3.4 Interval Estimation of a Population Parameter

The interval estimator of a parameter is a range of values within which the parameter is thought to lie. A confidence interval is a range of values that is about $100(1 - \alpha)\%$ of the times. It includes the true value of the unknown parameter, where $0 \leq \alpha \leq 1$. Suppose a confidence interval for parameter θ is a region (θ_1, θ_2) such that

$$\Pr\left(\theta_1 < \theta < \theta_2\right) = 1 - \alpha. \qquad (3.49)$$

θ_1 and θ_2 are called confidence limits. Regions $\theta \leq \theta_1$ and $\theta \leq \theta_2$ are regions of rejection. Both the confidence limit and the rejection region are determined by the probability $1 - \alpha$. This probability is known as the *level of confidence*, and the probability α represents a significance level.

Suppose that an outcome X of a random experiment is a random variable with a normal distribution $N(\mu, \sigma^2)$. Given the value of the significance level, confidence intervals for the mean and the variance can be determined based on outcomes of a random sample $X_1, X_2, \ldots, X_n$. If there are two random samples with different means and variances, a confidence interval for the mean difference can also be determined.

3.3.4   TESTS OF STATISTICAL HYPOTHESES

3.3.4.1   General Procedure for the Hypothesis Test

A statistical hypothesis is an assertion concerning the distribution of one or more random variables, or about one or more parameters to the distribution of random variable(s). A test of the statistical hypothesis is a decision rule for determining whether the hypothesis under consideration is accepted or rejected. This decision rule is determined using a power function of the test, which is a function that yields the probability of rejecting the hypothesis. The region leading to the rejection of the hypothesis is called the *critical region* of the test. The size of the critical region is called the *significance level* of the test. This is the maximum value of the power function when the hypothesis is true.

A test of the statistical hypothesis is performed with the following steps:

1. Construct a null hypothesis H_0 and its alternative hypothesis H_1.
2. Select a sample of data.
3. Choose the level of significance α.
4. Determine the critical values $K_{\alpha/2}$.
5. Calculate a test statistic u.
6. Statistical inference: in a two-tailed test, if $|u| < K_{\alpha/2}$, we accept H_0. Otherwise, we reject H_0. In a one-tailed test, if $u < K_\alpha$, (or $u > -K_\alpha$), we accept H_0 for a right (left) tailed H_1.

3.3.4.2   Statistical Hypothesis about Distribution Function

Let $F_0(X) = F_0(X; \theta_1, \theta_2, ..., \theta_k)$ be a given distribution function with parameters θ_1, θ_2, ..., θ_k. A random sample of the random variable X is $X_1, X_2, ..., X_n$. The distribution function of X is supposed to be $F(X)$. The following two tests of statistical hypotheses can be performed.

3.3.4.2.1   All Known Parameters

Suppose a real axis $(-\infty, \infty)$ is divided into m mutually exclusive regions: $(c_i, c_{i+1}]$ where $c_1 \to -\infty$, $c_{m+1} \to \infty$ and $i = 1, 2, ..., m$. A theoretical probability of $(c_i, c_{i+1}]$ is

$$r_i = F_0\left(c_{i+1}\right) - F_0\left(c_i\right) = P\left(c_i < X \le c_{i+1}\right). \tag{3.50}$$

The number of the sample $X_1, X_2, ..., X_n$ falling inside $(c_i, c_{i+1}]$ is denoted by ν_i. This value is called the *experimental frequency*. The following statistic has an approximately chi-square distribution with $m - 1$ degree of freedom,

$$\chi^2 = \sum_{i=1}^{m} \frac{\left(\nu_i - nr_i\right)^2}{nr_i}. \tag{3.51}$$

Hence, the chi-square test is applied to test the null hypothesis $H_0 : F(x) = F_0(X)$.

3.3.4.2.2 Unknown Parameters

Consider that $F_0(X)$ has l unknown parameters $\theta_{j_1}, \theta_{j_2}, \ldots, \theta_{j_l}$ where $l \leq k$. The values of the unknown parameters can be estimated based on the maximum likelihood. The estimators of $\theta_{j_1}, \theta_{j_2}, \ldots, \theta_{j_l}$ are assigned to be the corresponding parameters to $F_0(X)$. Similar to the case of known parameters, the theoretical frequency and the experimental frequency can be computed. The statistic

$$\chi^2 = \sum_{i=1}^{m} \frac{\left(v_i - nr_i\right)^2}{nr_i} \tag{3.52}$$

has a chi-square distribution with $m - l - 1$ degree of freedom when n is large. Applying the chi-square test, we can examine whether a null hypothesis $H_0 : F(x) = F_0(X)$ will be rejected.

3.3.5 STATISTICS FOR SPATIAL DATA

Spatial data can be classified into three categories (Cressie, 1993): geostatistical data, lattice data, and point patterns. Geostatistical data describe a statistical process in a continuous space. Lattice data relate to a spatial process (such as a spatio-temporal model) of lattice in the space. The term *point pattern* is used to describe the real world through a process of spatial points (including the process of representing a point or a random point set). Cressie (1993) introduced the following model that is applicable to geostatistical data, lattice data, point patterns, and objects.

Let S be a data set in an n-dimensional Euclidean space and $Z(s)$ be a random data at a location s in the space. Suppose that the location s belongs to a reference set D in the space. This yields a multidimensional random vector space (or random process), $\{Z(s) : s \in D\}$. Thus, spatial data are defined in accordance with a diversity definition of the reference set D.

3.3.5.1 Geostatistical Data

Consider that the reference set D is a fixed subset of R such that it is an N-dimensional rectangle containing the true value. For $s \in D$, $Z(s)$ is an N-dimensional random vector. Geostatistical data are continuously changing data of spatial reference s in a subset of R^n.

One of the important applications of geostatistics is to predict mine production, in that it enables the modeling of the spatial tendency and correlation synchronously. This spatial correlation has an impact on the estimation, prediction, and design of the spatial model.

3.3.5.2 Lattice Data

Suppose D is a fixed (either regular or irregular) countable set in R^n, and $Z(s)$ is an N-dimensional random vector for $s \in D$. Lattice is a term to describe regular points in the space R^n. A relation between two lattices can be studied by considering concepts of the nearest neighbor and the second-nearest neighbor, for instance, as follows.

3.3.5.3   Point Patterns

Let D be a point process in space R^n or be a subset of R^n. For $s \in D$, let $Z(s)$ be an N-dimensional random vector. In a general format, a lemma for the spatial point process is produced. If there is no definition of Z, a general point process will be obtained.

3.3.5.4   Objects

D is a point process in space R^n, and $Z(s)$ is a random set for $s \in D$; this leads to a process like Boolean model.

A conventional aspatial model is a special spatial model case. In this sense, spatial statistics is a more generic theory than aspatial statistics. Therefore, spatial statistics can be used for the further development of aspatial variables. Meanwhile, a statistics for spatio-temporal data is an even more generic development.

3.4   THE DEMPSTER–SHAFER THEORY OF EVIDENCE

Shannon (1948) first studied uncertainty problems using probability theory. In the 1980s, Klir and Folger applied the Dempster–Shafer theory of evidence (abbreviated as D-S evidence theory) to investigate uncertainty problems (Klir and Folger, 1988). Compared with the probability theory that deals with one kind of uncertainty (randomness), the evidence theory provides a way to assess two distinct types of uncertainty (discord and nonspecificity). The D-S evidence theory for dealing with an uncertainty problem is briefly introduced below. Further details are found in the work of Yager et al. (1994).

3.4.1   TERMINOLOGIES AND SYMBOLS

Suppose X is a finite set and $P(X)$ is a power set of X (the correction of all the subsets of X). A function m is defined as

$$m : P(X) \rightarrow [0,1], \tag{3.53}$$

which satisfies two conditions:

$$m(\varnothing) = 0 \tag{3.54}$$

and

$$\sum_{A \subset X} m(A) = 1. \tag{3.55}$$

The above is a function of basic probability assignments. The value $m(A)$ is a basic probability assignment for predicate A. It represents a level of support for A containing one element in X.

For any $A \in P(X)$, A is said to be a focal element if $m(A) \neq 0$. Let F be a collection of all focal elements. A pair (F,m) is called the *body of evidence*.

Every basic probability assignment m has a belief measure and a plausibility measure, given as

$$Bel(A) = \sum_{B \subset A} m(B) \tag{3.56}$$

and

$$Pl(A) = \sum_{B \cap A \neq \phi} m(B), \tag{3.57}$$

respectively. The belief measure represents the belief that elements belong to A as well as subsets of A. The plausibility measure represents the belief that elements belong to any set overlap A, including A and its subset. These two measures are interrelated; for all $A \in P(X)$,

$$Pl(A) = 1 - Bel(A). \tag{3.58}$$

When the focal element contains only one element, both the belief and the plausibility measures degenerate into a probability measure, Pr. This implies that, for all $A \in P(X)$,

$$Pr(A) = Bel(A) = Pl(A). \tag{3.59}$$

Any probability measure Pr, defined in a finite set, X can be derived from a probability density function. Let p be a probability density function, such that $p : X \rightarrow [0,1]$. The probability measure is then

$$Pr(A) = \sum_{x \in A} p(x). \tag{3.60}$$

From the evidence theory, when m is a set of single elements, function p is equal to function m. If all focal elements are nested, the plausibility measure and the belief measure will be mathematically identical to a possibility measure and a necessity measure, respectively. The possibility and the necessity measures provide the basics of the possibility theory. The possibility theory also has a close relation with fuzzy theory because any alpha cut fuzzy set forms a nested group.

A possibility measure Pos on a finite power set $P(X)$ is determined by a possibility distribution function

$$r : X \rightarrow [0,1] \tag{3.61}$$

via the formula

$$Pos(A) = \max_{x \in A} r(x) \tag{3.62}$$

for all $A \in P(X)$. A necessity measure is then given by

$$Nec(A) = 1 - Pos(A^c) \tag{3.63}$$

for all $A \in P(X)$ where $\bar{A}$ is a crisp complement of A.

A useful idea regarding possibility theory is now introduced. Suppose $X = \{x_1, x_2, \ldots, x_n\}$ is a completed sequence of nested subsets, such that $A_1 \subset A_2 \subset \ldots \subset A_n$ where $A_i = \{x_1, x_2, \ldots, x_i\}$ and $i = 1, 2, \ldots, n$. These nested subsets contain all focal elements of the possibility measure. That is, $A \in \{A_1, A_2, \ldots, A_n\}$ if $m(A) \neq 0$. Let $m_i = m(A_i)$ and $r_i = r(x_i)$ for all $i = 1, 2, \ldots, n$. A basic instruction and a possibility distribution can then be described from $m = (m_1, m_2, \ldots, m_n)$ and $r = (r_1, r_2, \ldots, r_n)$ completely. A nested structure has the following properties: for $i = 1, 2, \ldots, n - 1$ and $j = 1, 2, \ldots, n$,

$$r_i \geq r_{i+1} \tag{3.64}$$

and

$$r_j = \sum_{k=j}^{n} m_k , \tag{3.65}$$

where $m_j = r_j - r_{j+1}$ and $r_{n+1} = 0$.

3.4.2 Nonspecificity Measure

Evidence theory provides two measures to assess uncertainty: the nonspecificity measure and the discord measure. Nonspecificity measures are discussed in the text following.

A nonspecificity measure represents a measure of imprecision associated with a basic probability assignment and is expressed as follows:

$$N(m) = \sum_{A \subseteq X} m(A) \log_2 |A| , \tag{3.66}$$

where $|A|$ is a cardinality of A and X represents a set of all focal elements. The minimum value of this measure is zero when m is a probability distribution. It attains its maximum, $\log_2 |X|$, when $m(X) = 1$ and $m(A) = 0$ for all $A \subset X$. The function N is a generalization of Hartley's measure conceived in the classical set theory. Hartley (1928) measured the uncertainty of a finite set A with the amount of information $I(A)$,

$$I(A) = \log_2 |A| . \tag{3.67}$$

The nonspecificity measure is, obviously, a weighted sum of Hartley's measure.

When m is a probability distribution, focal elements of probability measures are singletons. Equation 3.66 then becomes

$$N(m) = \sum_{A \subseteq X} m(A) \log_2 1 = 0. \tag{3.68}$$

This shows that probability theory does not quantify discord.

When m depicts the possibility and the necessity measures, the function N is rewritten as

$$N(m) = \sum_{i=1}^{n} m_i \log_2 |A_i|, \tag{3.69}$$

where $m_i = m(A_i)$ and $A_i = \{x_1, x_2, \ldots, x_i\}$ for all $i = 1, 2, \ldots, n$. If $|A_i| = i$, this equation will be revised as

$$N(m) = \sum_{i=1}^{n} m_i \log_2 i. \tag{3.70}$$

Using Equation 3.65, possible discord having possibility distribution r and m can be expressed as

$$N(r) = \sum_{i=2}^{n} r_i \log_2 \frac{i}{i-1}. \tag{3.71}$$

Although the function N and Shannon's entropy are two distinct measures for uncertainty, the relationships between the entropy can be accurately described by the function N. For example, let $N(X)$ (or $N(Y)$), $N(X,Y)$ and $N(X|Y)$ (or $N(Y|X)$) be basic, joint and conditional forms of discord. The following relations are attained:

$$\begin{cases} N(X|Y) = N(X,Y) - N(Y) \\ N(Y|X) = N(X,Y) - N(X) \\ N(X,Y) \geq N(X) + N(Y) \\ N(X|Y) \leq N(X) \\ N(Y|X) \leq N(Y) \\ N(X) - N(Y) = N(X|Y) - N(Y|X) \\ N(Y) - N(X) = N(Y|X) - N(X|Y) \end{cases} \tag{3.72}$$

Also, an information conversion for discord can be defined as:

$$T_N(X,Y) = N(X) + N(Y) - N(X,Y).$$ (3.73)

Based on a simple algebra operation, for example, the following is obtained

$$T_N(X,Y) = N(X) - N(X|Y)$$ (3.74)

or

$$T_N(X,Y) = N(Y) - N(Y|X).$$ (3.75)

3.4.3 DISCORD MEASURE

Discord is the second type of uncertainty modeled by the evidence theory. Its measure is a generalized Shannon's entropy measure.

In evidence theory, Shannon's entropy is expressed as

$$H(m) = -\sum_{x \in X} m(\{x\}) \log_2 m(\{x\}).$$ (3.76)

For $x \in X$, the sum of $m[\{x\}]$ is equal to 1. Equation 3.76 is rewritten as

$$H(m) = -\sum_{x \in X} m[\{x\}] \log_2 \left[1 - \sum_{y \neq x} m(\{y\}) \right].$$ (3.77)

For all nonsingleton sets, $m(A) = 0$. The term

$$Con(\{x\}) = \sum_{y \neq x} m(\{y\})$$ (3.78)

in Equation 3.77 represents the total evidence claim pertaining to focal elements that are different from the focal element $\{x\}$. Here, for all $x \in X$, $0 \leq Con(\{x\}) \leq 1$. The function $-\log_2[1 - Con(\{x\})]$ is monotonic increasing with $Con(\{x\})$. Note that for two focal elements with probabilities 0.5, $H(m) = 1$. This shows that Shannon entropy is a measure of uncertainty and ranges in the interval $[0, \log_2 |X|]$.

Similar definitions of Shannon entropy were given in the 1980s: the dissonance measure and confusion measures are

$$\begin{cases} E(m) = -\sum_{A \in X} m(A) \log_2 Pl(A) \\ C(m) = -\sum_{A \in X} m(A) \log_2 Bel(A) \end{cases}$$ (3.79)

for any basic probability assignment m on X, and its corresponding plausibility and belief functions, respectively. These two measures are unable to depict discord of the evidence theory. Therefore, the following discord measure is introduced:

$$D(m) = -\sum_{A \in X} m(A) \log_2 \left[1 - \sum_{B \in X} m(B) \frac{|B - A|}{|B|} \right]. \tag{3.80}$$

Using Equation 3.65, the above equation can be expressed as

$$D(m) = -\sum_{i=1}^{n-1} (r_i - r_{i+1}) \log_2 \left[1 - i \sum_{j=i+1}^{n} \frac{r_j}{(j-1)j} \right]. \tag{3.81}$$

3.4.4 Measure of Total Uncertainty

Both nonspecificity and discord may exist in a given body of evidence. Therefore, a measure of total uncertainty $T(m)$ can be defined using the sum of these two uncertainty measures:

$$T(m) = N(m) + D(m) = \sum_{A \in X} m(A) \frac{|A|}{\sum_{B \in X} m(B) \frac{|A \cap B|}{|B|}}. \tag{3.82}$$

This function is in the range of $[0, \log_2|X|]$. The minimum is obtained when $m(\{x\}) = 1$ for some particular $x \in X$. The maximum is not unique. It may occur for $m(X) = 1$ (which is the unique maximum of N) and for the uniform probability distribution on X (which is the unique maximum of D). It may occur when other bodies of evidence possess certain symmetries.

The additive property of both functions N and D also leads to that of T. However, more work is needed to define various relations among the basic, joint and conditional forms of T.

3.4.5 Application of the Evidence Theory for Spatial Data

An example to illustrate how to apply the evidence theory for uncertainty modeling in spatial data is given below. A similar discussion is found in the work of in Shi (1994). Let Y be a series of hypotheses,

$$Y = \{P(H_1), P(H_2), ..., P(H_n), \Theta\}, \tag{3.83}$$

where $P(H_i)$ denotes the probability of hypothesis H_i, n is the number of the hypotheses, and Θ is a unmentioned part. We define, for $\Theta > 0$,

$$\sum_{i=1}^{n} P(H_i) + \Theta = 1. \tag{3.84}$$

For example, in the application of the maximum likelihood classification method for classifying a remote sensing image, a pixel is classified and its probability belongs to the three classes H_1, H_2, and H_3, which are $P(H_1) = 0.3$, $P(H_2) = 0.2$, $P(H_3) = 0.2$; the probability of unsupported type is 0.3. The belief function and the possibility are given as

$$Bel(H_i) = P(H_i) = 0.7 \tag{3.85}$$

and

$$Pl(H_i) = 1 - \sum_{i \neq j=1}^{n} P(H_j) = Bel(H_i) + \Theta = 0.7 + 0.3. \tag{3.86}$$

The interval $[Bel(H_i), Pl(H_i)]$ describes the uncertainty of the hypotheses H_i caused by the incomplete evidence.

The evidence theory is an extension of the probability theory. The probability function in the probability theory is a subset of the reliability function in the evidence theory. In addition, the conditional rule in the probability theory is a special case of the combine rule in the evidence theory. This leads to the combined theory, which is an important concept in the evidence theory. A detailed description can be found in Kramosil (2001).

3.5   FUZZY THEORY

3.5.1   INTRODUCTION

The representations of objects in the natural world may be imprecise and even vague. Most concepts in the human brain are vague, and any judgment or reasoning regarding these concepts is also vague. Zadeh (1965) introduced fuzzy set, which formed the basis of fuzzy mathematics.

Information about spatial objects is multidimensional, multisource, multiscale, and multitemporal. Because of the properties of diversity, complexity, and imprecision, a representation of these spatial objects can be vague. Fuzzy set theory can be applied to represent spatial objects with vague properties. Fuzzy set theory has also been adopted in the field of spatial information science for dealing with vague information.

3.5.2   FUZZY SET AND ITS OPERATIONS

3.5.2.1   Fuzzy Set

A fuzzy set A in a universe U means that, for any $u \in U$, there is a function $\mu_A(u)$ mapping u into the unit interval [0, 1]. The function is expressed as

$$\mu_A : U \to [0,1]$$

(3.87)

$$u \to \mu_A(u).$$

This is called the *membership function* of A. The value of the membership function gives the degree of membership of u belonging to A. If $\mu_A(u) = 0$, element u will not belong to A. The value 1 describes that u is definitely a member of A.

Similar to a classical set, characterized by its characteristic function, a fuzzy set is characterized by its membership function. The special case that $\mu_A(U) = \{0,1\}$, the function μ_A degenerates to a characteristic function of a classical set. Therefore, classical sets can be regarded as a special case of fuzzy sets.

Notation for a fuzzy set can be expressed in three ways. Below is an example for three different expressions of a fuzzy set. Let $U = \{1, 2, 3, 4, 5, 6, 7\}$ and A be a natural number close to 4. The membership function of A is given in Table 3.3.

Using Zadeh's method, A can be expressed as

$$A = \frac{0}{1} + \frac{0.25}{2} + \frac{0.7}{3} + \frac{1}{4} + \frac{0.7}{5} + \frac{0.25}{6} + \frac{0}{7}$$

(3.88)

or

$$A = \frac{0.25}{2} + \frac{0.7}{3} + \frac{1}{4} + \frac{0.7}{5} + \frac{0.25}{6}.$$

(3.89)

The "divisor" symbol at the right side is used to represent an element in a universe and its membership value: the "denominator" is the element, and the "numerator" is the corresponding membership value. When the membership value is zero, the relevant divisor is removed.

In a second way of the expression, A is represented as a set of order pairs:

$$A = \{(1,0),(2,0.25),(3,0.7),(4,1),(5,0.7),(6,0.25),(7,0)\}.$$

(3.90)

The third way of expression is that A is written in the form of vector:

$$A = (0,0.25,0.7,1,0.7,0.25,0).$$

(3.91)

TABLE 3.3

A Membership Function of a Fuzzy Set A

$\mu_A(u)$	0	0.25	0.7	1	0.7	0.25	0
u	1	2	3	4	5	6	7

These expressions given by Zadeh can be extended to different universes U:

$$A = \int_{u \in U} \left(\mu_A(u) / u \right). \tag{3.92}$$

The symbol of "integration" here does not mean integration, but rather the representation of the relationship between an element and its corresponding membership value.

3.5.2.2 Operations of Fuzzy Sets

Let A and B be two fuzzy sets on a universe U with membership functions $\mu_A(u)$ and $\mu_B(u)$, respectively, for all $u \in U$. Five fundamental operators on fuzzy sets are defined as follows:

1. Inclusion: $A \supseteq B$ if and only if (iff) $\mu_A(u) \geq \mu_B(u)$.
2. Equivalence: $A = B$ iff $A \supseteq B$ and $B \supseteq A$.
3. Union: $C = A \cup B$ iff $\mu_C(u) = \max\{\mu_A(u), \mu_B(u)\}$.
4. Intersection: $D = A \cap B$ iff $\mu_D(u) = \min\{\mu_A(u), \mu_B(u)\}$.
5. Complement: $E = A^C$ iff $\mu_E(u) = 1 - \mu_A(u)$.

In accordance with the notion of a fuzzy set given in Equation 3.92, the union, the intersection, and the complement operators are expressed as

1. Union: $\displaystyle A \cup B = \int_{u \in U} \left(\max\{\mu_A(u), \mu_B(u)\} / u \right).$

2. Intersection: $\displaystyle A \cap B = \int_{u \in U} \left(\min\{\mu_A(u), \mu_B(u)\} / u \right).$

3. Complement: $\displaystyle A^C = \int_{u \in U} \left((1 - \mu_A(u)) / u \right).$

If there are n fuzzy sets (A_1, A_2, ..., A_n with membership functions μ_{A_1}, μ_{A_2}, ..., μ_{A_n}, respectively), their union and intersection are given as, for all $u \in U$ and $T = \{1, 2, ..., n\}$,

$$A = \bigcup_{t \in T} A_t \text{ iff } \mu_A(u) = \max_{t \in T} \mu_{A(t)}(u) \tag{3.93}$$

and

$$B = \bigcap_{t \in T} A_t \text{ iff } \mu_B(u) = \min_{t \in T} \mu_{A(t)}(u). \tag{3.94}$$

For any nonempty fuzzy set A, B, and C, the following propositions are true:

1. Indempotency: $A \cup A = A$ and $A \cap A = A$.
2. Commutation: $A \cup B = B \cup A$ and $A \cap B = B \cap A$.
3. Association: $(A \cup B) \cup C = A \cup (B \cup C)$ and $(A \cap B) \cap C = A \cap (B \cap C)$.

4. Absorption: $(A \cap B) \cup A = A$ and $(A \cup B) \cap A = A$.
5. Distributivity: $A \cap (B \cup C) = (A \cap B) \cup (A \cap C)$ and $A \cup (B \cap C) = (A \cup B) \cap (A \cup C)$.
6. Identity: $A \cup U = U$, $A \cup \varnothing = A$, $A \cap U = A$ and $A \cap \varnothing = \varnothing$.
7. Involution: $(A^C)^C = A$.
8. De Morgan's law: $(A \cup B)^C = A^C \cap B^C$ and $(A \cap B)^C = A^C \cup B^C$.

These propositions are also true on sets. However, note that

$$A \cup A^C \neq U \tag{3.95}$$

and

$$A \cap A^C \neq \varnothing. \tag{3.96}$$

3.5.3 Relationships Between the Fuzzy Set and the Classical Set

Fuzzy sets cannot be represented graphically on a plane, and thus are different from classical sets. Instead, they are visualized by plotting their membership functions. In practical applications, a spatial decision is made based sometimes on classical sets and sometimes on fuzzy sets. What is a relationship between these two types of set?

As mentioned above, a classical set is a special case of a fuzzy set. For a classical set A in a universe U and $u \in U$, element u is said to be an element in A if the characteristic function is an identical one

$$C_A(u) = 1. \tag{3.97}$$

This definition is not applicable to fuzzy sets: the boundary of a fuzzy set is vague. The degree of membership in fuzzy sets is defined as follows: Given a variable $\lambda \in [0,1]$, when A is a fuzzy set with membership function $\mu_A(u)$ whose value is larger than λ at u (i.e., $\mu_A(u) \geq \lambda$), element u is said to be an element in A. This means that the constant λ is used to determine the degree of membership in that fuzzy set.

3.5.3.1 Definition of Level Set

For any fuzzy set A in a universe U and for all $\lambda \in [0,1]$, a subset of A defined by

$$A_{[\lambda]} = \left\{ u \in U \middle| \mu_A(u) \geq \lambda \right\} \tag{3.98}$$

is called the λ-level set where λ is the confidence level. This subset (the λ-level set of a fuzzy set) is the crisp set that contains all elements of a universe whose membership functions have values greater than or equal to the specified value of λ. A subset of A is said to be an open λ-level set if it is given as

$$A_{(\lambda)} = \left\{ u \in U \middle| \mu_A(u) > \lambda \right\}. \tag{3.99}$$

Both $A_{[\lambda]}$ and $A_{(\lambda)}$ are two classical sets in U.

The level set has the following properties:

1. $\left(A \cup B\right)_{[\lambda]} = A_{[\lambda]} \cup B_{[\lambda]}$ and $\left(A \cap B\right)_{[\lambda]} = A_{[\lambda]} \cap B_{[\lambda]}$.

2. $\left(A \cup B\right)_{(\lambda)} = A_{(\lambda)} \cup B_{(\lambda)}$ and $\left(A \cap B\right)_{(\lambda)} = A_{(\lambda)} \cap B_{(\lambda)}$.

3. $\left(\bigcup_{t \in T} A_t\right)_{[\lambda]} \supseteq \bigcup_{t \in T} \left(A_t\right)_{[\lambda]}$.

4. $\left(\bigcup_{t \in T} A_t\right)_{(\lambda)} = \bigcup_{t \in T} \left(A_t\right)_{(\lambda)}$.

5. $\left(\bigcap_{t \in T} A_t\right)_{[\lambda]} = \bigcap_{t \in T} \left(A_t\right)_{[\lambda]}$.

6. $\left(\bigcap_{t \in T} A_t\right)_{(\lambda)} \subseteq \bigcap_{t \in T} \left(A_t\right)_{(\lambda)}$.

7. $A_{(\lambda)} \subseteq A_{[\lambda]}$.

8. If $\lambda \leq \eta$, we have $A_{[\lambda]} \supseteq A_{[\eta]}$ and $A_{(\lambda)} \supset A_{(\eta)}$.

9. $A\left(\max_{t \in T} \lambda_t\right) = \bigcap_{t \in T} A_{[\lambda(t)]}$.

10. $A_{(\wp)} = \bigcup_{t \in T} A_{[\lambda(t)]}$ where $\wp = \max_{t \in T} \lambda_t$.

11. $\left(A^C\right)_{[\lambda]} = \left(A_{(1-\lambda)}\right)^C$.

12. $\left(A^C\right)_{(\lambda)} = \left(A_{[1-\lambda]}\right)^C$.

13. $A_{[0]} = U$ and $A_{(1)} = \varnothing$.

3.5.4   Fuzzy Sets in a Real Universe

A real set R is the common universe on which a fuzzy set is defined. When A is a fuzzy set defined on R, a function $\mu = \mu_A(x)$ is called the *membership function*. Some common fuzzy distributions are given below.

3.5.4.1   Left-Shouldered Membership Function

1. A left-shouldered rectangular distribution:

$$\mu(x) = \begin{cases} 1, x \leq a \\ 0, x > a \end{cases}. \tag{3.100}$$

2. A left-shouldered gamma distribution: for $k > 0$,

$$\mu(x) = \begin{cases} 1, & x \le a \\ e^{-k(x-a)}, & x > a \end{cases}.$$
(3.101)

3. A left-shouldered normal distribution: for $k > 0$,

$$\mu(x) = \begin{cases} 1, & x \le a \\ e^{-k(x-a)^2}, & x > a \end{cases}.$$
(3.102)

4. A left-shouldered Cauchy distribution: for $\alpha, \beta > 0$,

$$\mu(x) = \begin{cases} 1, & x \le a \\ \dfrac{1}{1 + a(x-a)^{\beta}}, & x > a \end{cases}.$$
(3.103)

5. A left-shouldered trapezoid distribution: for $\alpha, \beta > 0$,

$$\mu(x) = \begin{cases} 1, & x \le a_1 \\ \dfrac{a_2 - x}{a_2 - a_1}, & a_1 < x \le a_2 \\ 0, & a_2 < x \end{cases}.$$
(3.104)

6. A left-shouldered mountain distribution: for $\alpha, \beta > 0$,

$$\mu(x) = \begin{cases} 1, & x \le a_1 \\ \dfrac{1}{2} - \dfrac{1}{2} \sin \dfrac{\pi}{a_2 - a_1} \left(x - \dfrac{a_1 + a_2}{2} \right), & a_1 < x \le a_2 \\ 0, & a_2 < x \end{cases}.$$
(3.105)

3.5.4.2 Right-Shouldered Membership Function

1. A right-shouldered rectangular distribution:

$$\mu(x) = \begin{cases} 0, & x \le a \\ 1, & x > a \end{cases}.$$
(3.106)

2. A right-shouldered gamma distribution: for $k > 0$,

$$\mu(x) = \begin{cases} 0, & x \le a \\ 1 - e^{-k(x-a)}, & x > a \end{cases}.$$

(3.107)

3. A right-shouldered normal distribution: for $k > 0$,

$$\mu(x) = \begin{cases} 0, & x \le a \\ 1 - e^{-k(x-a)^2}, & x > a \end{cases}.$$

(3.108)

4. A right-shouldered Cauchy distribution: for $\alpha, \beta > 0$,

$$\mu(x) = \begin{cases} 0, & x \le a \\ \dfrac{1}{1 + a(x-a)^{-\beta}}, & x > a \end{cases}.$$

(3.109)

5. A right-shouldered trapezoid distribution: for $\alpha, \beta > 0$,

$$\mu(x) = \begin{cases} 0, & x \le a_1 \\ \dfrac{x - a_1}{a_2 - a_1}, & a_1 < x \le a_2 \\ 1, & a_2 < x \end{cases}.$$

(3.110)

6. A right-shouldered Mountain distribution: for $\alpha, \beta > 0$,

$$\mu(x) = \begin{cases} 0, & x \le a_1 \\ \dfrac{1}{2} + \dfrac{1}{2} \sin \dfrac{\pi}{a_2 - a_1} \left(x - \dfrac{a_1 + a_2}{2} \right), & a_1 < x \le a_2 \\ 1, & a_2 < x \end{cases}.$$

(3.111)

3.5.4.3 Middle Membership Function

1. A rectangular distribution:

$$\mu(x) = \begin{cases} 0, & x \le a - b \\ 1, & a - b < x \le a + b \\ 0, & a + b < x \end{cases}.$$

(3.112)

2. A Gamma distribution: for $k > 0$,

$$\mu(x) = \begin{cases} e^{k(x-a)}, & x \le a \\ e^{-k(x-a)}, & x > a \end{cases}. \tag{3.113}$$

3. A normal distribution: for $k > 0$,

$$\mu(x) = e^{-k(x-a)^2}. \tag{3.114}$$

4. A Cauchy distribution: for $\alpha, \beta > 0$,

$$\mu(x) = \frac{1}{1 + a(x-a)^\beta}. \tag{3.115}$$

5. A trapezoid distribution: for $\alpha, \beta > 0$,

$$\mu(x) = \begin{cases} 0, & x \le a - a_2 \\ \dfrac{a_2 + x - a}{a_2 - a_1}, & a - a_2 < x \le a - a_1 \\ 1, & a - a_1 < x \le a + a_1. \\ \dfrac{a_2 - x + a}{a_2 - a_1}, & a + a_1 < x \le a + a_2 \\ 0, & a + a_2 < x \end{cases} \tag{3.116}$$

6. A Mountain distribution: for $\alpha, \beta > 0$,

$$\mu(x) = \begin{cases} 0, & x \le -a_2 \\ \dfrac{1}{2} + \dfrac{1}{2} \sin \dfrac{\pi}{a_2 - a_1}\left(x - \dfrac{a_2 + a_1}{2}\right), & -a_2 < x \le -a_1 \\ 1, & -a_1 < x \le a_1 . \\ \dfrac{1}{2} - \dfrac{1}{2} \sin \dfrac{\pi}{a_2 - a_1}\left(x - \dfrac{a_2 + a_1}{2}\right), & a_1 < x \le a_2 \\ 0, & a_2 < x \end{cases} \tag{3.117}$$

3.5.5 FUZZY TOPOLOGY

Fuzzy topology is a combination of topology elements and fuzzy sets. A brief introduction to the foundations of fuzzy topology is given below.

3.5.5.1 Fuzzy Projection and Its Properties

3.5.5.1.1 Fuzzy Projection

The class of fuzzy sets in a universe X is denoted by I^X. Suppose I^X and I^Y are two fuzzy spaces, and the function $f: X \to Y$ is a universal projection. A fuzzy projection $f^\to: I^X \to I^Y$ and its inverse projection $f^\leftarrow: I^Y \to I^X$ are defined for all $A \in I^X$, for all $y \in Y$, for all $B \in I^Y$, and for all $x \in X$,

$$f^\to(A)(y) = \cup \{A(x) : x \in X, f(x) = y\} \tag{3.118}$$

and

$$f^\leftarrow(B)(x) = B(f(x)). \tag{3.119}$$

Proposition 3.5.5.1: Let I^X and I^Y be two fuzzy spaces and $f: X \to Y$ be a universal projection. For $A \in I^X$ and $B \in I^Y$, we have

1. $f^\leftarrow f^\to(A) \supseteq A$ and $f^\to f^\leftarrow(B) \supseteq B$.
2. $f^\leftarrow(B^C) = (f^\leftarrow(B))^C$.
3. $f^\to(A)(y) = 1$ if $A(x) = 1$.
4. $f^\leftarrow(B)(x) = \alpha$ if $B(y) = \alpha$.

Let I^X and I^Y be two fuzzy spaces and $f: X \to Y$ be a universal projection. For any $a \in I$ and any $A \in I^X$, $f^\to(aA) = af^\to(A)$. Therefore, $f^\to(aA) = af^\to(A)$.

Let I^X and I^Y be two fuzzy spaces and $f: X \to Y$ be a universal projection. For $A \subset X$ and $B \subset X$, the following two statements are true:

1. $f^\to(\chi_A) = \chi_{f[A]}$.
2. $f^\leftarrow(\chi_B) = \chi_{f^{-1}(B)}$.

3.5.5.2 Fuzzy Topological Space

Fuzzy topology theory is a generalization of classical topology, and is one of the theories used to model spatial data uncertainties. It enables the fusion of ordered structure with topological structure.

Let X be a nonempty classical set, I be the closed unit interval. A family δ of fuzzy sets in X is called a *fuzzy topology* for X iff

1. $\emptyset, X \in \delta$
2. $A \cap B \in \delta$ if $A, B \in \delta$
3. for all $A_i \in \delta$ ($i \in I$), $\cup A_i \in \delta$.

The pair (X, δ) is called a *fuzzy topological space*. Every member of δ is called an *open* fuzzy set, while its complement, denoted by, δ' is a *closed* fuzzy set.

3.5.5.2.1 Interior and closure

For each fuzzy set A,

1. Set A^o is said to be the interior of A, if it is an open subset of A; i.e., $A^o = \bigcup \{B \in \delta : B \subseteq A\}$.
2. The closure $\overline{A}$ of A is the intersection of all closed subsets of A: $\overline{A} = \bigcap \{B \in \delta' : B \supseteq A\}$.

Boundary

The boundary of a fuzzy set A is defined as $\partial A = \overline{A} \cup \overline{(A^C)}$.

Definition 3.5.5.5: Let (I^X, δ) be a fuzzy topological space. The following properties are true:

1. $0^o = 0$, $1^o = 0$, $\overline{0} = 0$, and $\overline{1} = 1$. This means that both the empty set and the entire set are open and also closed.
2. $A^0 \subseteq A$ and $A \subseteq \overline{A}$.
3. $A^{oo} = A^o$ and $\overline{\overline{A}} = \overline{A}$.
4. $A \subseteq B$ implies $A^o \subseteq B^o$ and $A \subseteq B$ implies $\overline{A} \subseteq \overline{B}$.
5. $(A \cap B)^o = A^o \cap B^o$ and $\overline{A \cup B} = \overline{A} \cup \overline{B}$.

3.5.5.2.2 Fuzzy Injection

Let (I^X, δ) and (I^Y, μ) be two fuzzy topological spaces. A function $f^\rightarrow : (I^X, \delta) \rightarrow (I^Y, \mu)$ is said to be a fuzzy injection if $A \neq B, f^\rightarrow(A) \neq f^\rightarrow(B)$.

3.5.5.2.3 Fuzzy Surjection

Let (I^X, δ) and (I^Y, μ) be two fuzzy topological spaces. A function $f^\rightarrow : (I^X, \delta) \rightarrow (I^Y, \mu)$ is said to be a *surjection* if for every $B \in \mu$, there exists, $A \in \delta$, such that $B = f^\rightarrow(A)$.

3.5.5.2.4 Fuzzy Bijection

Let (I^X, δ) and (I^Y, μ) be two fuzzy topological spaces. A function $f^\rightarrow : (I^X, \delta) \rightarrow (I^Y, \mu)$ is said to be a *bijection* if it is both an injection and a surjection.

Proposition 3.5.5.4: Let I^X and I^Y be two fuzzy spaces and $f : X \rightarrow Y$ be a universal projection. Function $f^\rightarrow : (I^X, \delta) \rightarrow (I^Y, \mu)$ is a bijection only if $f : X \rightarrow Y$ is a bijection.

3.5.5.2.5 Fuzzy Continuous Projection

Let (I^X, δ) and (I^Y, μ) be two fuzzy topological spaces. A function $f^\rightarrow : (I^X, \delta) \rightarrow (I^Y, \mu)$ is said to be a *fuzzy continuous projection* if for every $A \in I^X$ and for every fuzzy open set B containing $f^\rightarrow(A)$, there exists a fuzzy open set C containing A, such that $f^\rightarrow(C) \subseteq B$.

3.5.5.2.6   *Fuzzy Homogeneous Projection*

Let (I^X, δ) and (I^Y, μ) be two fuzzy topological spaces. A function $f^{\rightarrow} : (I^X, \delta) \rightarrow (I^Y, \mu)$ is said to be a *homogenous projection* if it is a continuous and open bijection.

Topological properties that do not change in a fuzzy projection are said to be topologically invariant. The existence of the topological invariant is subject to a fuzzy projection. A homogeneous projection, for example, can guarantee that the result of the spatial data transformation is a topological invariant.

3.6   ROUGH SETS

3.6.1   DEFINITIONS AND PROPERTIES

Suppose U is a nonempty set of objects (called the universe), and R is a U equivalence relation. An ordered pair, $K = (U, R)$, is called an *approximation space*. The equivalence relation partitions the universe into disjoint subsets, each of which is denoted by U/R. An equivalence class induced by R and the empty set $\varnothing$ is called an R-elementary set in the approximation space. Any finite union of R-elementary sets in $K = (U, R)$ is said to be R-definable.

For a set $X \subseteq U$ and an equivalence relation R, X can be characterized by a pair of R-lower and R-upper approximations:

$$R_-(X) = \bigcup \{Y \in U/R : Y \subseteq X\} \tag{3.120}$$

and

$$R^-(X) = \bigcup \{Y \in U/R : Y \cap X \neq \varnothing\}. \tag{3.121}$$

They can also be expressed as

$$R_-(X) = \bigcup \{x \in U : [x]_R \subseteq X\} \tag{3.122}$$

and

$$R^-(X) = \bigcup \{x \in U : [x]_R \cap X \neq \varnothing\}, \tag{3.123}$$

where $[x]_R$ is the equivalence class of R, containing x. The R-lower approximation is the union of those elementary sets in $K = (U, R)$ that are subsets of X. Each element in the R-lower approximation necessarily belongs to X. The R-upper approximation is the union of those elementary sets that intersect X. All elements in the R-upper approximation possibly belong to X.

The R-lower and the R-upper approximations are used to divide the universe into three disjoint regions: the positive region $pos_R(X)$, the negative region $neg_R(X)$, and the boundary region $bn_R(X)$. These three regions are defined as

$$pos_R\left(X\right)=R_-\left(X\right),\tag{3.124}$$

$$neg_R\left(X\right)=U-R_-\left(X\right),\tag{3.125}$$

and

$$bn_R\left(X\right)=R^-\left(X\right)-R_-\left(X\right).\tag{3.126}$$

The positive region is a collection of all elements that definitely belong to X, which is the lower approximation of set X. The negative region is a collection of all elements that do not belong to X. The boundary region is considered as the uncertainty region, in which each element, whether or not it is an element in X, cannot be determined definitely. The union of the positive region and the boundary region is the upper approximation of the set X.

For any two subsets $X, Y \subseteq U$, the lower and the upper approximations satisfy the following properties:

1. $R^-(X) = R_-(X)$ if X is a R-definable set
2. $R^-(X) \neq R_-(X)$ if the set X is referred to as rough with respect to R
3. $R_-(X) \subseteq X \subseteq R^-(X)$
4. $R_-(\emptyset) = R^-(\emptyset) = \emptyset, R_-(U) = R^-(U) = U$
5. $R^-(X \cup Y) = R^-(X) \cup R^-(Y)$
6. $R_-(X \cap Y) = R_-(X) \cap R_-(Y)$
7. $X \subseteq Y \Rightarrow \begin{cases} R_-(X) \subseteq R_-(Y) \\ R^-(X) \subseteq R_-(Y) \end{cases}$
8. $R_-(X \cup Y) \supset R_-(X) \cup R_-(Y)$
9. $R^-(X \cap Y) \subseteq R^-(X) \cap R^-(Y)$
10. $R_-(-X) = -R^-(X)$
11. $R^-(-X) = -R_-(X)$
12. $R_-(R_-(-X)) = R^-(R_-(X)) = R_-(X)$
13. $R^-(R^-(X)) = R_-(R^-(X)) = R^-(X)$

where $-X$ denotes $U - X$.

Every approximation space $K = (U, R)$ can define a unique topological space $T_A = (U, dis(R))$, where $dis(R)$ is the family of all unions of one or more R-elementary sets and all closed sets in T_A.

3.6.2 APPROXIMATION AND MEMBERSHIP RELATION

Let $x \in_{-R}(X)$ denote that x definitely belongs to X, which is simplified as $x \in_-(X)$. In addition, let $x \in_{\bar{R}}(X)$ denote that x possibly belongs to X, which is simplified as $x \in^-(X)$. The membership of x in the collection X satisfies only when (a) $x \in R_-(X)$ or $x \in_-(X)$, or (b) $x \in R^-(X)$ or $x \in^-(X)$. X in case (a) is called the lower membership, whereas the one in case (b) is called the upper membership.

Using the membership relationship, the following propositions are true:

1. $x \in_{-} (X)$ iff $\begin{cases} x \in X \\ x \in^{-} (X) \end{cases}$

2. $X \subseteq Y \Rightarrow \begin{cases} if\ x \in_{-} (X) \Rightarrow x \in_{-} (Y) \\ if\ x \in^{-} (X) \Rightarrow x \in^{-} (Y) \end{cases}$

3. $x \in^{-} (X)$ or $x \in^{-} (Y)$ iff $x \in^{-} (X \cup Y)$
4. $x \in_{-} (X)$ and $x \in_{-} (Y)$ iff $x \in_{-} (X \cup Y)$
5. $x \in_{-} (X)$ or $x \in_{-} (Y) \Rightarrow x \in_{-} (X \cup Y)$
6. $x \in^{-} (X \cap Y) \Rightarrow x \in^{-} (X)$ and $x \in^{-} (Y)$
7. $x \in^{-} (X)$ not if $x \in_{-} (-X)$
8. $x \in_{-} (X)$ not if $x \in^{-} (-X)$

The membership relationship is an elementary and important definition in set theory. Each element in a universe has a strict membership relationship in a given set: it belongs to either the given set or the complement of that set. However, this section considers the problem that the membership relationship cannot be precisely determined, based on the available information or knowledge of the objects under consideration. The uncertain regions of these objects cannot be determined from available information or knowledge. Therefore, classical membership relationships given in set theory are only applicable to a region with certainty. However, rough set theory can be considered as an extension of classical set theory if membership relationships in an approximation space are completely consistent with classical membership relations.

3.6.3   CHARACTERISTIC OF INACCURACY

3.6.3.1   Numerical Characteristic

Uncertainty in a set is due to the set boundary. The larger the boundary, the lower the accuracy. The concept of the accuracy of approximation to describe the uncertainty in a set is introduced as follows. The accuracy of approximation is defined as

$$d_R(X) = \frac{\left| \left(R_{-}(X) \right) \right|}{\left| \left(R^{-}(X) \right) \right|}. \tag{3.127}$$

where $|\cdot|$ is a cardinality and $X \neq \varnothing$.

This accuracy measuring $d_R(X)$ is used to reflect the understanding level of the available information or knowledge about X. Let R be an equivalence relation and X be the subset of universe U, $0 \leq d_R(X) \leq 1$. When $d_R(X) = 1$, the boundary region of X is an empty set and set X is R-definable in the approximation space (U,R); i.e., X is crisp with respect to R. When $d_R(X) < 1$, the boundary region is nonempty, and X is not R-definable; X is rough compared with to R.

Inaccuracy in X can be assessed with another measure—a rough measure—that has an opposite meaning from the accuracy measure. The rough measure reflects the incompleteness degree of the available information or knowledge. It is defined as

$$P_R(X) = 1 - d_R(X).$$ (3.128)

The above discussion shows that rough set theory differs from probability theory and fuzzy theory. The inaccuracy measure is not derived from a predefined hypothesis. The accuracy measure is computed by expressing the concept of inaccuracy of the available information or knowledge mathematically. The value of the accuracy measure shows the result of incomplete information or knowledge.

3.6.3.2 Topological Characteristic

Based on the R-lower and the R-upper approximations, four basic classes of rough sets are given as follows:

1. Set X is roughly R-definable if $R\_(X) \neq \emptyset$ and $R^-(X) \neq U$,
2. Set X is internally R-indefinable if $R\_(X) = \emptyset$ and $R^-(X) \neq U$,
3. Set X is externally R-indefinable if $R\_(X) \neq \emptyset$ and $R^-(X) = U$,
4. Set X is totally R-indefinable if $R\_(X) = \emptyset$ and $R^-(X) = U$.

The meaning of this classification is obvious. If X is roughly R-definable, it can be decided on which elements of U belong to X or $-X$. An internal $R-$ in a definable set X means a decision can be made regarding which elements of U belong to $-X$ but which elements belong to X cannot be decided. When X is externally R-indefinable, the elements of U that belong to X can be decided, but the elements which belong to $-X$ cannot be decided. If X is totally R-indefinable, the elements of U that belong to X cannot be decided

The above classification yields the following two propositions:

1. The three statements: "X is roughly R-definable," "X is totally R-indefinable," and "$-X$ is R-definable" are equivalent to the statement "X is R-definable."
2. "$-X$ is internally (or externally) R-indefinable" is equivalent to "X is externally (or internally) R-indefinable."

The accuracy of approximation and the classification of rough set are two ways by which rough sets can be depicted, each having different meanings yet with an interrelationship. The numerical characteristic of a rough set represents the set's boundary size, but does not account for the boundary structure. Topological characteristics of the rough set do not provide information about the boundary size of the rough set but provide information about the boundary structure. As indicated above, the ways of depicting a rough set are interrelated. When a set is internally or totally indistinguishable, its accuracy measure is 0. When a set is externally or totally indistinguishable, the accuracy measure of the complement of the set is 0. From these two cases, it is clear that the accuracy measure of a rough set cannot entirely determine its topological structure. Knowledge of the topological set of a rough set also does not contain accuracy information. Therefore, in a practical application, a combination of the numerical and the topological characteristics is necessary.

3.6.4 APPROXIMATION FOR CLASSIFICATION

It is quite possible and even likely that not all information or knowledge for a certain set, is given when required. Hence, partitioning the set based on the available information or knowledge F results in an approximate classification. Inaccuracy of the classification approximation can be characterized by the concept of set approximation.

Let $P = \{x_1, x_2, \ldots, x_n\}$ denote a family of sets. The R-lower approximation and the R-upper approximation are defined as

$$R_-\left(F\right)=\left\{R_-\left(X_1\right), R_-\left(X_2\right), \ldots, R_-\left(X_n\right)\right\} \tag{3.129}$$

and

$$R^-\left(F\right)=\left\{R^-\left(X_1\right), R^-\left(X_2\right), \ldots, R^-\left(X_n\right)\right\}. \tag{3.130}$$

The classification approximation is a simple generalization of the set approximation. Its inaccuracy can be assessed by the accuracy of the approximation and the quality of the approximation. The accuracy of the approximation is defined as

$$D_R\left(F\right)=\frac{\sum\left|\left(R_-\left(X_i\right)\right)\right|}{\sum\left|\left(R^-\left(X_i\right)\right)\right|}. \tag{3.131}$$

Classification accuracy is used to describe the case when P is classified based on the available information and the class (among all possible categories) that has the correct determined percentage.

The classification quality is defined as

$$r_R\left(F\right)=\sum\frac{\left|\left(R_-\left(X_i\right)\right)\right|}{\left|\left(U\right)\right|}. \tag{3.132}$$

It represents the percentage of elements P, which can be classified as F based on the available information or knowledge R.

Let $F = \{x_1, x_2, \ldots, x_n\}$ (for $n > 1$) denote a class in U and R denote an equivalence relation. For $i \in \{1, 2, \ldots, n\}, j \in \{1, 2, \ldots, n\}$ and $i \neq j$. The following propositions are true when set F is R definable:

1. If $R_-\left(X_i\right) \neq \emptyset, R^-\left(X_j\right) \neq U$.
2. If $R^-\left(X_i\right) = U, R^-\left(X_j\right) = \emptyset$.
3. If $R_-\left(X_i\right) \neq \emptyset, R^-\left(X_i\right) \neq U$.
4. If $R^-\left(X_i\right) = U, R_-\left(X_i\right) = \emptyset$.

3.6.5 ROUGH EQUALITY AND ROUGH INCLUSION

3.6.5.1 Rough Equality of a Set

A distinction between a rough set and classical set is drawn from the definition of equality. Equality in the classical sets refers to two sets, the elements of which are *completely* the same, while in a rough set, two sets are said to be equal if elements in these two sets are *approximately* the same. The definition in a rough set is sometimes more realistic. This can be explained by the fact that it is always difficult to guarantee that the available information or knowledge about the two sets is complete. Thus, the set equality is derived based on the information or knowledge available, regarding the relevant objects.

Suppose $K = (U, R)$ is an approximation space, X and Y belong to a universe U, and $R \in ind(K)$. Three approximate rough equalities are defined as

1. When $R_-(X) = R_-(Y)$, sets X and Y have lower R-equality, abbreviated as $(X)_-R(Y)$.
2. When $R^-(X) = R^-(Y)$, sets X and Y have upper R-equality, abbreviated as $(X)^-R(Y)$.
3. When $R_-(X) = R_-(Y)$ and $R^-(X) = R^-(Y)$, sets X and Y have R-equality, abbreviated as $(X)R(Y)$.

For any equivalence relation, the following properties are true:

1. When $(X \cap Y)_- R(X)$ and $(X \cap Y)_- R(Y)$, $(X)_- R(Y)$.
2. When $(X \cup Y)^- R(X)$ and $(X \cup Y)^- R(Y)$, $(X)_- R(Z)$.
3. When $(X)^- R(X')$ and $(Y)^- R(Y')$, $(X \cup Y)^- R(X' \cup Y')$.
4. When $(X)_- R(X')$ and $(Y)_- R(Y')$, $(X \cap Y)_- R(X' \cap Y')$.
5. When $(X)^- R(Y)$, $(X \cup -Y)^- R(U)$.
6. When $(X)_- R(X')$, $(X \cap -Y)_- R(\emptyset)$.
7. When $X \subseteq Y$ and $(Y)^- R(\emptyset)$, $(X)^- R(\emptyset)$.
8. When $X \subseteq Y$ and $(Y)^- R(U)$, $(X)^- R(U)$
9. $(-X)_- R(-Y)$ iff $(X) R(Y)$.
10. When $(X)_- R(\emptyset)$ or $(Y)_- R(\emptyset)$, $(X \cap Y)_- R(\emptyset)$.
11. When $(X)^- R(U)$ or $(Y)^- R(U)$, $(X \cup Y)^- R(U)$.

For any equivalence relation R, we have $(X)_- R(Y)$ if $R_-(X)$ is the intersection of all $Y \subseteq U$. If $R^-(X)$ is the union of all $Y \subseteq U$, we have $(X)^- R(Y)$.

Rough equality is based on the upper approximation and the lower approximation of the topological characteristic. This means that rough equality relies on the available information or knowledge about the sets: a pair of sets is equal in one approximate space, but may not be equal in another.

3.6.5.2 Rough Inclusion of a Set

Inclusion in set theory is another important concept. Rough inclusion can also be defined in a similar way to the definition of rough equality.

Suppose $K = (U, R)$ is an approximate space, X and Y belong to a universe U, and $R \in ind(K)$. The rough inclusion is defined as follows:

If $R_-(X) \subseteq R_-(Y)$, set X is said to be the lower R-inclusion of Y and is abbreviated as $(X)C_-R(Y)$. If $R^-(X) \subseteq R^-(Y)$, X is said to be the upper R-inclusion of Y and is abbreviated as $(X)C^-R(Y)$. If both $R_-(X) \subseteq R_-(Y)$ and, $R^-(X) \subseteq R^-(Y)$ satisfy, X is then said to be the R-inclusion of Y and is denoted by $(X)CR(Y)$. These definitions give the following properties:

1. $(X)C_-R(Y)$, $(X)C^-R(Y)$ and $(X)CR(Y)$ if $X \subseteq Y$.
2. $(X)C_-R(Y)$ if $(X)C_-R(Y)$ and $(Y)C^-R(X)$.
3. $(X)C^-R(Y)$ if $(X)C^-R(Y)$ and $(Y)C^-R(X)$.
4. $(X)CR(Y)$ if $(X)CR(Y)$ and $(Y)CR(X)$.
5. $(X \cup Y)^-R(Y)$ iff $(X)C^-R(Y)$.
6. $(X \cap Y)_-R(Y)$ if $(X)C_-R(Y)$.
7. $(X')C_-R(Y')$ if $X \subseteq Y$, $(X)_-R(X')$ and $(Y)_-R(Y')$.
8. $(X')C^-R(Y')$ if $X \subseteq Y$, $(X)^-R(X')$ and $(Y)^-R(Y')$.
9. $(X')CR(Y')$ if $X \subseteq Y$, $(X)^-R(X')$ and $(Y)R(Y')$.
10. $(X' \cup Y')C^-R(X \cup Y)$ if $(X')C^-R(X)$ and $(Y')C^-R(Y)$.
11. $(X' \cap Y')C^-R(X \cap Y)$ if $(X')C_-R(X)$ and $(Y')C_-R(Y)$.
12. $(Z)C_-R(Y)$ if $(X)C_-R(Y)$ and $(X)_-R(Z)$.
13. $(Z)C^-R(Y)$ if $(X)C^-R(Y)$ and $(X)^-R(Z)$.
14. $(Z)CR(Y)$ if $(X)CR(Y)$ and $(X)R(Z)$.

3.7 INFORMATION THEORY AND ENTROPY

Information refers to the interchanging content of the states of different objects, which is the state and rule of dynamic objects. Information can be expressed in the form of texts, graphs, images, or sounds. To give an information definition and provide an effective measure of the information amount, Shannon (1948) defined entropy as an uncertainty associated with a random variable. Section 3.7 introduces information theory and entropy. More details of information theory can be found in Jiang (2001) and Yeung (2002).

3.7.1 MATHEMATICAL INFORMATION SOURCE MODEL

The information source, as its name suggests is a carrier of information. It provides, information modeled as a random variable or process, and a set of probabilities assigned to a set of outcomes. The simplest information source is a single symbol.

Suppose an information source yields r outcomes: $a_1, a_2, ..., a_r$ with probabilities $p(a_1), p(a_2), ..., p(a_r)$, respectively. Let a random variable X denote this information source. Its information space is given by

$$[X \bullet P]: \begin{cases} X: & a_1 & a_2 & ... & a_r \\ P(X): & p(a_1) & p(a_2) & ... & p(a_r) \end{cases}, \tag{3.133}$$

where for $i = 1, 2, \ldots, r,\ 0 \le p(a_i) \le 1$ and

$$\sum_{i=1}^{r} p(a_i) = 1.$$

An information source in an information space is bijective (a one-to-one relationship): different information sources are mapped into different information spaces. To present an information space as a mathematical information source model, it is necessary to provide predefined or premeasurable probabilities of all information source outcomes. This means that the probability space is vital in the formation of the information space. The starting point of describing the information source mathematically represents, for example, a binary discrete information source with a discrete random variable X. The physical and the probability spaces of this random variable are then two elementary parts of the creation of the information space. Probability space provides the foundation of the information theory of Shannon.

3.7.2 AMOUNT OF SELF-INFORMATION

From the definition of the information amount, the received information amount equates to the uncertainty amount removed upon the revelation of the outcome in magnitude. Suppose the outcome of the information source emitted is a_i, for example, due to noise in data transmission, the information received may be different from a_i. Let b_j be the information received by a_i. The information amount of a_i extracted from b_j (or the amount of uncertainty removed upon revealing the outcome) is expressed as

$$I\left(a_i; b_j\right) = \left[\text{uncertainty in } a_i \text{ before } b_j \text{ is received}\right]$$
$$-\left[\text{uncertainty in } a_i \text{ after } b_j \text{ is received}\right]. \tag{3.134}$$

Without noise in the process of data transmission, an outcome a_i of the information source will be propagated through the process without uncertainty. The amount of self-information is then given by

$$I\left(a_i\right) = I\left(a_i; a_i\right)$$
$$= \left[\text{uncertainty in } a_i \text{ before } a_i \text{ is received}\right]. \tag{3.135}$$

This expression shows that the issue of assessing information contained in a_i is converted to assessing the uncertainty in a_i. In Section 3.5, it is mentioned, that uncertainty and possibility are interrelated when possibility can be described by probability. It is thus reasonable to consider self-information $I(a_i)$ contained in an outcome a_i as a prior probability function $p(a_i)$ of a_i: for $i = 1, 2, \ldots, r$,

$$I\left(a_i\right) = f\left[p\left(a_i\right)\right]. \tag{3.136}$$

Generally, this function satisfies the following four conditions:

1. Suppose two outcomes of information source are a_i and a_j with prior probabilities $p(a_i)$ and $p(a_j)$, respectively, where $0 < p(a_i), p(a_j) < 1$. If $p(a_i) > p(a_j)$, $I(a_i) = f[p(a_i)] < I(a_j) = f[p(a_j)]$.
2. If an outcome a_i of information source has a prior probability $p(a_i) = 0$, $I(a_i) = f[p(a_i)] \rightarrow \infty$.
3. If an outcome a_i of information source has a prior probability $p(a_i) = 1$, $I(a_i) = f[p(a_i)] = 0$.
4. Let X and Y be two statistical-independent information sources. Suppose an outcome of X is a_i with a prior probability $p(a_i)$, an outcome of Y is b_j with a prior probability $p(b_j)$. The joint probability of outcomes a_i and b_j is $p(a_ib_j)$. Then,

$$I\left(a_ib_j\right) = f\left[p\left(a_ib_j\right)\right] = I\left(a_i\right) + I\left(b_j\right). \qquad (3.137)$$

If function $I(a_i)$ satisfies the above four conditions, it will be the logarithm of the reciprocal of $p(a_i)$,

$$I\left(a_i\right) = \log\left[\frac{1}{p\left(a_i\right)}\right] = -\log\left[p\left(a_i\right)\right], \qquad (3.138)$$

where $i = 1, 2, \ldots, r$. This is called the information function. It is used to assess uncertainty in an outcome a_i of information source

$$\left[X \bullet P\right]: \begin{cases} X: & a_1 & a_2 & \cdots & a_r \\ P(X): & p\left(a_1\right) & p\left(a_2\right) & \cdots & p\left(a_r\right) \end{cases}.$$

Here, the prior probability of each a_i is the only certain information. Therefore, the function $I(a_i)$ in Equation 3.139 is also called the probability information. The unit of the amount of self-information is subject to the base of the logarithm. If the base is 2, the unit will be bit. That is,

$$I\left(a_i\right) = \log_2\left[\frac{1}{p\left(a_i\right)}\right] \text{ bit.} \qquad (3.139)$$

The information function provides a measure of the information amount.

3.7.3 ENTROPY OF INFORMATION

The information function has two limitations. The first is that the information function is a tool to measure information based on the prior probability of each information source outcome. However, the probability may not be always known. Second, information function $I(a_i)$ cannot be used to measure the overall information amount obtained in all information source outcomes, which assesses the amount of information obtained in one outcome a_i only.

Suppose information source X has r outcomes $\{a_1, a_2, ..., a_r\}$ in the probability space $\{p(a_1), p(a_2), ..., p(a_r)\}$. An overall information measure is expressed as a linear combination of all $I(a_i)$. This is called entropy of X:

$$H(X) = p(a_1)I(a_1) + p(a_2)I(a_2) + ... + p(a_r)I(a_r) \tag{3.140}$$

$$= -\sum_{i=1}^{r} p(a_i) \log p(a_i).$$

This formula measures the average information amount contained in X.

The information function $I(a_i)$ describes the amount of information in a single outcome a_i. It is either a measure of the unexpectedness of outcome a_i or a measure of the information yielded by the outcome. The entropy can be interpreted not only as the average amount of information in all $I(a_i)$ in the probability space of the information source, but also as an average uncertainty in the information source X.

3.7.4 Algebraic and Analytical Properties of Entropy

Let $p_i = p(a_i)$. The entropy is then rewritten as

$$H(X) = -\sum_{i=1}^{r} p_i \log[p_i]. \tag{3.141}$$

This expression shows that $H(X)$ is a function of p_i. That is,

$$H(p_1, p_2, ..., p_r) = -\sum_{i=1}^{r} p_i \log[p_i]. \tag{3.142}$$

If the probabilities of all outcomes a_i ($i = 1, 2, ..., r$) of the information source X is expressed in the form of vector,

$$P = (p_1, p_2, ... p_r). \tag{3.143}$$

The entropy in Equation 3.143 is then denoted by $H(P) = H(p_1, p_2, ..., p_r)$.

The entropy has the following algebraic properties:

1. *Symmetry:* $H(p_1, p_2, ..., p_r) = H(p_2, p_1, ..., p_r) = H(p_r, p_2, ..., p_1)$. Swapping locations of all p_i does not change the value of the entropy. That is, the entropy is invariant with the order in which the outcomes occur.
2. *Nonnegativity:* $H(p_1, p_2, ..., p_r) \geq 0$. This property states that in any experiment, we either gain some information or do not gain any information.

3. *Nullity:* If the probability of one outcome of information source X is equal to 1 and the remaining probabilities of other outcomes are 0, the entropy of X must be 0. This property means that a certain outcome almost occurs and the other outcomes almost do not occur. This certain outcome is the certain information source.

4. *Continuity:*

$$\lim_{\varepsilon \to 0, \varepsilon_j \to 0, i \neq j} H\{(p_1 - \varepsilon_1), (p_2 - \varepsilon_2), ..., (p_i + \varepsilon), ..., (p_r - \varepsilon_r)\}$$

$$= H(p_1, p_2, ..., p_r),$$

where

$$\varepsilon_j \geq 0 \ (j \neq i), \sum_{j \neq i} \varepsilon_j = \varepsilon, \ 0 \leq (p_j - \varepsilon_j) \leq 1 \ (j \neq i) \text{ and } 0 \leq (p_i + \varepsilon) \leq 1.$$

5. *Expansibility:* If the information space is given as

$$[X' \bullet P] : \begin{cases} X: & a_1 \quad a_2 \quad ... \quad a_i \quad ... \quad a_r \quad a_{r+1} \quad ... \quad a_{r+k} \\ P(X): & p_1 \quad p_2 \quad ... \quad p_i - \varepsilon \quad ... \quad p_r \quad \varepsilon_1 \quad ... \quad \varepsilon_k \end{cases},$$

where

$$0 < \varepsilon_l < 1 \ (l = 1, 2, ..., k), \sum_{l=1}^{k} \varepsilon_l = \varepsilon \text{ and } 0 < (p_i - \varepsilon) < 1,$$

$$\lim_{\substack{\varepsilon \to 0, \varepsilon_l \to 0 \\ (l=1, 2, ..., k)}} H\{p_1, p_2, ..., p_i - \varepsilon, ..., p_r; \varepsilon_1, \varepsilon_2, ..., \varepsilon_r\}$$

$$= H(p_1, p_2, ..., p_i, ..., p_r).$$

This is the extension of the entropy function. It states that addition of an outcome occurring with nearly zero probability does not contribute anything to the entropy (or the amount of information).

6. *Additivity:* Suppose X and Y are two statistical independent information sources. The joint entropy of X and Y is

$$H(XY) = H(X) + H(Y). \tag{3.144}$$

This additive property simply states that the joint entropy is the sum of the entropies of the individual information sources.

7. *Recursivity:* Let an information space of information source X be

$$[X \bullet P]: \begin{cases} X: & a_1 \quad a_2 \quad \cdots \quad a_{k-1} \quad a_k \quad a_{k+1} \quad \cdots \quad a_r \\ P(X): & p_1 \quad p_2 \quad \cdots \quad p_{k-1} \quad p_k \quad p_{k+1} \quad \cdots \quad p_r \end{cases}$$

and an information space of another information source X' be

$$[X' \bullet P]: \begin{cases} X': & a_1 \quad \cdots \quad a_{k-1}; \quad a_1' \quad \cdots \quad a_k'; \quad a_{k+1} \quad \cdots \quad a_r \\ P(X'): & p_1 \quad \cdots \quad p_{k-1}; \quad p_1' \quad \cdots \quad p_k'; \quad p_{k+1} \quad \cdots \quad p_r \end{cases}.$$

We have

$$H(X') = H(X) + p_k H\left(\frac{p_1'}{p_k}, \frac{p_2'}{p_k}, \ldots, \frac{p_k'}{p_k}\right). \tag{3.145}$$

This progress entropy property means that if k of r outcomes are combined, the average amount of information will be lower.

Let X and Y be two independent experiments with r values. Let $P = (p_1, p_2, \ldots, p_r)$ be a probability distribution associated with X and $Q = (q_1, q_2, \ldots, q_r)$ be a probability distribution associated with Y. The entropy has the following two analytical properties.

1. Shannon-Gibbs inequality and maximality: Inequality

$$-\sum_{i=1}^r p_i \log p_i = H(p_1, p_2, \ldots, p_r) \le -\sum_{i=1}^r p_i \log q_i \tag{3.146}$$

has an equality iff $p_i = q_i$ for all i. The entropy $H(p_2, p_1, \ldots, p_r)$ is maximum when all the probabilities are equal,

$$H(p_1, p_2, \ldots, p_r) \le H\left(\frac{1}{r}, \frac{1}{r}, \ldots, \frac{1}{r}\right) = \log r. \tag{3.147}$$

If the average of the outcomes satisfies the condition

$$\sum_{i=1}^r a_i p_i = m, \tag{3.148}$$

the maximum value of the entropy is expressed as

$$H\left(p_1, p_2, \ldots, p_r; m\right) = -m\lambda_2 + \log\left(\sum_{j=1}^{r} 2^{\lambda_2 a_j}\right), \tag{3.149}$$

where λ_2 is obtained by solving

$$\sum_{i=1}^{r} a_i 2^{\lambda_2 a_i} = m \sum_{i=1}^{r} 2^{\lambda_2 a_i}. \tag{3.150}$$

This is the maximum entropy of the average bound.

2. For $0 < \alpha < 1$, we have

$$H\left[\alpha P + \left(1 - \alpha\right) Q\right] \geq \alpha H\left(P\right) + \left(1 - \alpha\right) H\left(Q\right). \tag{3.151}$$

This property states that the average of the entropies of X and Y is not greater than the entropy of the average of X and Y.

3.8   SUMMARY

Because of the complex nature of uncertainties in spatial data such as those connected with imprecision, ambiguity, and vagueness, a number of mathematical theories have been identified and adopted for modeling uncertainties in spatial data and spatial analyses. These theories are described briefly in this chapter, including probability theory, statistics, information theory and entropy, rough set, evidence theory, fuzzy set, and fuzzy topology. For instance, probability theory and statistics have been applied for handling imprecision uncertainty, such as modeling position uncertainty in spatial data (as described in Chapter 4). Fuzzy set theories provide a natural framework for handling vagueness uncertainty. Fuzzy topology has been developed for modeling uncertain relationships between spatial objects (as described in Chapter 7).

REFERENCES

Cressie, N., 1993. *Statistics for Spatial Data*. New York: John Wiley.

Hartley, R. V. L., 1928. Transmission of information. *Bell System Tech. J.* 7: 463–535.

Jaynes, E. T., 2003. *Probability Theory: The Logic of Science*. Cambridge; New York: Cambridge University Press.

Jiang, D., 2001. *Information Theory and Coding*. Hefei, Anhui: University of Science and Technology of China Press.

Klir, G. J. and Folger, T. A., 1988. *Fuzzy Sets, Uuncertainty, and Information.* Singapore: Prentice Hall International. 355.

Kramosil, I., 2001. *Probabilistic Analysis of Belief Functions.* New York: Kluwer Academic/ Plenum Publishers.

Shannon, C. E., 1948. The mathematical theory of communication. *Bell System Tech. J.* 27:379–423, 623–656.

Yager, R. R., Fedrizzi, M., and Kacprzyk, J., 1994. *Advances in the Dempster-Shafer Theory of Evidence.* New York: John Wiley.

Yeung, R. W., 2002. *A First Course in Information Theory.* New York: Kluwer Academic/ Plenum Publishers.

Zadeh, L. A. 1965. Fuzzy sets. *Information and Control.* 8(3): 338–353.

Section II

Modeling Uncertainties in Spatial Data

4 Modeling Positional Uncertainty in Spatial Data

Section II of this book, including Chapters 4 to 7, addresses uncertainty modeling for spatial data. Spatial features in GISci are classified as points, line segments, polylines, curves, and area objects. The methods for modeling the positional uncertainties of these features are introduced in this chapter. These methods are a further development of currently used methods for point positional error modeling, the latter being the focus of error modeling in the field of surveying data processing. The new methods are motivated by the need to model spatial objects in GISci. Hence, in addition to points, modeling of polylines, curves, and area objects are included. The theory of positional uncertainty modeling, described in this chapter, is based on statistics and probability theory.

4.1 INTRODUCTION

A geographic object on the Earth's surface is described as a spatial feature and further identified in GISci by its position, its attributes associated with the position, and relationships between this object and other surrounding objects. The position of a geographic object can be represented either by an object-based model, by a field-based model, or by an analytical function with parameters. Positional uncertainty modeling for spatial features in an object-based model is addressed in this chapter. The object-based model in which the coordinates of points are used to describe the locations of a spatial feature is, currently, the most commonly used model in GISci.

Positional uncertainty (or error) can be determined by the differences between the measured position of the spatial feature and its "true" value. Because of the comparatively low level of cognition of the real world and limitation of the measurement technologies, positional uncertainty exists in spatial features in GISci. Such uncertainties affect the reliability of the results in GISci, applications. The quality of the spatial data used similarly affects the quality of decision making.

In geometric terms, geographic objects in the real world can be classified into point objects (such as a lamppost or a tree), linear objects (such as road network or coastlines), and area objects (such as a land parcel or a water body). In a two-dimensional GIS, these objects are represented as point features, a curved feature, a polyline feature (with line segments as the basic element), and an area feature.

Positional uncertainty models for points, line segments, polylines, curve lines, and areas, therefore, are the basic models for spatial feature positional uncertainty modeling in GISci. These uncertainty models are introduced and described below.

The positional uncertainty model for a point has been extensively studied in the fields of geodesy and surveying data processing. Point positional uncertainty is usually assumed to be distributed in the proximity of an error ellipse centered at the "true" location of that point. Positional uncertainty models for line segment studies are mainly researched by using the error propagation law, statistical theories, and analytical approaches. These commonly used methods have been further extended to include the development of models for assessing positional uncertainty for a curved line. Positional uncertainty for a polygon can be modeled based on either the positional error of the component vertices or the component line segments. Polygon uncertainty is indicated by parameters such as area, perimeter, or gravity.

The details of modeling positional uncertainty of spatial features in GISci is organized as follows.

A brief review of the existing error models for spatial features (spatial features are defined in Section 4.3) is given in Section 4.2. Positional uncertainty models for points are introduced in Section 4.4. Uncertainty models for polylines with the basic element of line segments are introduced in Section 4.5. A solution for the correlated uncertainty of the line segment is provided by a G-band model in Section 4.7. In addition to describing polylines, curved lines also describe. The positional uncertainty model of curves is presented in Section 4.8. Positional uncertainty models for polygons are introduced in Section 4.9. The above uncertainty models are mainly derived based on statistics and probability theory.

4.2   A REVIEW OF EXISTING ERROR MODELS

Several error models for spatial features have been proposed in the literature for describing the positional error of spatial features due to random error sources in spatial data. These error models can be classified into three categories: (a) positional error models for points on a line segment, (b) positional error distribution models for spatial features, and (c) error-band models for line features.

4.2.1   Positional Error for a Point on a Line Segment

A line segment is composed of an infinite number of implicit points between any two locations, those locations being known as explicit points. All intermediate points can be derived from a linear function of the two known explicit points. For example, let the two known explicit points be $A = (x_A, y_A)$ and $B = (x_B, y_B)$. Any implicit point between endpoints A and B of a line segment is given as

$$(x, y) = (1 - r) \cdot (x_A, y_A) + r \cdot (x_B, y_B) = f(x_A, y_A, x_B, y_B, r), \qquad (4.1)$$

where $r \in [0, 1]$. By varying the value of r, intermediate points on the line segment are obtained.

Measurement of the variables x_A, y_A, x_B, y_B may contain positional error. However, a question arises: How can the positional error in the measurement be propagated into the result, through the linear function in Equation 4.1? To answer this question, the positional error at each intermediate point on the line segment is derived by using the statistics error propagation law.

Based on the error propagation law, the error at the middle point of the line segment is smaller than that at the two endpoints, given that positional errors of the endpoints are identical and independently distributed.

4.2.2 Positional Error Distribution of Spatial Features

Positional error distribution of a spatial feature depends on data capture methods. The most common data-entry method is map digitization. Map digitization error distribution has been investigated by researchers including Bolstad et al. (1990). It was found that the digitizing error was approximately normally distributed. Meng et al. (1998) and Tong et al. (2000) found that in some circumstances, the digitizing error was approximate to the NL distribution and the least P-norm distribution, respectively. However, the normal distribution is still widely used to describe the point positional error, since it is relatively easier for follow-up processing, such as auto-correlation and cross-correlation modeling errors at the vertices of spatial features

Under the assumption of normally distributed point positional errors, the positional error distribution of a line or a polygon has been studied using mainly two approaches: simulation and analytical. Dutton (1992) modeled the error distribution of a line segment using a simulation approach. This model considered that the true location a line segment, is inside an error region (or error ellipse in general) centered at the middle of the line. If there is only one observation, the mean location will be equivalent to the observation. The simulation of the endpoints, based on this assumption, will obtain a series of line segments. This could outline the line segment error distribution.

Another method to derive the error distribution is based on an analytical approach (Shi, 1994, 1998). The early work on positional error of a line from one- to n-dimensional space was carried out under the assumption that the positional error at each endpoint of the line was independently normally distributed. Later, Shi and Liu (2000) introduced a model to study the positional error distribution of a line, based on a stochastic process, where the error of the line segment composite points was correlated.

4.2.3 Error-Band Model for Line Features

An error-band model is a band around the true location of a spatial feature. The first error-band model was proposed by Perkal (1966). Perkal's ε-band can be created by rolling a circle along the true location of a line segment or a polygon's boundary. The positional error for the line or the polygon's boundary is then expressed as the area of the ε-band.

Chrisman (1982) further developed this theoretical ε-band model for practical applications. He applied ε-band measurements to model positional error in a GIRAS digital file of the U.S. Geological Surveys land-use and cover data. For practical purposes, the bandwidth was determined by a statistical function of those positional errors on the line. The errors accumulated from the first stage to the final stage of spatial data capture (Chapman et al., 1997). However, this model is sensitive to any outlier of a line, because in this model, the true location of a line is supposed to be definitely located inside the error band.

An alternative approach to determine the bandwidth was suggested by Goodchild and Hunter (1997). The bandwidth was estimated by calculating the proportion of a measured location of the line being within the ε-band, such that the proportion must be equal to, or larger than, the predefined tolerance (say, 0.95). This modified ε-band is a more appropriate error-band model than the previous version in terms of the outlier sensitivity.

In addition to the above limitations, the existing error-band models should be further developed by taking the following cases into consideration: the endpoints of a line may not have the same error distribution. Shi (1994) proposed a confidence region error model. In this error model, the positional error at the endpoints of a line may follow normal distributions (Shi, 1997, 1998). This confidence region is a band around a measured location of the line, within which the true location of the line is located, with a probability larger than a predefined confidence level. The confidence region error model is described in detail in Section 4.5.

The error bands, mentioned above, provide a feasible region that encloses the true location of a spatial feature. The width of each error band is subject to positional uncertainties in the vertices of the spatial feature and a probability of the true location of the spatial feature outside the error band.

To obtain an error band invariant, Fan and Guo (2001), Li (2003) proposed an entropy error band (H-band) for a line, using Shannon information theory. The bandwidth of the H-band is determined by the entropy coefficient, being a ratio of error entropy, to the standard deviation of a point on the line, or by the average error entropy.

4.3   DEFINITION OF SPATIAL FEATURES IN GEOGRAPHIC INFORMATION SCIENCE

Spatial features in GISci can have different dimensions. A one-dimensional spatial feature is a profile of a two-dimensional spatial feature; this may be a road section for example. Spatial features in a two-dimensional space are commonly used in the GISci community: most geographic objects in GIS are stored and presented in a two-dimensional space. To describe geographic objects in the real world, a three-dimensional system is required. These three-dimensional spatial features contain horizontal (x- and y-coordinates) and elevation information (z values) about geographic objects. In addition to the spatial dimensions (including the x-, y-, and/or z-coordinates), time can be of interest. In other words, both space and time factors of a spatial object may be acquired, maintained, analyzed, and presented. This leads to spatio-temporal GISci which thus deals with four-dimensional spatial features. As a generalization, spatial features can be represented in an n-dimensional space.

Consequently, error models for one-, two-, three- and n-dimensional spatial features are required.

4.3.1 One-Dimensional Spatial Features

A point in one-dimensional space is defined as a one-dimensional stochastic vector: $Q_{11} = [x]$. This vector may follow a univariate normal distribution (N_1) as given in Equation 4.2.

$$Q_{11} \sim N_1\left[\mu_x, \sigma_x^2\right], \tag{4.2}$$

where μ_x and σ_x^2 are the parameters to define the statistical properties of the point (i.e., its mean and variance). Coordinates in a GIS are thus derived from a probability distribution, where it is assumed that the expectations are equal to the population values.

Joining two points forms a line segment, called $Q_{11}Q_{12}$, in one-dimension. This line segment is derived from two one-dimensional stochastic vectors $Q_{11} = [x_1]$ and $Q_{12} = [x_2]$. Let $\Sigma_{Q_{11}Q_{12}}$ denote the covariance matrix of the two vertices, that is,

$$\Sigma_{Q_{11}Q_{12}} = \begin{bmatrix} \sigma_{x_1}^2 & \sigma_{x_1 x_2} \\ \sigma_{x_2 x_1} & \sigma_{x_2}^2 \end{bmatrix}. \tag{4.3}$$

Any point on the line segment $Q_{11}Q_{12}$ is then represented by

$$Q_{1r} = (1-r)Q_{11} + rQ_{12} = (1-r)\mu_{x_1} + r\mu_{x_2}, \, r \in [0, 1], \tag{4.4}$$

where μ_{x1} and μ_{x2} are the expectation of x_1 and x_2.

Joining several one-dimensional line segments forms a polyline. This polyline is derived from one-dimensional stochastic vectors of its composite points: $Q_{11} = [x_1]$, $Q_{12} = [x_2]$, ..., $Q_{1i} = [x_i]$, ..., $Q_{1m} = [x_m]$, where $i \in [1, m]$ and m is the number of composition points. These stochastic vectors are used to represent the polyline with the line segments $Q_{11}Q_{12}$, $Q_{12}Q_{13}$, ..., $Q_{1(m-1)}Q_{1m}$.

Let $\Sigma_{Q_{1i}Q_{1,i+1}}$ denote the covariance matrix of two succeeding vertices, that is,

$$\Sigma_{Q_{1i}Q_{1,i+1}} = \begin{bmatrix} \sigma_{x_i}^2 & \sigma_{x_i x_{i+1}} \\ \sigma_{x_{i+1} x_i} & \sigma_{x_{i+1}}^2 \end{bmatrix} (i = 1, 2, \cdots, m-1). \tag{4.5}$$

Any point on the line segment $Q_{1i}Q_{1,i+1}$ is then represented by

$$Q_{1ir} = (1-r)Q_{1i} + rQ_{1,i+1} = (1-r)\mu_{x_i} + r\mu_{x_{i+1}} \, (i = 1, 2, \cdots, m-1), \, r \in [0, 1], \tag{4.6}$$

where μ_{x_i} and $\mu_{x_{i+1}}$ are the expectation of x_i and x_{i+1}, respectively.

4.3.2 TWO-DIMENSIONAL SPATIAL FEATURES

A point in a two-dimensional space is presented as a two-dimensional stochastic vector $Q_{21} = [x, y]^T$. It is supposed that this stochastic vector follows a bivariate normal distribution (N_2). This implies that

$$Q_{21} = \begin{bmatrix} x \\ y \end{bmatrix} \sim N_2 \left(\begin{bmatrix} \mu_x \\ \mu_y \end{bmatrix}, \begin{bmatrix} \sigma_x^2 & \sigma_{xy} \\ \sigma_{yx} & \sigma_y^2 \end{bmatrix} \right), \tag{4.7}$$

where μ_x and μ_y are the expectation of x and y; and σ_x^2, σ_y^2 are variance of x and y, and σ_{xy} and σ_{yx} are covariance coefficients of x and y, and y and x, respectively.

A line segment in two-dimensional space is defined by two stochastic vectors of its endpoints: $Q_{21} = [x_1, y_1]^T$ and $Q_{22} = [x_2, y_2]^T$. Let $\Sigma_{Q_{21}Q_{22}}$ denote the covariance matrix of the two vertices, i.e.,

$$\Sigma_{Q_{21}Q_{22}} = \begin{bmatrix} \sigma_{x_1}^2 & \sigma_{x_1 x_2} & \sigma_{x_1 y_1} & \sigma_{x_1 y_2} \\ \sigma_{x_2 x_1} & \sigma_{x_2}^2 & \sigma_{x_2 y_1} & \sigma_{x_2 y_2} \\ \sigma_{y_1 x_1} & \sigma_{y_1 x_2} & \sigma_{y_1}^2 & \sigma_{y_1 y_2} \\ \sigma_{y_2 x_1} & \sigma_{y_2 x_2} & \sigma_{y_2 y_1} & \sigma_{y_2}^2 \end{bmatrix}. \tag{4.8}$$

Any point on the line segment $Q_{21}Q_{22}$ is then represented by

$$Q_{2r} = (1-r)Q_{21} + rQ_{22} = \begin{bmatrix} (1-r)\mu_{x_1} + r\mu_{x_2} \\ (1-r)\mu_{y_1} + r\mu_{y_2} \end{bmatrix}, r \in [0, 1], \tag{4.9}$$

where μ_{x_1}, μ_{y_1}, μ_{x_2}, and μ_{y_2} are the expectation of x_1 and y_1, x_2 and y_2, respectively.

Joining several two-dimensional line segments forms a polyline. A special case of a polyline with identical beginning and end vertices is a closed polygon (A). This polygon is derived from two-dimensional stochastic vectors of its composite points: $Q_{21} = [x_1, y_1]^T$, $Q_{22} = [x_2, y_2]^T$, ..., $Q_{2i} = [x_i, y_i]^T$, ..., $Q_{2m} = [x_m, y_m]^T$, where $i \in [1, m]$ and m is the number of composition points. These stochastic vectors are used to represent the boundary of polygon A with the line segments $Q_{21}Q_{22}$, $Q_{22}Q_{23}$, ..., $Q_{2(m-1)}Q_{2m}$. Let $\Sigma_{Q_{2i}Q_{2,i+1}}$ denote the covariance matrix of two succeeding vertices, that is,

$$\Sigma_{Q_{2i}Q_{2,i+1}} = \begin{bmatrix} \sigma_{x_i}^2 & \sigma_{x_i x_{i+1}} & \sigma_{x_i y_{i+1}} & \sigma_{x_i y_{i+1}} \\ \sigma_{x_{i+1} x_i} & \sigma_{x_{i+1}}^2 & \sigma_{x_{i+1} y_i} & \sigma_{x_{i+1} y_{i+1}} \\ \sigma_{y_i x_i} & \sigma_{y_i x_{i+1}} & \sigma_{y_i}^2 & \sigma_{y_i y_{i+1}} \\ \sigma_{y_{i+1} x_i} & \sigma_{y_{i+1} x_{i+1}} & \sigma_{y_{i+1} y_i} & \sigma_{y_{i+1}}^2 \end{bmatrix} (i = 1, 2, \cdots, m-1). \tag{4.10}$$

Any point on the line segment $Q_{2i}Q_{2,i+1}$ is then represented by

$$Q_{2ir} = (1-r)Q_{2i} + rQ_{2,i+1} = \begin{bmatrix} (1-r)\mu_{x_i} + r\mu_{x_{i+1}} \\ (1-r)\mu_{y_i} + r\mu_{y_{i+1}} \end{bmatrix} \quad (i = 1, 2, \cdots, m-1) \,, \, r \in [0, 1], \quad (4.11)$$

where μ_{x_i}, μ_{y_i}, $\mu_{x_{i+1}}$, and $\mu_{y_{i+1}}$ are the expectation of x_i and y_i, x_{i+1} and y_{i+1}, respectively.

4.3.3 N-DIMENSIONAL SPATIAL FEATURES

In an n-dimensional space, a point is represented by an n-dimensional stochastic vector: $Q_{n1} = [x_1, x_2, ..., x_i, ..., x_n]^T$, where $i \in [1, n]$. If the error at this point follows a normal distribution, the stochastic vector can be expressed as

$$Q_{n1} \sim N_n \left(\begin{bmatrix} \mu_{x_1} \\ \mu_{x_2} \\ \vdots \\ \mu_{x_n} \end{bmatrix}, \begin{bmatrix} \sigma_{x_1}^2 & \sigma_{x_1 x_2} & \cdots & \sigma_{x_1 x_n} \\ \sigma_{x_2 x_1} & \sigma_{x_2}^2 & \cdots & \sigma_{x_2 x_n} \\ \vdots & \cdots & & \\ \sigma_{x_n x_1} & \sigma_{x_n x_2} & \cdots & \sigma_{x_n}^2 \end{bmatrix} \right), \quad (4.12)$$

where μ_{xi} is the expectation of x_i, and $\sigma_{xixj} = \sigma_{xjxi}$ is the covariance coefficient of x_i and x_j, where $i, j \in [1, n]$.

A line segment in n-dimensional space is defined by two stochastic vectors of its endpoints: $Q_{n1} = [x_{11}, x_{12}, ..., x_{1n}]^T$ and $Q_{n2} = [x_{21}, x_{22}, ..., x_{2n}]^T$. Let Σ_{Qn1Qn2} denote the covariance matrix of two succeeding vertices, that is,

$$\Sigma_{Qn1Qn2} =$$

$$\begin{bmatrix} \sigma_{x_{11}}^2 & \sigma_{x_{11}x_{12}} & \cdots & \sigma_{x_{11}x_{1n}} & \sigma_{x_{11}x_{21}} & \sigma_{x_{11}x_{22}} & \cdots & \sigma_{x_{11}x_{2n}} \\ \sigma_{x_{12}x_{11}} & \sigma_{x_{12}}^2 & \cdots & \sigma_{x_{12}x_{1n}} & \sigma_{x_{12}x_{21}} & \sigma_{x_{12}x_{22}} & \cdots & \sigma_{x_{12}x_{2n}} \\ \vdots & \vdots & \vdots & \vdots & \vdots & \vdots & \vdots & \vdots \\ \sigma_{x_{1n}x_{11}} & \sigma_{x_{1n}x_{12}} & \cdots & \sigma_{x_{1n}}^2 & \sigma_{x_{1n}x_{21}} & \sigma_{x_{1n}x_{22}} & \cdots & \sigma_{x_{1n}x_{2n}} \\ \sigma_{x_{21}x_{11}} & \sigma_{x_{21}x_{12}} & \cdots & \sigma_{x_{21}x_{1n}} & \sigma_{x_{21}}^2 & \sigma_{x_{21}x_{22}} & \cdots & \sigma_{x_{21}x_{2n}} \\ \sigma_{x_{22}x_{11}} & \sigma_{x_{22}x_{12}} & \cdots & \sigma_{x_{22}x_{1n}} & \sigma_{x_{22}x_{21}} & \sigma_{x_{22}}^2 & \cdots & \sigma_{x_{22}x_{2n}} \\ \vdots & \vdots & \cdots & \vdots & \vdots & \vdots & \cdots & \vdots \\ \sigma_{x_{2n}x_{11}} & \sigma_{x_{2n}x_{12}} & \cdots & \sigma_{x_{2n}x_{1n}} & \sigma_{x_{2n}x_{21}} & \sigma_{x_{2n}x_{22}} & \cdots & \sigma_{x_{2n}}^2 \end{bmatrix}. \quad (4.13)$$

Any point on the line segment $Q_{n1}Q_{n2}$ is then represented by

$$Q_{nr} = \begin{bmatrix} x_{1r} \\ x_{2r} \\ \vdots \\ x_{nr} \end{bmatrix} = (1-r)Q_{n1} + rQ_{n2} = \begin{bmatrix} (1-r)\mu_{x_{11}} + r\mu_{x_{21}} \\ (1-r)\mu_{x_{12}} + r\mu_{x_{22}} \\ \vdots \\ (1-r)\mu_{x_{1n}} + r\mu_{x_{2n}} \end{bmatrix} \quad r \in [0,1], \quad (4.14)$$

where $\mu_{x_{11}}, \mu_{x_{12}}, \cdots \mu_{x_{1n}}, \mu_{x_{21}}, \mu_{x_{22}}, \cdots \mu_{x_{2n}}$, are the expectation of $x_{11}, x_{12}, \cdots x_{1n}, x_{21}, x_{22}, \cdots x_{2n}$, respectively.

Joining several n-dimensional line segments forms an n-dimensional polyline. A special case of a polyline, with identical beginning and end vertices is a closed polygon (A). This polygon is derived from n-dimensional stochastic vectors of its composite points: $Q_{n1} = [x_{11}, x_{12}, \ldots x_{1n}]^T$, $Q_{n2} = [x_{21}, x_{22}, \ldots, x_{2n}]^T$, $\ldots$, $Q_{ni} = [x_{i1}, x_{i2}, \ldots x_{in}]^T$, $\ldots$, $Q_{nm} = [x_{m1}, x_{m2}, \ldots x_{mn}]^T$, where $i \in [1, m]$ and m is the number of composition points. These stochastic vectors are used to represent the boundary of polygon A with the n-dimensional line segments $Q_{n1}Q_{n2}$, $Q_{n2}Q_{n3}$, $\ldots$, $Q_{n(m-1)}Q_{nm}$. Let $\Sigma_{Q_{ni}Q_{n,i+1}}$ denote the covariance matrix of two succeeding vertices, where $i = 1, 2, \ldots, m-1$, that is,

$$\Sigma_{Q_{ni}Q_{n,i+1}} = \tag{4.15}$$

$$
\begin{bmatrix}
\sigma^2_{x_{i1}} & \sigma_{x_{i1}x_{i2}} & \cdots & \sigma_{x_{i1}x_{in}} & \sigma_{x_{i1}x_{i+1,1}} & \sigma_{x_{i1}x_{i+1,2}} & \cdots & \sigma_{x_{i1}x_{i+1,n}} \\
\sigma_{x_{i2}x_{i1}} & \sigma^2_{x_{i2}} & \cdots & \sigma_{x_{i2}x_{in}} & \sigma_{x_{i2}x_{i+1,1}} & \sigma_{x_{i2}x_{i+1,2}} & \cdots & \sigma_{x_{i2}x_{i+1,n}} \\
\vdots & \vdots & \vdots & \vdots & \vdots & \vdots & \vdots & \vdots \\
\sigma_{x_{in}x_{i1}} & \sigma_{x_{in}x_{i2}} & \cdots & \sigma^2_{x_{in}} & \sigma_{x_{in}x_{i+1,1}} & \sigma_{x_{in}x_{i+1,2}} & \cdots & \sigma_{x_{in}x_{i+1,n}} \\
\sigma_{x_{i+1,1}x_{i1}} & \sigma_{x_{i+1,1}x_{i2}} & \cdots & \sigma_{x_{i+1,1}x_{in}} & \sigma^2_{x_{i+1,1}} & \sigma_{x_{i+1,1}x_{i+1,2}} & \cdots & \sigma_{x_{i+1,1}x_{i+1,n}} \\
\sigma_{x_{i+1,2}x_{i1}} & \sigma_{x_{i+1,2}x_{i2}} & \cdots & \sigma_{x_{i+1,2}x_{in}} & \sigma_{x_{i+1,2}x_{i+1,1}} & \sigma^2_{x_{i+1,2}} & \cdots & \sigma_{x_{i+1,2}x_{i+1,n}} \\
\vdots & \vdots & \cdots & \vdots & \vdots & \vdots & \cdots & \vdots \\
\sigma_{x_{i+1,n}x_{i1}} & \sigma_{x_{i+1,n}x_{i2}} & \cdots & \sigma_{x_{i+1,n}x_{in}} & \sigma_{x_{i+1,n}x_{i+1,1}} & \sigma_{x_{i+1,n}x_{i+1,2}} & \cdots & \sigma^2_{x_{i+1,n}}
\end{bmatrix}.
$$

Any point on the line segment $Q_{ni}Q_{n,i+1}$ is then represented by

$$
Q_{nir} = \begin{bmatrix} x_{i1r} \\ x_{i2r} \\ \vdots \\ x_{inr} \end{bmatrix} = (1-r)Q_{ni} + rQ_{n,i+1}
$$

$$
= \begin{bmatrix} (1-r)\mu_{x_{i1}} + r\mu_{x_{i+1,1}} \\ (1-r)\mu_{x_{i2}} + r\mu_{x_{i+1,2}} \\ \vdots \\ (1-r)\mu_{x_{in}} + r\mu_{x_{i+1,n}} \end{bmatrix} \quad (i = 1, 2, \cdots, m-1),\ r \in [0,1], \tag{4.16}
$$

where $\mu_{x_{i1}}, \mu_{x_{i2}}, \cdots \mu_{x_{in}}, \mu_{x_{i+1,1}}, \mu_{x_{i+1,2}} \cdots \mu_{x_{i+1,n}}$, are the expectation of $x_{i1}, x_{i2}, \cdots x_{in}, x_{i+1,1}, x_{i+1,2} \cdots x_{i+1,n}$, respectively.

4.4 MODELING POSITIONAL UNCERTAINTY FOR A POINT

A point feature is one of the fundamental elements of geographic objects. The point feature is also a composite of a line or a polygon in GISci. Positional error modeling of the point feature is thus a basis for positional error modeling of other spatial features in GISci.

Positional error at a point can be described by a possibility region. The region may be in the form of an error ellipse or an error rectangle. An error ellipse is commonly used to describe the error at the point. Let Q_{21} denote a point with the x-directional variance of σ_x^2, the y-directional variance of σ_y^2, and the covariance of σ_{xy}. The correlation of this random point in the x- and y-directions is then denoted by ρ_{xy}.

If the error at point Q_{21} follows a normal distribution, its probability density function is given as

$$f(x,y) = \exp\left\{-\left[(x-\mu_x)^2/\sigma_x^2 + (y-\mu_y)^2/\sigma_y^2 - 2\rho_{xy}(x-\mu_x)(y-\mu_y)/(\sigma_x\sigma_y)\right]\right.$$
$$\left./2(1-\rho_{xy}^2)\right\}/\left(2\pi\sigma_x\sigma_y\sqrt{1-\rho_{xy}^2}\right). \tag{4.17}$$

The major and minor semiaxes (E and F) as well as the orientation θ of the corresponding error ellipse are

$$E = \frac{\sqrt{\sigma_x^2+\sigma_y^2+\left[\left(\sigma_x^2-\sigma_y^2\right)^2+4\sigma_{xy}^2\right]}}{2},$$

$$F = \frac{\sqrt{\sigma_x^2+\sigma_y^2-\left[\left(\sigma_x^2-\sigma_y^2\right)^2+4\sigma_{xy}^2\right]}}{2}, \tag{4.18}$$

$$\theta = \frac{1}{2}\arctan\left(\frac{2\sigma_{xy}}{\sigma_x^2-\sigma_y^2}\right).$$

The probability P that the point is inside the error ellipse can then be indicated by the volume of the two-dimensional error curved surface over the error ellipse (as the case shown in Figure 4.1):

$$P(x,y \subset \Omega) = \iint_\Omega f(x,y)\,dxdy = 4\int_0^r \int_0^{\frac{\pi}{2}} f(r,\theta)\,drd\theta = 0.393. \tag{4.19}$$

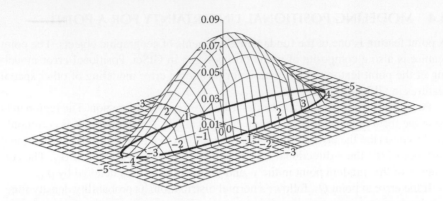

FIGURE 4.1   Two-dimensional error curved surface over the error ellipse.

This probability increases with the size of the error ellipse. For example, the error ellipse in double size will yield a probability equal to 0.893. When the shape of the error ellipse triples, the probability of the point inside the error ellipse is 0.989.

Rectangles can also be considered as possible regions to describe positional errors at point Q_{21}, when the correlation ρ_{xy}, within the point, is zero. First, the coordinate system is translated and rotated such that the expected location of point Q_{21} in the new coordinate system is the origin, and the two semi-axes of the error ellipse are the x- and y-axes. Suppose that the coordinate of Q_{21} in the new coordinate system is (x', y'), probability density function in Equation 4.17 is then expressed as

$$f(x', y') = f_u(x') f_v(y')$$

$$= \left\{ \exp\left[-x'^2/(2E^2)\right] / \left(\sqrt{2\pi}E\right) \right\} \left\{ \exp\left[-y'^2/(2F^2)\right] / \left(\sqrt{2\pi}F\right) \right\} \quad (4.20)$$

$$= \exp\left[-x'^2/(2E^2) - y'^2/(2F^2)\right] / (2\pi EF).$$

The probability that point Q_{21} is within the intervals $-E < x' < E$ and $-F < y' < F$ is equal to the volume of the two-dimensional error curved surface over the rectangles circumscribed around the error ellipse as shown in Figure 4.2. It is given mathematically by

$$P\left(-E < x' < E \text{ and } -F < y' < F\right)$$

$$= 4 \int_{u=0}^{E} \int_{v=0}^{F} \exp\left[-x'^2/(2E^2) - y'^2/(2F^2)\right] / (2\pi EF) \, dx' dy' \quad (4.21)$$

$$= 0.683 \times 0.683$$

$$= 0.466.$$

The probabilities that the point is within the doubled and tripled rectangles are computed in a similar way:

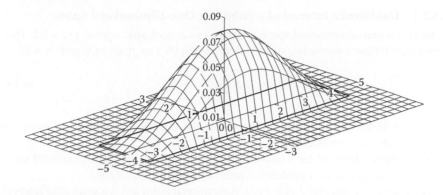

FIGURE 4.2 Two-dimensional error curved surface over the rectangle circumscribed around the error ellipse.

$$P\left(-2E < x' < 2E \text{ and } -2F < y' < 2F\right) = 0.954^2 = 0.9111,$$

$$P\left(-3E < x' < 3E \text{ and } -3F < y' < 3F\right) = 0.997^2 = 0.9946. \tag{4.22}$$

The above results show that the probability of a point within a possibility region varies with the approach used to define that region. The probability that the point is within the standard error ellipse equals 0.393, whereas the probability that it is within the rectangle conscribed around the error ellipse equals 0.466. These two fixed values of the probability are widely used to describe point positional errors.

4.5 CONFIDENCE REGION MODELS

A polyline is composed of a number of line segments. With the assumption of uncorrelated positional errors for the endpoints (or vertices) of a line, two types of error models are proposed for the polyline: (a) confidence region models and (b) positional error distribution models. A confidence region for a two-dimensional line was first introduced Shi (1994) and was further developed to assess the positional error for a line from a one-dimensional to an *n*-dimensional geometrical space (Shi, 1998). In these models, the confidence region is derived based on the independent positional error of the composite points and a predefined confidence level. Its shape is different from an ε-model, which assumes identical positional errors for any point of a line segment. The confidence region model considers that positional errors for all points on the line can be dissimilar.

4.5.1 CONFIDENCE INTERVAL MODEL FOR SPATIAL FEATURES IN ONE-DIMENSIONAL SPACE

A confidence interval for one-dimensional features includes the confidence interval of a point, the confidence interval of a line segment, and the confidence interval of a line. These confidence intervals are as follows.

4.5.1.1   Confidence Interval of a Point in a One-Dimensional Space

A point in a one-dimensional space is defined as a stochastic vector: $Q_{11} = [x]$. This vector may follow a univariate normal distribution (N_1) as given in Equation 4.23.

$$Q_{11} \sim N_1\left[\mu_x, \sigma_x^2\right], \tag{4.23}$$

where μ_x and σ_x^2 are the parameters to define the statistical properties of the point (i.e., its mean and variance).

A confidence interval for a one-dimensional point describes the positional error for a measured point. This confidence interval is derived as follows.

The confidence interval J_{11} for a one-dimensional point Q_{11} is a stochastic interval around a measured location of the point (X). This stochastic interval contains the true location of the point with a probability larger than or equal to a predefined confidence level α. This is expressed as:

$$P(Q_{11} \in J_{11}) > \alpha. \tag{4.24}$$

J_{11} is a set of points $\{x\}$ that satisfy

$$X - a_{11} \leq x \leq X + a_{11}, \tag{4.25}$$

where $a_{11} = k^{1/2}\sigma_x$, and $k = \chi^2_{1:\alpha}$.

The parameter k depends upon the predefined confidence level α, and can be found in a chi-square distribution table. When the confidence level α is equal to 0.97, for example, the chi-square parameter k is equal to 4.789.

The corresponding confidence interval is shown in Figure 4.3a, where the confidence interval for the measured point Q_{11} is indicated by the interval J_{11}; this confidence interval contains the true location $\varnothing_{11}$ of the point, with the probability larger than the 97% confidence level.

4.5.1.2   Confidence Interval of a Line Segment in a One-Dimensional Space

Joining two points forms a line segment, called $Q_{11}Q_{12}$, in one-dimension. This line segment is derived from two one-dimensional stochastic vectors $Q_{11} = [x_1]$ and $Q_{12} = [x_2]$. It is supposed that these stochastic vectors follow univariate normal distributions

$$Q_{11} = \left[x_1\right] \sim N_1\left[\mu_{x_1}, \sigma_x^2\right],$$

$$Q_{12} = \left[x_2\right] \sim N_1\left[\mu_{x_2}, \sigma_x^2\right], \tag{4.26}$$

where μ_{x_1}, μ_{x_2}, σ_x^2 are the parameters to define the statistical properties of the point; μ_{x_1} and μ_{x_2} are the expectation of x_1 and x_2.

The confidence interval J_1 for the one-dimensional line segment $Q_{11}Q_{12}$ is a stochastic interval, such that all points on the true location $\varnothing_{11}\varnothing_{12}$ of the line segment

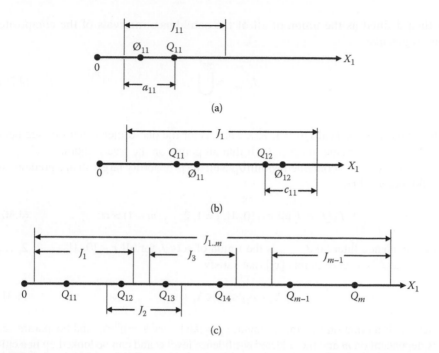

(a)

(b)

(c)

FIGURE 4.3 (a) The confidence interval for a one-dimensional point, (b) the confidence interval for a one-dimensional line segment, and (c) the confidence interval for a one-dimensional polyline. (After Shi, W. Z., 1998. *Int. J. Geog. Inform. Sci.* 12: 131–143. With permission.)

are contained within J_r with the probability larger than a predefined confidence level α:

$$P(Q_{1r} \in J_{1r}, \text{all } r \in [0, 1]) > \alpha. \tag{4.27}$$

The confidence interval J_1 is the union of sets J_r for all $r \in [0, 1]$, and J_r is a set of points $[x_r]$, which satisfy

$$X_r - c_{11} \le x_r \le X_r + c_{11}, \tag{4.28}$$

where X_r *is a point on the line segment*, $c_{11} = [k((1 - r)^2 + r^2)]^{1/2}\sigma_x$, and $k = \chi^2_{2;\alpha}$.

If the confidence level α equals 0.97, the chi-square parameter k is equal to 7.174. Figure 4.3b shows the confidence interval for the measured location of the line segment $Q_{11}Q_{12}$ indicated by the interval J_1. This confidence interval contains the true location of the line segment with the probability larger than the 97% confidence level.

4.5.1.3 Confidence Interval of a One-Dimensional Polyline

In order to create a confidence interval for a one-dimensional polyline, a polyline (Q_1Q_m) is supposed to be composed of a sequence of line segments: Q_1Q_2, Q_2Q_3, ..., $Q_{m-1}Q_m$; the confidence intervals for these line segments are denoted by J_1, J_2, ..., J_{m-1}, respectively. The confidence interval for the one-dimensional polyline is then defined as the union of all of the confidence intervals of the composite line segments:

$$J_{1...m-1} = \bigcup_{i=1}^{m-1} J_i. \tag{4.29}$$

The confidence interval J_i ($i = 1, 2, ..., m - 1$) for the one-dimensional line segment $Q_{1i}Q_{1,i+1}$ is a stochastic interval, such that all points on the true location $\varnothing_{1i}\varnothing_{1,i+1}$ of the line segment are contained within J_{ir} with the probability larger than a predefined confidence level α:

$$P(Q_{ir} \in J_i \text{ all } r \in [0, 1], i = 1, 2, ..., m - 1) > \alpha. \tag{4.30}$$

The confidence interval $J_{1...m-1}$ is the union of sets J_i for all $r \in [0, 1]$, $i = 1, 2, ...,$ $m - 1$, and J_i is a set of points $[x_{ir}]$ that satisfy

$$X_{ir} - c_{11} \leq x_{ir} \leq X_{ir} + c_{11}, \tag{4.31}$$

where X_{ir} is a point on the line segment, $c_{11} = [k((1 - r)^2 + r^2)]^{1/2}\sigma_x$, and the parameter k is dependent on m and the selected confidence level α and can be looked up in a chi-square distribution table, $k = \chi^2_{2;1-(m-1)(1-\alpha)}$. Figure 4.3c shows the confidence interval $J_{1...m-1}$ for the polyline, which is constructed by the union of the confidence intervals for the composite line segments. The confidence interval $J_{1...m-1}$ contains the true location of the line with the probability larger than a predefined confidence level.

The confidence interval for a one-dimensional point can be used to describe its positional error. A confidence interval for a one-dimensional line can be applied to describe, for example, the positional error in a part of a road section that is determined by two road section points in a one-dimensional geometrical space. A confidence interval for a line can thus be used to describe the positional error in a road. All three descriptions give the intervals within which the true locations of points, line segments, and lines are contained with the probability larger than a predefined confidence level.

4.5.2 Confidence Region Model for Spatial Features in Two-Dimensional Space

4.5.2.1 Confidence Region Model for a Point in Two-Dimensional Space

A point in a two-dimensional space is represented as a two-dimensional stochastic vector $Q_{21} = [x, y]^T$. It is supposed that this stochastic vector follows a bivariate normal distribution (N_2). This implies that

$$Q_{21} = \begin{bmatrix} x \\ y \end{bmatrix} \sim N_2 \left(\begin{bmatrix} \mu_x \\ \mu_y \end{bmatrix}, \begin{bmatrix} \sigma_x^2 & \sigma_{xy} \\ \sigma_{yx} & \sigma_y^2 \end{bmatrix} \right), \tag{4.32}$$

where μ_x and μ_y are the expectation of x and y; σ_x^2, σ_y^2 are a variance of x and y; and σ_{xy} and σ_{yx} ($\sigma_{xy} = \sigma_{yx}$) are covariance coefficients of x and y and that of y and x, respectively.

The confidence region J_{21} for the measured two-dimensional point Q_{21} is defined as a region within which the true location $\varnothing_{21}$ of the point is contained with a probability larger than a predefined confidence level α:

$$P(Q_{21} \in J_{21}) > \alpha. \tag{4.33}$$

The confidence region J_{21} is determined by a set of points $(x,y)^T$ with x and y satisfying

$$X - a_{21} \le x \le X + a_{21}, \tag{4.34}$$

$$Y - a_{22} \le y \le Y + a_{22}, \tag{4.35}$$

where $a_{21} = k^{1/2}\sigma_x$, $a_{22} = k^{1/2}\sigma_y$, and $k = \chi^2_{1;(1+\alpha)/2}$.

If the confidence level α is 0.97, the chi-square parameter k is 6.024. The corresponding confidence region is then illustrated in Figure 4.4a, where Q_{21} and $\varnothing_{21}$ represent a measured location and the true location of a two-dimensional point, respectively; the confidence region for the measured location of the point is a rectangle of size $2x$ by $2y$. The rectangle contains the true location of the point with the probability larger than the predefined confidence level, 97%.

4.5.2.2 Confidence Region Model for a Line Segment in Two-Dimensional Space

A line segment in two-dimensional space is defined by two stochastic vectors of its endpoints: $Q_{21} = [x_1, y_1]^T$ and $Q_{22} = [x_2, y_2]^T$. Bivariate normally distributed stochastic vectors are given as

$$Q_{21} = \begin{bmatrix} x_1 \\ y_1 \end{bmatrix} \sim N_2\left(\begin{bmatrix} \mu_{x_1} \\ \mu_{y_1} \end{bmatrix}, \begin{bmatrix} \sigma_x^2 & \sigma_{xy} \\ \sigma_{yx} & \sigma_y^2 \end{bmatrix} \right),$$

$$Q_{22} = \begin{bmatrix} x_2 \\ y_2 \end{bmatrix} \sim N_2\left(\begin{bmatrix} \mu_{x_2} \\ \mu_{y_2} \end{bmatrix}, \begin{bmatrix} \sigma_x^2 & \sigma_{xy} \\ \sigma_{yx} & \sigma_y^2 \end{bmatrix} \right), \tag{4.36}$$

where μ_{x_1}, μ_{y_1}, μ_{x_2}, and μ_{y_2} are the expectation of x_1 and y_1, x_2 and y_2, respectively; σ_x^2, σ_y^2 are a variance of x_1 and y_1, σ_{xy} and σ_{yx} ($\sigma_{xy} = \sigma_{yx}$) are covariance coefficients of x_1 and y_1 and that of y_1 and x_1, respectively; σ_x^2 and σ_y^2 are also a variance of x_2 and y_2; σ_{xy} and σ_{yx} ($\sigma_{xy} = \sigma_{yx}$) are also covariance coefficients of x_2 and y_2, and y_2 and x_2, respectively.

A confidence region for a line segment in two-dimensional space is defined as a region containing the true location of the line segment with a predefined confidence

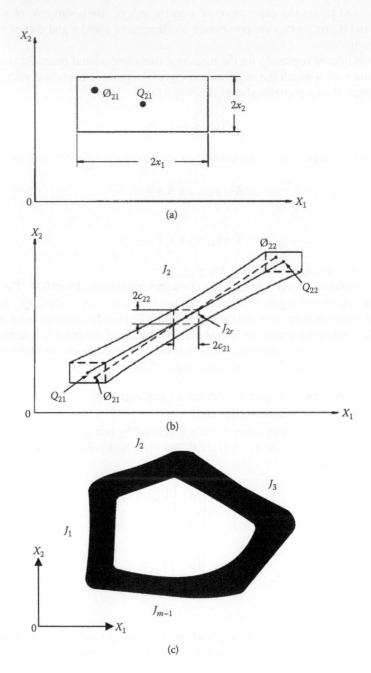

FIGURE 4.4 The confidence regions of (a) a point, (b) a line segment, and (c) a polyline in two-dimensional space. (After Shi, W. Z., 1998. *Int. J. Geog. Inform. Sci.* 12: 131–143. With permission.)

level. This confidence region J_2, is the union of the confidence regions for all points Q_{2r} on the line segment for $r \in [0, 1]$, so that the true locations $\varnothing_{2r}$ of all points on the line segment are contained within J_2, with the probability larger than a predefined confidence level:

$$P(\varnothing_{2r} \in J_2 \text{ for all } r \in [0, 1]) > \alpha. \tag{4.37}$$

The confidence region J_2 is a set of points $(x_r, y_r)^T$ with x_r and y_r satisfying

$$X_r - c_{21} \le x_r \le X_r + c_{21}, \tag{4.38}$$

$$Y_r - c_{22} \le y_r \le Y_r + c_{22}, \tag{4.39}$$

where $c_{21} = [k((1 - r)^2 + r^2)]^{1/2}\sigma_x$, $c_{22} = [k((1 - r)^2 + r^2)]^{1/2}\sigma_y$, and $k = \chi^2_{2;(1 + \alpha)/2}$.

The chi-square parameter k is 8.517, $\alpha = 0.97$. Figure 4.4b shows the confidence region J_2 for a measured location $Q_{21}Q_{22}$ of a line segment with a predefined confidence level $\alpha = 0.97$. Points Q_{21} and Q_{22} are measured locations of the endpoints of the line segment, while points $\varnothing_{21}$ and $\varnothing_{22}$ are their corresponding true locations. The confidence region for the measured location $Q_{21}Q_{22}$ of the line segment is composed of the rectangles with the individual measured points Q_{2r} on the line segment as their centers. The details of the approval of the confidence region model for a two-dimensional line segment can be found in Shi (1994).

To analyze the shape of the confidence region of a two-dimensional line segment, the confidence regions for all of the points on the line segments are studied. Parameters c_{21} and c_{22} in Equations 4.38 and 4.39 are then differentiated by r. With the conditions $\partial c_{21}/\partial r = 0$ and $\partial c_{22}/\partial r = 0$, parameters c_{21} and c_{22} get their minimum value at $r = 1/2$. The other two extreme values of c_{21} and c_{22} occur when $r = 0$ and 1. Therefore, the shape of the confidence region for a two-dimensional lines segment is similar to that in Figure 4.4b. This result is obtained under the assumption that error in the two endpoints is identical and follows the normal distribution. The above analytical form of the confidence region for a line segment shows that the size of the confidence region is subject to the positional error at the endpoints of that segment and the predefined confidence level. The larger the positional error at the endpoints and the predefined confidence level, the wider the line segment confidence region.

Furthermore, the shape of the confidence region is narrower at the middle point and wider at the two endpoints. Details regarding the approval of the confidence region of a two-dimensional line segment are given in Shi (1994).

4.5.2.3 A Confidence Region Model for a Polyline in Two-Dimensional Space

Joining several two-dimensional line segments forms a polyline. A special case of a polyline, with identical beginning and end vertices is a closed polygon (A). This polygon is derived from two-dimensional stochastic vectors of its composite points: $Q_{21} = [x_1, y_1]^T$, $Q_{22} = [x_2, y_2]^T$, ..., $Q_{2i} = [x_i, y_i]^T$, ..., $Q_{2m} = [x_m, y_m]^T$, where $i \in [1, m]$ and

m is the number of composition points. These stochastic vectors are used to represent the boundary of polygon A with the line segments $Q_{21}Q_{22}$, $Q_{22}Q_{23}$, ..., $Q_{2(m-1)}Q_{2m}$. Let $\Sigma_{Q_{2i}Q_{2,i+1}}$ denote the covariance matrix between two succeeding vertices, i.e.,

$$\Sigma_{Q_{2i}Q_{2,i+1}} = \begin{bmatrix} \sigma_{x_i}^2 & \sigma_{x_i x_{i+1}} & \sigma_{x_i y_{i+1}} & \sigma_{x_i y_{i+1}} \\ \sigma_{x_{i+1} x_i} & \sigma_{x_{i+1}}^2 & \sigma_{x_{i+1} y_i} & \sigma_{x_{i+1} y_{i+1}} \\ \sigma_{y_i x_i} & \sigma_{y_i x_{i+1}} & \sigma_{y_i}^2 & \sigma_{y_i y_{i+1}} \\ \sigma_{y_{i+1} x_i} & \sigma_{y_{i+1} x_{i+1}} & \sigma_{y_{i+1} y_i} & \sigma_{y_{i+1}}^2 \end{bmatrix} \quad (i=1, 2, \cdots, m-1). \quad (4.40)$$

Any point on the line segment $Q_{2i}Q_{2,i+1}$ is then represented by

$$Q_{2ir} = (1-r)Q_{2i} + rQ_{2,i+1} = \begin{bmatrix} (1-r)\mu_{x_i} + r\mu_{x_{i+1}} \\ (1-r)\mu_{y_i} + r\mu_{y_{i+1}} \end{bmatrix} \quad (4.41)$$

$$(i=1, 2, \cdots, m-1), \, r \in [0,1],$$

where μ_{x_i}, μ_{y_i}, $\mu_{x_{i+1}}$, and $\mu_{y_{i+1}}$ are the expectations of x_i and y_i, x_{i+1}, and y_{i+1}, respectively.

The variance of x_{ir} and y_{ir}, and the covariance coefficients of x_{ir} and y_{ir}, are then,

$$\begin{cases} \sigma_{x_{ir}}^2 = (1-r)^2 \sigma_{x_i}^2 + 2r(1-r)\sigma_{x_i x_{i+1}} + r^2 \sigma_{x_{i+1}}^2 \\ \sigma_{x_{ir} y_{ir}} = (1-r)^2 \sigma_{x_i y_i} + r^2 \sigma_{x_{i+1} y_{i+1}} \\ \qquad\qquad + r(1-r)(\sigma_{x_{i+1} y_i} + \sigma_{x_i y_{i+1}}) \\ \sigma_{y_{ir}}^2 = (1-r)^2 \sigma_{y_i}^2 + 2r(1-r)\sigma_{y_i y_{i+1}} + r^2 \sigma_{y_{i+1}}^2 \end{cases} \quad (i=1, 2, \cdots, m-1, r \in [0,1]). \quad (4.42)$$

A region J that contains the true location of the polyline $(\phi_{21}\phi_{22} \cdots \phi_{2m})$ is constructed around a "measured" polyline, that is, all the points ϕ_{ir} ($i = 1, 2, \cdots$, m − 1, $r \in [0,1]$) on the polyline are contained in this region with a probability that is larger than a predefined confidence level $1 - \alpha (0 < \alpha < 1)$, that is,

$$P\left(\phi_{ir} \in J, i=1, 2, \cdots, m-1, r \in [0,1]\right) \geq 1-\alpha. \quad (4.43)$$

The confidence region for the expectations ϕ_{ir} ($i = 1, \cdots$, m − 1, $r \in [0,1]$) is now constructed. We are interested in a simultaneous confidence region for ϕ_{ir} ($i = 1, 2, \cdots$, m − 1, $r \in [0,1]$), not only in the confidence region of a particular point of the polyline. If every point of the polyline is located within a region simultaneously, the whole polyline is located within the region. Therefore, we construct a region such that all points are located within the region simultaneously with a probability larger than a predefined confidence level.

$$\left(Q_{2ir}-\phi_{2ir}\right)^{T}\Sigma_{ir}^{-1}\left(Q_{2ir}-\phi_{2ir}\right)=\left\|\Sigma_{ir}^{-1/2}\left(Q_{2ir}-\phi_{2ir}\right)\right\|^{2}$$

$$=\left\|\left(1-r\right)\Sigma_{ir}^{-1/2}\left(Q_{2i}-\phi_{2i}\right)+r\Sigma_{ir}^{-1/2}\left(Q_{2,i+1}-\phi_{2,i+1}\right)\right\|^{2}.$$

ϕ_{2ir} can be computed by formula (Equation 4.41), and Σ_{ir} can be computed by (Equation 4.42). $\Sigma_{ir}^{-1/2}$ is the square root matrix of Σ_{ir}^{-1}. Based on norm inequality, we have

$$\left(Q_{2ir}-\phi_{2ir}\right)^{T}\Sigma_{ir}^{-1}\left(Q_{2ir}-\phi_{2ir}\right)\leq\left[\left(1-r\right)\left\|\Sigma_{ir}^{-1/2}\left(Q_{2i}-\phi_{2i}\right)\right\|+r\left\|\Sigma_{ir}^{-1/2}\left(Q_{2,i+1}-\phi_{2,i+1}\right)\right\|\right]^{2}.$$

Based on the Cauchy-Schwarz inequality, we further have

$$\left(Q_{2ir}-\phi_{2ir}\right)^{T}\Sigma_{ir}^{-1}\left(Q_{2ir}-\phi_{2ir}\right)$$

$$\leq\left[\left(1-r\right)^{2}+r^{2}\right]\left[\left(Q_{2i}-\phi_{2i}\right)^{T}\Sigma_{ir}^{-1}\left(Q_{2i}-\phi_{2i}\right)+\left(Q_{2,i+1}-\phi_{2,i+1}\right)^{T}\Sigma_{ir}^{-1}\left(Q_{2,i+1}-\phi_{2,i+1}\right)\right]$$

$$\leq\left[\left(1-r\right)^{2}+r^{2}\right]\left[\left(Q_{2i}-\phi_{2i}\right)^{T}\Sigma_{\min}^{-1}\left(Q_{2i}-\phi_{2i}\right)+\left(Q_{2,i+1}-\phi_{2,i+1}\right)^{T}\Sigma_{\min}^{-1}\left(Q_{2,i+1}-\phi_{2,i+1}\right)\right].$$

Σ_{min} is the variance and covariance matrix of the point with the minimum error among all the points on the polyline $Q_{21}Q_{22}$, ..., Q_{2i}, ..., Q_{2m}, and Σ_{min} can be computed by numerical analysis method.

Dividing each side of this inequality by $[(1 - r^2 + r^2]$, we have

$$\frac{\left(Q_{2ir}-\phi_{2ir}\right)^{T}\Sigma_{ir}^{-1}\left(Q_{2ir}-\phi_{2ir}\right)}{\left[\left(1-r\right)^{2}+r^{2}\right]}\leq$$

$$\left[\left(Q_{2i}-\phi_{2i}\right)^{T}\Sigma_{\min}^{-1}\left(Q_{2i}-\phi_{2i}\right)+\left(Q_{2,i+1}-\phi_{2,i+1}\right)^{T}\Sigma_{\min}^{-1}\left(Q_{2,i+1}-\phi_{2,i+1}\right)\right]. \quad (4.44)$$

The right side of Equation 4.44 does not depend on m and r, thus it is true for the entire polyline.

Let

$$Y_{i}=\left[\left(Q_{2i}-\phi_{2i}\right)^{T}\Sigma_{\min}^{-1}\left(Q_{2i}-\phi_{2i}\right)+\left(Q_{2,i+1}-\phi_{2,i+1}\right)^{T}\Sigma_{\min}^{-1}\left(Q_{2,i+1}-\phi_{2,i+1}\right)\right]$$

$$(i=1, 2, \cdots, m-1).$$

We determined the probability distribution of Y_i by means of Monte Carlo simulations. This is more tractable than using an analytical method.

Working with a probability of $1 - \alpha (0 < \alpha < 1)$, because $P[Y_i \leq Y_{i;\alpha}] \geq 1 - \alpha$, from (4.44), it is concluded that

$$P\left[\frac{\left(Q_{2ir} - \phi_{2ir}\right)^T \Sigma_{ir}^{-1}\left(Q_{2ir} - \phi_{2ir}\right)}{\left[\left(1-r\right)^2 + r^2\right]} \leq Y_{i;\alpha}\right] \geq 1 - \alpha, \tag{4.45}$$

which is equivalent to

$$P\left[\left(\phi_{2ir} - Q_{2ir}\right)^T \Sigma_{ir}^{-1}\left(\phi_{2ir} - Q_{2ir}\right) \leq Y_{i;\alpha}\left[\left(\left(1-r\right)^2 + r^2\right)\right], i = 1, 2, \cdots m-1, r \in \left[0,1\right]\right]$$

$$\geq 1 - \alpha.$$

Thus we identify an area within which the true location of the polyline $Q_{21}Q_{22}, \cdots,$ $Q_{2i}, \cdots, Q_{2m}$ is simultaneously included with a probability larger than a predefined confidence level. The size and shape of the region are determined by the error of the composite points and the predefined confidence level. The confidence region for a polyline is illustrated in Figure 4.4c.

4.5.3   CONFIDENCE SPACE MODEL FOR SPATIAL FEATURE
IN AN N-DIMENSIONAL SPACE

4.5.3.1   Confidence Space for a Point in an n-Dimensional Space

In an n-dimensional space, a point is represented by an n-dimensional stochastic vector: $Q_{n1} = [x_1, x_2, \ldots, x_i, \ldots, x_n]^T$, where $i \in [1, n]$. If the error at this point follows a normal distribution, the stochastic vector can be expressed as

$$Q_{n1} \sim N_n\left[\begin{bmatrix}\mu_{x_1} \\ \mu_{x_2} \\ \vdots \\ \mu_{x_n}\end{bmatrix}, \begin{bmatrix}\sigma_{x_1}^2 & \sigma_{x_1 x_2} & \cdots & \sigma_{x_1 x_n} \\ \sigma_{x_2 x_1} & \sigma_{x_2}^2 & \cdots & \sigma_{x_2 x_n} \\ \cdots & & & \\ \sigma_{x_n x_1} & \sigma_{x_n x_2} & \cdots & \sigma_{x_n}^2\end{bmatrix}\right], \tag{4.46}$$

where μ_{x_i} is the expectation of x_i, $\sigma_{x_i}^2$ is the variance of x_i, and $\sigma x_i x_j = \sigma x_j x_i$ is the covariance coefficient of x_i and x_j, where $i, j \in [1, n]$.

A confidence space of a measured n-dimensional point is used to describe positional uncertainty of the position measurement in an n-dimensional space. The confidence space J_{n1} for a point is a space containing the true location $\varnothing_{n1}$ of the point, with a probability that is larger than a predefined confidence level α:

$$P(\varnothing_{n1} \in J_{n1}) > \alpha. \tag{4.47}$$

The vector J_{n1} is a set $[x_1, x_2, ..., x_n]^T$ with x_i satisfying

$$X_{i1} - a_{ni} \leq x_i \leq X_{i1} + a_{ni}, i = 1, 2, ..., n \qquad (4.48)$$

where $a_{ni} = k^{1/2}\sigma_{x_i}$, and $k = \chi^2_{1;(n-1+\alpha)/n}$.

The parameter k is dependent on the predefined confidence level α and can be found in a chi-square distribution table.

4.5.3.2. Confidence Space for a Line Segment in an n-Dimensional Space

A line segment in n-dimensional space is defined by two stochastic vectors of its endpoints: $Q_{n1} = [x_{11}, x_{12}, ..., x_{1n}]^T$ and $Q_{n2} = [x_{21}, x_{22}, ..., x_{2n}]^T$. Bivariate normally distributed stochastic vectors are given as

$$
Q_{n1} \sim N_n \left[\begin{bmatrix} \mu_{x_{11}} \\ \mu_{x_{12}} \\ \vdots \\ \mu_{x_{1n}} \end{bmatrix}, \begin{bmatrix} \sigma^2_{x_1} & \sigma_{x_1 x_2} & \cdots & \sigma_{x_1 x_n} \\ \sigma_{x_2 x_1} & \sigma^2_{x_2} & \cdots & \sigma_{x_2 x_n} \\ & \cdots & & \\ \sigma_{x_n x_1} & \sigma_{x_n x_2} & \cdots & \sigma^2_{x_n} \end{bmatrix} \right],
$$

$$
Q_{n2} \sim N_n \left[\begin{bmatrix} \mu_{x_{21}} \\ \mu_{x_{22}} \\ \vdots \\ \mu_{x_{2n}} \end{bmatrix}, \begin{bmatrix} \sigma^2_{x_1} & \sigma_{x_1 x_2} & \cdots & \sigma_{x_1 x_n} \\ \sigma_{x_2 x_1} & \sigma^2_{x_2} & \cdots & \sigma_{x_2 x_n} \\ & \cdots & & \\ \sigma_{x_n x_1} & \sigma_{x_n x_2} & \cdots & \sigma^2_{x_n} \end{bmatrix} \right],
$$

$$(4.49)$$

where $\mu_{x_{11}}, \mu_{x_{12}}, \cdots \mu_{x_{1n}}, \mu_{x_{21}}, \mu_{x_{22}}, \cdots \mu_{x_{2n}}$, are the expectation of $x_{11}, x_{12}, \cdots x_{1n}, x_{21}, x_{22}, \cdots x_{2n}$, respectively.

A confidence space J_n for a line segment in n-dimensional space is a space around a measured location of the n-dimensional line segment. It is defined as the union of confidence space J_{nr} for all points on the line segment as given in Equation 4.50. This confidence space contains the true location of the line segment with the probability larger than a predefined confidence level α:

$$P(\varnothing_{nr} \in J_{nr}, \text{ for all } r \in [0, 1]) > \alpha, \qquad (4.50)$$

where $\varnothing_{nr}$ is the true location of any point on the line segment; J_{nr} is a set $[x_1, x_2, ..., x_n]^T$ with x_i satisfying

$$X_{ir} - c_{ni} \leq x_i \leq X_{ir} + c_{ni}, \qquad (4.51)$$

where $c_{ni} = [k((1 - r)^2 + r^2)]^{1/2}\sigma_{x_i}$, $k = \chi^2_{2;(n-1+\alpha)/n}$, and $i = 1, 2, ..., n$.

4.5.3.3 Confidence Space for an n-Dimensional Polyline

Joining several n-dimensional line segments forms a polyline. A special case of a polyline, with identical beginning and end vertices is a closed polygon (A). This polygon is derived from n-dimensional stochastic vectors of its composite points: $Q_{n1} = [x_{11}, x_{12}, \ldots, x_{1n}]^T$, $Q_{n2} = [x_{21}, x_{22}, \ldots, x_{2n}]^T$, $\ldots$, $Q_{ni} = [x_{i1}, x_{i2}, \ldots, x_{in}]^T$, $\ldots$, $Q_{nm} = [x_{m1}, x_{m2}, \ldots x_{mn}]^T$, where $i \in [1, m]$ and m is the number of composition points. These stochastic vectors are used to represent the boundary of polygon A with the line segments $Q_{n1}Q_{n2}$, $Q_{n2}Q_{n3}$, $\ldots$, $Q_{n(m-1)}Q_{nm}$. Let $\Sigma_{Q_{ni}Q_{n,i+1}}$ denote the covariance matrix of the two vertices, i.e.,

$$\Sigma_{Q_{ni}Q_{n,i+1}} = \tag{4.52}$$

$$\begin{bmatrix} \sigma^2_{x_{i1}} & \sigma_{x_{i1}x_{i2}} & \cdots & \sigma_{x_{i1}x_{in}} & \sigma_{x_{i1}x_{i+1,1}} & \sigma_{x_{i1}x_{i+1,2}} & \cdots & \sigma_{x_{i1}x_{i+1,n}} \\ \sigma_{x_{i2}x_{i1}} & \sigma^2_{x_{i2}} & \cdots & \sigma_{x_{i2}x_{in}} & \sigma_{x_{i2}x_{i+1,1}} & \sigma_{x_{i2}x_{i+1,2}} & \cdots & \sigma_{x_{i2}x_{i+1,n}} \\ \vdots & \vdots & \vdots & \vdots & \vdots & \vdots & \vdots & \vdots \\ \sigma_{x_{in}x_{i1}} & \sigma_{x_{in}x_{i2}} & \cdots & \sigma^2_{x_{in}} & \sigma_{x_{in}x_{i+1,1}} & \sigma_{x_{in}x_{i+1,2}} & \cdots & \sigma_{x_{in}x_{i+1,n}} \\ \sigma_{x_{i+1,1}x_{i1}} & \sigma_{x_{i+1,1}x_{i2}} & \cdots & \sigma_{x_{i+1,1}x_{in}} & \sigma^2_{x_{i+1,1}} & \sigma_{x_{i+1,1}x_{i+1,2}} & \cdots & \sigma_{x_{i+1,1}x_{i+1,n}} \\ \sigma_{x_{i+1,2}x_{i1}} & \sigma_{x_{i+1,2}x_{i2}} & \cdots & \sigma_{x_{i+1,2}x_{in}} & \sigma_{x_{i+1,2}x_{i+1,1}} & \sigma^2_{x_{i+1,2}} & \cdots & \sigma_{x_{i+1,2}x_{i+1,n}} \\ \vdots & \vdots & \cdots & \vdots & \vdots & \vdots & \cdots & \vdots \\ \sigma_{x_{i+1,n}x_{i1}} & \sigma_{x_{i+1,n}x_{i2}} & \cdots & \sigma_{x_{i+1,n}x_{in}} & \sigma_{x_{i+1,n}x_{i+1,1}} & \sigma_{x_{i+1,n}x_{i+1,2}} & \cdots & \sigma^2_{x_{i+1,n}} \end{bmatrix}.$$

$$\left(i = 1, 2, \cdots, m-1 \right)$$

Any point on the line segment $Q_{ni}Q_{n,i+1}$ is then, represented by

$$Q_{nir} = \begin{bmatrix} x_{i1r} \\ x_{i2r} \\ \vdots \\ x_{inr} \end{bmatrix} = \left(1-r \right) Q_{ni} + r Q_{n,i+1}$$

$$= \begin{bmatrix} \left(1-r\right)\mu_{x_{i1}} + r\mu_{x_{i+1,1}} \\ \left(1-r\right)\mu_{x_{i2}} + r\mu_{x_{i+1,2}} \\ \vdots \\ \left(1-r\right)\mu_{x_{in}} + r\mu_{x_{i+1,n}} \end{bmatrix} \left(i = 1, 2, \cdots, m-1 \right), r \in [0, 1], \tag{4.53}$$

where $\mu_{x_{i1}}, \mu_{x_{i2}}, \cdots \mu_{x_{in}}, \mu_{x_{i+1,1}}, \mu_{x_{i+1,2}}, \cdots \mu_{x_{i+1,n}}$ are the expectation of $x_{i1}, x_{i2}, \cdots x_{in}, x_{i+1,1}, x_{i+1,2}, \cdots x_{i+1,n}$, respectively.

A space J that contains the true location of the polyline $(\phi_{n1}\phi_{n2} \cdots \phi_{nm})$ is constructed around a "measured" polyline; that is, all the points ϕ_{ir} ($i = 1, 2, \cdots, m - 1$, $r \in [0,1]$) are contained in this space with a probability that is larger than a predefined confidence level $1 - \alpha (0 < \alpha < 1)$, that is,

$$P\left(\phi_{ir} \in J, i = 1, 2, \cdots, m-1, r \in [0,1] \right) \geq 1-\alpha. \tag{4.54}$$

Let

$$
Y_i = \left[\left(Q_{ni} - \phi_{ni} \right)^T \Sigma_{\min}^{-1} \left(Q_{ni} - \phi_{ni} \right) + \left(Q_{n,i+1} - \phi_{n,i+1} \right)^T \Sigma_{\min}^{-1} \left(Q_{n,i+1} - \phi_{n,i+1} \right) \right]
$$

$$
\left(i = 1, 2, \cdots, m-1 \right)
$$

$\Sigma_{\min}$ is the variance and covariance matrix of the point with the minimum error among all the points on the polyline $Q_{n1} Q_{n2}, \cdots, Q_{ni}, \cdots, Q_{nm}$, and $\Sigma_{\min}$ can be computed by numerical analysis method.

The probability distribution of Y_i can be determined by means of Monte Carlo simulations.

Working with a probability of $1 - \alpha (0 < \alpha < 1)$, because $P[Y_i \le Y_{i;\alpha}] \ge 1 - \alpha$, it is concluded that

$$
P\left[\frac{\left(Q_{nir} - \phi_{nir} \right)^T \Sigma_{ir}^{-1} \left(Q_{nir} - \phi_{nir} \right)}{\left[\left(1 - r \right)^2 + r^2 \right]} \le Y_{i;\alpha} \right] \ge 1 - \alpha,
$$

which is equivalent to

$$
P\left[\left(\phi_{nir} - Q_{nir} \right)^T \Sigma_{ir}^{-1} \left(\phi_{nir} - Q_{nir} \right) \le Y_{i;\alpha} \left[\left(1 - r \right)^2 + r^2 \right] \right], \quad i = 1, 2, \cdots m-1, r \in [0,1]
$$

$$
\ge 1 - \alpha.
$$

Therefore, we identify a space within which the true location of the polyline $Q_{n1} Q_{n2}$, $\cdots, Q_{ni}, \cdots, Q_{nm}$ is simultaneously included with a probability larger than a predefined confidence level. The size and shape of the space are determined by the error of the composite points and the predefined confidence level.

4.6 PROBABILITY DISTRIBUTION MODEL FOR LINE SEGMENT

The confidence region models given from Section 4.5.1 to Section 4.5.3 provide a region that potentially contains the true location of a line. However, probability analysis can be applied to depict the error distribution of that line. Based on a probability analysis, the positional error on any straight line is modeled from a joint probability function of those points on the straight line. This joint probability function gives the probability distribution for the line within a particular region. The probability of the line falling inside the corresponding region is defined as the volume of the line's error curved surface formed by integrating the error curved surfaces of individual points on the line.

In Figure 4.5, random points $Q_{21} = (x_1, y_1)^T$ and $Q_{22} = (x_2, y_2)^T$ are two vertices (also called explicit points) of a straight-line segment $Q_{21} Q_{22}$; l_r and l are distances between Q_{21} and Q_{2r}, and between Q_{21} and Q_{22}, respectively. For $r = l_r/l$, the random point $Q_{2r} = (x_r, y_r)^T$ is a point on the line segment, derived from the linear combination of the two explicit points.

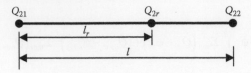

FIGURE 4.5 A point Q_{2r} on a straight-line segment $Q_{21}Q_{22}$.

Let $\Sigma_{Q_{21}Q_{22}}$ denote the covariance matrix of the two vertices, i.e.,

$$\Sigma_{Q_{21}Q_{22}} = \begin{bmatrix} \sigma_{x_1}^2 & \sigma_{x_1x_2} & \sigma_{x_1y_1} & \sigma_{x_1y_2} \\ \sigma_{x_2x_1} & \sigma_{x_2}^2 & \sigma_{x_2y_1} & \sigma_{x_2y_2} \\ \sigma_{y_1x_1} & \sigma_{y_1x_2} & \sigma_{y_1}^2 & \sigma_{y_1y_2} \\ \sigma_{y_2x_1} & \sigma_{y_2x_2} & \sigma_{y_2y_1} & \sigma_{y_2}^2 \end{bmatrix}. \tag{4.55}$$

The variance of x_r and y_r, and the covariance coefficients of x_r and y_r are then,

$$\begin{cases} \sigma_{x_r}^2 = (1-r)^2 \sigma_{x_1}^2 + 2r(1-r)\sigma_{x_1x_2} + r^2\sigma_{x_2}^2 \\ \sigma_{x_ry_r} = (1-r)^2 \sigma_{x_1y_1} + r^2\sigma_{x_2y_2} \\ \qquad\qquad + r(1-r)(\sigma_{x_2y_1} + \sigma_{x_1y_2}) \\ \sigma_{y_r}^2 = (1-r)^2 \sigma_{y_1}^2 + 2r(1-r)\sigma_{y_1y_2} + r^2\sigma_{y_2}^2 \end{cases}. \tag{4.56}$$

In the special case that $\Sigma_{Q_{21}Q_{22}}$ is a diagonal matrix, the probability density function of Q_{2r} will be

$$f(x_r, y_r) = \frac{\exp\left\{-x_r^2/(2\sigma_{x_r}^2) - (y_r - l_r)^2/(2\sigma_{y_r}^2)\right\}}{2\pi\sigma_{x_r}\sigma_{y_r}}. \tag{4.57}$$

When n is an even number, $n/2$ points are considered, including two vertexes and $(n/2 - 2)$ intermediate points, on the line segment. These points are then denoted by $z_{t1} = [x_1\ y_1]^T, \ldots, z_{ti} = [x_i\ y_i]^T, \ldots, z_{tn/2} = [x_{n/2}\ y_{n/2}]^T$. Their united probability density function is

$$f\left(x_1, y_1, x_2, y_2, \ldots, x_{\frac{n}{2}}, y_{\frac{n}{2}}\right) = \frac{\exp\left\{-(z-\mu_z)^T \Sigma_{zz}^{-1}(z-\mu_z)/2\right\}}{(2\pi)^n |\Sigma_{zz}|^{\frac{1}{2}}}, \tag{4.58}$$

where $Z = (x_1, y_1, x_2, y_2, \ldots, x_{n/2}, y_{n/2})^T$; $\mu_z = (\mu_{x_1}, \mu_{y_1}, \mu_{x_2}, \mu_{y_2}, \cdots, \mu_{x_{n/2}}, \mu_{x_{n/2}})^T$; and

$$\Sigma_{ZZ} = \begin{bmatrix} \sigma^2_{x_1} & \sigma_{x_1 y_1} & \cdots & \sigma_{x_1 y_{n/2}} \\ \sigma_{y_1 x_1} & \sigma^2_{y_1} & \cdots & \sigma_{y_1 y_{n/2}} \\ \cdots & & & \\ \sigma_{y_{n/2} x_1} & \sigma_{y_{n/2} y_1} & \cdots & \sigma^2_{y_{n/2}} \end{bmatrix}.$$

The probability that the true locations of the individual points on the line segments are simultaneously inside the corresponding error ellipses is given as

$$P = \int_\Omega f(z) dz$$

$$= \int_\Omega \frac{\exp\left\{-(z-\mu_z)^T \Sigma_{zz}^{-1} (z-\mu_z)/2\right\}}{(2\pi)^n |\Sigma_{zz}|^{\frac{1}{2}}} dz \tag{4.59}$$

where Ω is the region in a two-dimensional plane composed of the error ellipses of the $n/2$ points of the line segment (Figure 4.6).

When n tends to infinity, the probability given in Equation 4.59 becomes

$$P = \frac{T_r + T_0}{T_l + 1}, \tag{4.60}$$

where T_r is the volume of the error curved surface, which is the union of profiles of the intermediate point error curved surface in the direction, perpendicular to the line segment over domain Ω_r ($-\varepsilon < x_r < \varepsilon$ and $0 < y_r < l$); T_l is the volume over domain Ω_r ($-\infty < x_r < \infty$ and $0 < y_r < l$); and T_0 is the volume of the error curved surface, which

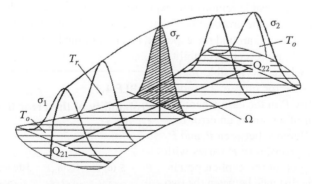

FIGURE 4.6 Probability distribution of the line segment $Q_{21}Q_{22}$.

is formed by a union of the regions of the two explicit points at the end of the line segment.

If

$$\varepsilon = \sigma_{x_r} = \sqrt{(1-r)^2 \sigma_{x_1}^2 + r^2 \sigma_{x_2}^2}\,, \tag{4.61}$$

the following equation will be obtained:

$$T_r = 2 \int_{x_r=0}^{\sigma_{x_r}} \int_{y_r=0}^{l} f(x_r)\,dx_r dy_r$$

$$= 2 \int_{x_r=0}^{\sigma_{x_r}} \int_{r=0}^{1} f(x_r)\,dx_r l dr \tag{4.62}$$

$$= 2 \int_{x_r=0}^{\sigma_{x_r}} \int_{r=0}^{1} \frac{\left[l \cdot \exp\left(\frac{-x_r^2}{2}\right)\right]}{\sqrt{2\pi}}\,dx_r dr$$

$$= 0.6826 l$$

and

$$T_l = 2 \int_{x_r=0}^{\infty} \int_{r=0}^{1} \frac{\left[l \cdot \exp\left(-\frac{x_r^2}{2\sigma_{x_r}^2}\right)\right]}{\sqrt{2\pi}\sigma_{x_r}}\,dx_r dr \tag{4.63}$$

$$= l.$$

When the feasible regions of the vertices are expressed in terms of error ellipse, T_0 is 0.393; when the feasible regions are expressed in terms of rectangle, T_0 is 0.466. Table 4.1 summarizes the probability P that a measured location of the line segment is inside the error band, around the expected location of the line segment.

In Table 4.1, P_1 represents probability P in the case where the feasible region of a point is an error ellipse; P_2 represents probability P in the case where the feasible region of a point is an error rectangle. The difference between P_1 and P_2 shows that the probability P varies with the feasible region of the explicit points. In addition, the distance between the two explicit points affects the probability of the error band for the line segment: when

TABLE 4.1
Probability of the Error Band for the Line Segment

		l		
P	5 (σ_{x_1})	10 (σ_{x_1})	100 (σ_{x_1})	500 (σ_{x_1})
P_1	0.6345	0.6565	0.6799	0.6822
P_2	0.6433	0.6613	0.6805	0.6823

the distance is five times as large as the standard error in the vertex points, the probability of the error band defined from the error ellipse is 0.6345 and that defined from the error rectangle is 0.6433. When the distance is 500 times as large as the standard error, the probability equals 0.6822 for the error ellipse and 0.6823 for the error rectangle. Therefore, when the distance l is much larger than the standard error σ_{x_i} at the vertex points, the probability P of the error band for the line segment approaches 68.3%.

The above confidence region models enable the description of the positional uncertainty in spatial features in two- and three-dimensional GIS, and even in a four-dimensional spatio-temporal GIS. The models are defined under assumptions that the positional error at composite points of a spatial feature is independent and normally distributed.

Introduced in the following section is an error model for lines where errors of composite points can be correlated, instead of assuming that the error of the composite points of a line segment feature is independent.

4.7 G-BAND ERROR MODEL FOR LINE SEGMENT

4.7.1 ERROR DISTRIBUTION AND PROBABILITY DENSITY FUNCTIONS

A line segment $Q_{21}Q_{22}$ composed of (explicit) points Q_{21} and Q_{22} has an infinite set of intermediate points Q_{1r} on the line segment, where Q_{1r} can be derived from Equation 4.9 with $r \in [0, 1]$. The cumulative distribution function of line segment $Q_{21}Q_{22}$ can be defined by the joint distribution function of the stochastic vectors of points Q_{21}, Q_{22}, and Q_{1r} for all r.

Let $Z_{t_1} = [X_{t_1}\ Y_{t_1}]^T = [X_{Q_{11}}\ X_{Q_{21}}]^T$ and $Z_{t_{n/2}} = [X_{t_{n/2}}\ Y_{t_{n/2}}]^T = [X_{Q_{12}}\ X_{Q_{22}}]^T$ denote the stochastic vectors of explicit points Q_{21} and Q_{22}; and let $Z_{t_i} = [X_{t_i}\ Y_{t_i}]^T$ denote an implicit point on the line segment for any positive integer $i \in (1, n/2)$ and any positive even integer n. The cumulative distribution function of line segment $Q_{21}Q_{22}$ is approximated by the joint distribution function of the stochastic vector of two explicit points and finite number of the implicit points:

$$F\left(x_1, y_1, x_2, y_2, \ldots, x_{n/2}, y_{n/2}\right) =$$
$$P\left\{X_{t_1} < x_1, Y_{t_1} < y_1, X_{t_2} < x_2, Y_{t_2} < y_2, \cdots, X_{t_{n/2}} < x_{n/2}, Y_{t_{n/2}} < y_{n/2}\right\}$$

$$(4.64)$$

This is a family of the finite dimensional distribution function of line segment $Q_{21}Q_{22}$.

The probability density function of the line segment is obtained by differentiating Equation 4.64 with $x_1, y_1, x_2, y_2, \ldots, x_{n/2}, y_{n/2}$.

$$f\left(x_1, y_1, x_2, y_2, \ldots, x_{n/2}, y_{n/2}\right) =$$
$$\partial^{2n} F\left(x_1, y_1, x_2, y_2, \ldots, x_{n/2}, y_{n/2}\right) / \partial x_1 \partial y_1 \partial x_2 \partial y_2 \ldots \partial x_{n/2} \partial y_{n/2} .$$

$$(4.65)$$

Equations 4.64 or 4.65 can not only describe the statistical characteristics of all stochastic variables $X_{t_1}, Y_{t_1}, X_{t_2}, Y_{t_2}, \cdots, X_{t_i}, Y_{t_i}, X_{t_{n/2}}, Y_{t_{n/2}}$ for all $i \in [1, n/2]$, but can also describe the statistical characteristics of individual stochastic variables or any of their combinations, as well as the correlation between any two of the stochastic variables, with a marginal distribution or density function.

If positional errors at points Q_{21} and Q_{22} follow bivariate normal distributions, the probability density function in Equation 4.65 will be expressed as the probability density function of an n-dimensional normal distribution from Equation 4.9:

$$f(Z) = \frac{\exp\left\{-(Z-\mu_z)^T \Sigma_{ZZ}^{-1}(Z-\mu_z)/2\right\}}{(2\pi)^n |\Sigma_{ZZ}|^{\frac{1}{2}}}, \tag{4.66}$$

where $Z = (z_1, z_2, z_3, z_4, \ldots, z_{n-1}, z_n)^T = (x_1, y_1, x_2, y_2, \ldots, x_{n/2}, y_{n/2})^T$; μ_z
$= (\mu_{z_1}, \mu_{z_2}, \mu_{z_3}, \mu_{z_4}, \ldots, \mu_{z_{n-1}}, \mu_{z_n})^T$
$= (\mu_{x_1}, \mu_{y_1}, \mu_{x_2}, \mu_{y_2}, \ldots, \mu_{x_{n/2}}, \mu_{y_{n/2}})^T$; and

$$\Sigma_{ZZ} = \begin{bmatrix} \sigma_{z_1}^2 & \sigma_{z_1 z_2} & \cdots & \sigma_{z_1 z_n} \\ \sigma_{z_2 z_1} & \sigma_{z_2}^2 & \cdots & \sigma_{z_2 z_n} \\ \cdots & & & \\ \sigma_{z_n z_1} & \sigma_{z_n z_2} & \cdots & \sigma_{z_n}^2 \end{bmatrix}.$$

The stochastic line segment $Q_{21}Q_{22}$ is regarded as a stochastic process. Figure 4.7 illustrates this stochastic process, which is an extension of the conventional probability density function of a point.

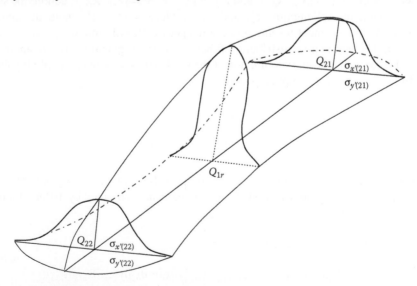

FIGURE 4.7   The probability density function of line segment $Q_{21}Q_{22}$. (After Shi, W. Z., 1994. ITC Publication, Enschede, Number 22: 147. With permission.)

4.7.2 ERROR INFORMATION OF A LINE SEGMENT

The statistical characteristics of a line segment can be described by Equations 4.64 and 4.65. However, they cannot be used to assess the positional error of the line segment directly. A symmetrical matrix is therefore defined as follows:

$$
\Sigma_{z_{t_1} z_{t_2}} = \begin{bmatrix} \sigma^2_{x_{t_1} x_{t_2}} & \sigma_{x_{t_1} y_{t_2}} \\ \sigma_{y_{t_2} x_{t_1}} & \sigma^2_{y_{t_1} y_{t_2}} \end{bmatrix}. \tag{4.67}
$$

This matrix provides the error information of a line segment. It is a conventional function of t_1 and t_2. The matrix (4.67) is used to describe the linear correlation of the (vertex and intermediate) points on the line segment.

Based on Equations 4.9 and 4.67, the elements of the matrix (4.67) are given as

$$
\begin{cases}
\sigma^2_{x_{t_1} x_{t_2}} = (1-t_1)(1-t_2)\sigma^2_{x_1} + \left[t_1(1-t_2) + t_2(1-t_1) \right]\sigma_{x_1 x_2} + t_1 t_2 \sigma^2_{x_2} \\
\sigma_{x_{t_1} y_{t_2}} = (1-t_1)(1-t_2)\sigma_{x_1 y_1} + t_1(1-t_2)\sigma_{x_2 y_1} + t_2(1-t_1)\sigma_{x_1 y_2} + t_1 t_2 \sigma_{x_2 y_2} \\
\sigma_{y_{t_2} x_{t_1}} = (1-t_1)(1-t_2)\sigma_{y_1 x_1} + t_2(1-t_1)\sigma_{y_2 x_1} + t_1(1-t_2)\sigma_{y_1 x_2} + t_1 t_2 \sigma_{y_2 x_2} \\
\sigma^2_{y_{t_1} y_{t_2}} = (1-t_1)(1-t_2)\sigma^2_{y_1} + \left[t_1(1-t_2) + t_2(1-t_1) \right]\sigma_{y_1 y_2} + t_1 t_2 \sigma^2_{y_2}
\end{cases} \tag{4.68}
$$

In the extreme case that $t_1 = t_2$, Equation 4.68 becomes

$$
\begin{cases}
\sigma^2_{x_{t_1} x_{t_2}} = (1-t)^2 \sigma^2_{x_1} + 2\left[t(1-t) \right]\sigma_{x_1 x_2} + t^2 \sigma^2_{x_2} \\
\sigma_{x_{t_1} y_{t_2}} = (1-t)^2 \sigma_{y_1 x_1} + t(1-t)(\sigma_{y_2 x_1} + \sigma_{y_1 x_2}) + t^2 \sigma_{y_2 x_2} \\
\sigma_{y_{t_2} x_{t_1}} = (1-t)^2 \sigma_{x_1 y_1} + t(1-t)(\sigma_{x_2 y_1} + \sigma_{x_1 y_2}) + t^2 \sigma_{x_2 y_2} \\
\sigma^2_{y_{t_1} y_{t_2}} = (1-t)^2 \sigma^2_{y_1} + 2\left[t(1-t) \right]\sigma_{y_1 y_2} + t^2 \sigma^2_{y_2}
\end{cases} \tag{4.69}
$$

The matrix (4.67) is, thus, an extension of the conventional covariance matrix of a point.

4.7.3 THE G-BAND MODEL

Refer to Figure 4.7, and let the surface for the probability density function of line segment $Q_{21}Q_{22}$ be cut by the plane that is parallel to the x-O-y plane and projected onto the x-O-y plane. A class of concentric bands is then obtained. The concentric band, which is cut by the plane passing through the error ellipses of the endpoints of the line segment, is defined as the generic error band, the *G-band* (Shi and Liu, 2000).

The shape and size of the G-band vary with the shape of the surface for the probability density function. The surface's shape is determined by the matrix $\Sigma_{z_{t_1} z_{t_2}}$. The visualization of the G-band is derived in the following.

Suppose that Q_{21} and Q_{22} follow a two-dimensional normal distribution. Any point $Z_t = [x_t\ y_t]^T$ on line segment $Q_{21}Q_{22}$ follows the two-dimensional normal distribution

$$Z_t = \begin{bmatrix} x_t & y_t \end{bmatrix}^T \sim N_2\left(\mu_{z_t}, \Sigma_{z_t z_t}\right), \tag{4.70}$$

where $\Sigma_{z_t z_t}$ is derived from Equations 4.67 and 4.69, and μ_{z_t} is given by

$$\mu_{Z_t} = \begin{bmatrix} \mu_{x_t}, \mu_{y_t} \end{bmatrix}^T = \begin{bmatrix} (1-t)\mu_{x_1} + t\mu_{x_2}, (1-t)\mu_{y_1} + t\mu_{y_2} \end{bmatrix}^T. \tag{4.71}$$

A spectral analysis of the matrix then equals

$$\Sigma_{z_t z_t} = S_t \Lambda_t S_t^T = \begin{bmatrix} S_{t_1} & S_{t_2} \end{bmatrix} \begin{bmatrix} \lambda_{t_1} & 0 \\ 0 & \lambda_{t_2} \end{bmatrix} \begin{bmatrix} S_{t_1}^T \\ S_{t_2}^T \end{bmatrix}, \tag{4.72}$$

where Λ_t is the spectral matrix constructed by the eigenvalues λ_{t_1} and λ_{t_2} of the matrix $\Sigma_{z_t z_t}$; S_t is a standardized orthogonal matrix being a composite of the eigenvectors S_{t_1} and S_{t_2} corresponding to the eigenvalues λ_{t_1} and λ_{t_2}, respectively.

The semiaxes and the direction to the major semi axis of the error ellipse for an arbitrary implicit point on the line segment are expressed as

$$\begin{cases} A_t^2 = \sigma_0^2 \lambda_{t_1} = \left[\sigma_{x_t}^2 + \sigma_{y_t}^2 + \omega \right]/2 \\ B_t^2 = \sigma_0^2 \lambda_{t_2} = \left[\sigma_{x_t}^2 + \sigma_{y_t}^2 - \omega \right]/2, \\ \tan\left(2\phi_t\right) = 2\sigma_{x_t y_t} / \left[\sigma_{x_t}^2 - \sigma_{y_t}^2 \right] \end{cases} \tag{4.73}$$

where

$$\omega = \sqrt{\left(\sigma_{x_t}^2 - \sigma_{y_t}^2\right)^2 + 4\sigma_{x_t y_t}^2}.$$

By varying $t \in (0, 1)$, infinitely many error ellipses of individual implicit points $[x_t\ y_t]^T$ on the line segment can be drawn. This set of the error ellipses describes the nature of the positional error in the line segment. The boundaries of the error ellipses of both the implicit points and the endpoints of the line segment construct the band-shaped region around the true or mean location of the line segment.

When λ_{t_1} and λ_{t_2} are equal at $t = \tau$, the error in Z_τ is minimal; that is,

$$\lambda_{\tau_i} = \min\left\{\lambda_{t_i}, \forall t \in [0,1]\right\}. \tag{4.74}$$

The error ellipse determined by λ_τ is defined as the critical error ellipse. Therefore, the shape and size of the G-band are determined by the error ellipses of the endpoints of the line segment and the critical error ellipse of Z_τ.

4.7.4 STATISTICAL CHARACTERISTICS OF G-BAND

4.7.4.1 Directional Independence

It is first supposed that the positional error in the x- and y-components of the two endpoints of the line segment are independent and equal: $\sigma_{x_1}^2 = \sigma_{y_1}^2 = \sigma_1^2$ and $\sigma_{x_2}^2 = \sigma_{y_2}^2 = \sigma_2^2$. The matrix given in Equation 4.67 then simplifies as

$$\Sigma_{z_{t_1} z_{t_2}} = \left[(1-t_1)(1-t_2)\sigma_1^2 + t_1 t_2 \sigma_2^2\right] I_2, \tag{4.75}$$

where I_2 is the two-dimensional identity matrix.

Let R denote the following two-dimensional rotation matrix,

$$R = \begin{bmatrix} \cos\gamma & \sin\gamma \\ -\sin\gamma & \cos\gamma \end{bmatrix}, \tag{4.76}$$

where γ is a rotation angle. Equation 4.75 then further simplifies as

$$R\Sigma_{z_{t_1} z_{t_2}} R^T = \Sigma_{z_{t_1} z_{t_2}}. \tag{4.77}$$

The matrix $\Sigma_{z_{t_1} z_{t_2}}$ is thus independent of the rotation angle. Therefore, it can be concluded that, under the assumption that the two endpoints of the line segment are independent and equal, the line segment error is directionally independent. Under the condition of directional independence, Equation 4.69 is simplified as

$$\Sigma_{Z_t Z_t} = \left[(1-t)^2 \sigma_1^2 + t^2 \sigma_2^2\right] I_2. \tag{4.78}$$

This indicates that the error ellipse at any implicit point on the line segment becomes an error circle. However, the radii of the error circles of the endpoints of the line segment are unequal. In order to resolve the extreme value, let

$$\left(d\Sigma_{z_t z_t}/dt\right)\big|_{t_\Delta} = 0. \tag{4.79}$$

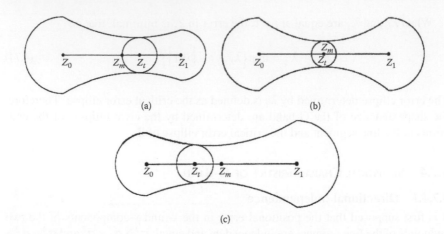

FIGURE 4.8 The directional independence characteristic of the G-band: (a) $t_\Delta < 1/2$, (b) $t_\Delta = 1/2$, and (c) $t_\Delta > 1/2$. (From Shi, W. Z. and Liu, W. B., 2000. *Int. J. Geog. Inform. Sci.* 14: 51–66. With permission.)

We then, have

$$t_\Delta = \sigma_1^2 \Big/ \left(\sigma_1^2 + \sigma_2^2 \right). \tag{4.80}$$

When $t = t_\Delta$, the diagonal elements of matrix Σ_{ztzt} will equal their minimum values. This corresponds to the minimum radii of the error circles.

The location of the minimum error circle on the line segment, according to Equation 4.80, is now discussed

- If $\sigma_1^2 > \sigma_2^2$, and thus $t_\Delta > 1/2$, then the minimum error circle on the line segment approaches to endpoint Q_{22}.
- If $\sigma_1^2 = \sigma_2^2$, and thus $t_\Delta = 1/2$, then the minimum error circle on the line segment is located at the middle point of line segment $Q_{21}Q_{22}$.
- If $\sigma_1^2 < \sigma_2^2$, and thus $t_\Delta < 1/2$, then the minimum error circle on the line segment approaches to endpoint Q_{21}.

From the above, it can be deduced that the minimum error on the line segment is always closer to the endpoint with the smallest error circle. Figure 4.8 shows the relationship between the error circles of the endpoints of a line segment and the minimum error circle.

4.7.4.1.1   Homogeneity and Directional Independence

Suppose that the positional error in the x- and y-directions of the two endpoints of the line segment is homogeneous, independent, and equal. Their covariance matrix $\Sigma_{z t1 z t2}$ meets the following conditions:

$$\Sigma_{z_{t1} z_{t2}} = \Phi_{ij} \left(t_1, t_2 \right) \left(e^i \otimes e^j \right) \tag{4.81}$$

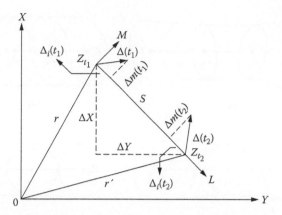

FIGURE 4.9 The geometric meaning of the correlation function in the parallel and perpendicular directions. (From Shi, W. Z. and Liu, W. B., 2000. *Int. J. Geog. Inform. Sci.* 14: 51–66. With permission.)

and

$$\Phi_{ij}\left(t_1, t_2\right) = \Phi_m\left(L_\tau\right)\delta_{ij} + \left[\Phi_l\left(L_\tau\right) - \Phi_m\left(L_\tau\right)\right] / L^2\left(\Delta x_i \Delta y_i / \tau^2\right), \qquad (4.82)$$

where e^i and e^j are spatial base vectors; τ is the normalized distance between the points Z_{t_1} and Z_{t_2} of line segment $Q_{21}Q_{22}$; L is the total length of the line segment; and $\Phi_l(L_\tau)$ and $\Phi_m(L_\tau)$ are the correlation coefficients in the parallel and perpendicular directions. They meet the boundary condition of $\Phi_l(0)$ and $\Phi_m(0)$. The δ_{ij} is the Kronecker δ-function, that is,

$$\delta_{ij} = \begin{cases} 1 \text{ when } i = j \\ 0 \text{ when } i \neq j \end{cases} \quad i, j \in 1, 2 \underline{\Delta} x(t), y(t), \qquad (4.83)$$

where $\Delta x = x(t_2) - x(t_1)$ and $\Delta y = y(t_2) - y(t_1)$. The subscripts in Δx_i and Δy_j $(i, j = 1, 2)$ indicate the two projected coordinate components.

The geometrical meaning of the correlation function in the parallel and perpendicular directions is illustrated in Figure 4.9. Vectors Δ_{t_1} and Δ_{t_2} represent the true value of the error at points Z_{t_1} and Z_{t_2}, respectively. Variables $\Delta_l(t_i)$ and $\Delta_m(t_i)$ are projections of Δ_{t_i} $(i = 1, 2)$ in the parallel and perpendicular directions, respectively.

The following definitions of $\Phi_l(L_\tau)$ and $\Phi_m(L_\tau)$ are needed:

$$\begin{bmatrix} \Phi_l\left(L_\tau\right) = E\left\{\Delta_l\left(t_1\right)\Delta_l\left(t_2\right)\right\} \\ \Phi_m\left(L_\tau\right) = E\left\{\Delta_m\left(t_1\right)\Delta_m\left(t_2\right)\right\} \\ \Phi_{lm}\left(L_\tau\right) = E\left\{\Delta_l\left(t_1\right)\Delta_m\left(t_2\right)\right\} \end{bmatrix} \qquad (4.84)$$

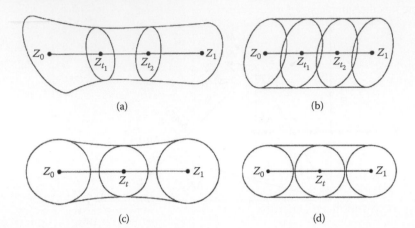

FIGURE 4.10   Error bands of stochastic process-based line segments with different statistical characteristics. (From Shi, W. Z. and Liu, W. B., 2000. *Int. J. Geog. Inform. Sci.* 14: 51–66. With permission.)

In Equation 4.82, if $t_1 = t_2 = t$, we have

$$\Phi_{ij}(t) = \Phi_m(L_\tau)I_2 + \left[\Phi_l(L_\tau) - \Phi_m(L_\tau)\right]\begin{bmatrix} \cos^2\gamma & \sin\gamma\cos\gamma \\ \sin\gamma\cos\gamma & \sin^2\gamma \end{bmatrix}, \qquad (4.85)$$

where γ is the coordinate-bearing angle of the line segment. According to procedures in Grafarend and Sanso (1985), values of $\Phi_l(L_\tau)$ and $\Phi_m(L_\tau)$ are obtained. In addition, the covariance matrix of any implicit point Z_t on the line segment $Q_{21}Q_{22}$ is an identity matrix. Therefore, error ellipses of all intermediate points on the line segment are error circles with identical radii. In this case, the G-band is a parallel band along the line segment.

Homogeneity of a line segment means that the error ellipse for any point on the line segment is translation independent. Geometrically, this indicates, on the one hand, that the size of the error ellipse and the direction of its major axis are the same for any two points on the line segment (Figure 4.10b). However, directional independence means that the error ellipse for any point is rotationally independent.

Geometrically, this means that all error ellipses become error circles (Figure 4.10c). Both homogeneity and directional independence mean that any error ellipse centered at a point on the line segment is simultaneously translation and rotationally independent. Therefore, all error ellipses on the line segment become error circles with the same radius (Figure 4.10d).

With the graphic interface of a commercial GIS, it is possible to implement the G-band model. The error, including the quantity and direction of error, of an arbitrary implicit point on the line segment can thus be visualized.

4.7.5 Note on the G-Band Error Model

In this section, the G-band model has been introduced. The model can handle positional error for a line segment when the two endpoints have correlated positional error. The model gives the analytical relationships between the band shape and size with the error at the two endpoints and also relationships between them.

For a directionally independent stochastic process, the G-band degenerates to the error-band models. If the line is simultaneously homogeneous and directionally independent, the G-band model degenerates to a parallel band along the line segment, such as the ε-band model.

Having introduced the error modeling methods for a straight line in Sections 4.5 to 4.7, in the following section, the focus now turns to the error modeling method for a curved line.

4.8 MODELING POSITIONAL UNCERTAINTY OF A CURVE

Spatial objects in GISci may be represented by another line type, the curve. For example, coastlines, roads, borderlines, and contour lines can be represented as curves. A curved line can be approximated by one of the following three methods: (a) a series of straight-line segments, (b) a regular curve model, or (c) an irregular curve model.

In method (a), a curve is represented as a series of straight-line segments. This method has been widely adopted in GISci. The regular curve model of the method (b) is to approximate a regular curve entity based on an analytical function. A circular or elliptical curve, for example, can be used to express curved lines. In method (c), an irregular curve is approximated by various kinds of interpolation functions, such as polynomial or spline functions.

Goodchild and Hunter (1997) proposed an error model to assess positional error in digitized linear features. The model was further applied to estimate the error in a coastline, which was approximated by a series of straight-line segments. Huang and Liu (1997) estimated the digitization errors in contours based on a line series model, where the contours were approximated by the irregular curve model.

In fact, modeling error of the method (a) is similar to the error modeling for a polyline, which has been described earlier in this chapter. Therefore, in this section, the concentration is on error modeling for curve lines that are approximated by using either methods (b) or (c). Two error indicators are used to describe curve line errors: the root mean square error in the normal direction of a curve and the maximum error at a point of the curve (Shi et al., 2000).

4.8.1 Positional Error at Any Point on a Curve

In this section, we will derive the variance of an intermediate point on the curve based on the covariance of the two endpoints and that of the measured distance between one of the two endpoints and the intermediate point on the curve.

FIGURE 4.11 A curve of two explicit points (Q_1 and Q_2).

A curve, such as in Figure 4.11 consists of two known (or explicit) points $Q_1 = (x_1, y_1)$ and $Q_2 = (x_2, y_2)$, an intermediate point $Q_i = (x_i, y_i)$ between the two explicit points. This curve is expressed as

$$f(x, y, \alpha, \beta) = 0,$$
(4.86)

where (x, y) is a point on the curve; and $\alpha = [\alpha_1, \alpha_2, ..., \alpha_n]^T$ and $\beta = [\beta_1, \beta_2, ..., \beta_m]^T$ are two parameter vectors of the function. This yields the following expressions

$$\begin{cases} f(x_1, y_1, \alpha, \beta) = 0 \\ f(x_2, y_2, \alpha, \beta) = 0. \\ f(x_i, y_i, \alpha, \beta) = 0 \end{cases}$$
(4.87)

The distance $s_{1,i}$ between explicit the point Q_1 and intermediate point Q_i is given as

$$s_{1,i}^2 = (x_i - x_1)^2 + (y_i - y_1)^2.$$
(4.88)

Assume that two parameter vectors in Equation 4.86 contain errors; by differentiating Equations 4.87 and 4.88, we can obtain the differential equation between the coordinates of the points on the curve and the parameter vectors of the curve as follows

$$\begin{cases} A_{\alpha_1}\partial\alpha + A_{\beta_1}\partial\beta + a_{x_1}dx_1 + a_{y_1}dy_1 = 0 \\ A_{\alpha_2}\partial\alpha + A_{\beta_2}\partial\beta + a_{x_2}dx_2 + a_{y_2}dy_2 = 0 \\ A_{\alpha_i}\partial\alpha + A_{\beta_i}\partial\beta + a_{x_i}dx_i + a_{y_i}dy_i = 0 \\ s_{1,i}ds_{1,i} = \Delta x_{1i}(dx_i - dx_1) + \Delta y_{1i}(dy_i - dy_1) \end{cases},$$
(4.89)

where $\partial\alpha = [\partial\alpha_1, \partial\alpha_2, ..., \partial\alpha_n]$, $\partial\alpha = [\partial\beta_1, \partial\beta_2, ..., \partial\beta_m]$, $A_{\alpha j}$ ($j = 1, 2, i$) and $A_{\beta j}$ ($j = 1, 2, i$) are the coefficient matrices, a_{xj} ($j = 1, 2, i$) and a_{yj} ($j = 1, 2, i$) are the coefficients, and $\Delta x_{1i} = x_i - x_1$, $\Delta y_{1i} = y_i - y_1$.

A matrix form of the first two expressions in Equation 4.89 is

$$\partial\beta = - \begin{bmatrix} A_{\beta_1} \\ A_{\beta_2} \end{bmatrix}^{-1} \begin{bmatrix} a_{x_1} & a_{y_1} & 0 & 0 & A_{\alpha_1} \\ 0 & 0 & a_{x_2} & a_{y_2} & A_{\alpha_2} \end{bmatrix} \begin{bmatrix} dx_1 \\ dy_1 \\ dx_2 \\ dy_2 \\ \partial\alpha \end{bmatrix}^T . \tag{4.90}$$

Equation 4.90 is put into the last two expressions of Equation 4.89 to obtain a derivative of (x_i, y_i),

$$\begin{bmatrix} dx_i' \\ dy_i' \end{bmatrix} = \begin{bmatrix} A_x & A_{xs} & 0 \\ A_y & A_{ys} & 0 \end{bmatrix} \begin{bmatrix} d\eta \\ ds_{1,i} \\ ds_{2,i} \end{bmatrix} = Ad\zeta , \tag{4.91}$$

where $d\eta = [dx_1, dy_1, dx_2, dy_2, \partial\alpha]$;

$$A = \begin{bmatrix} A_x & A_{xs} & 0 \\ A_y & A_{ys} & 0 \end{bmatrix} = \begin{bmatrix} \Delta x_{1i} & \Delta y_{1i} \\ a_{x_i} & a_{y_i} \end{bmatrix}^{-1} \begin{bmatrix} \alpha_{i1} & s_{1,i} & 0 \\ \alpha_{i2} & 0 & 0 \end{bmatrix} ;$$

$$\alpha_{i1} = [\Delta x_{1i}, \Delta y_{1i}, 0, 0, 0];$$

$$\alpha_{i2} = \begin{bmatrix} 0 \\ -A_{\alpha_i} \end{bmatrix} + A_{\beta_i} \begin{bmatrix} A_{\beta_1} \\ A_{\beta_2} \end{bmatrix}^{-1} \begin{bmatrix} a_{x_1} & a_{y_1} & 0 & 0 & A_{\alpha_1} \\ 0 & 0 & a_{x_2} & a_{y_2} & A_{\alpha_2} \end{bmatrix} ; \text{ and}$$

$$d\zeta = [dx_1, dy_1, dx_2, dy_2, \partial\alpha, ds_{1,i}, ds_{2,i}]^T.$$

Considering the distance $s_{2,i}$ between explicit point Q and implicit point Q_i in Equation 4.89, the obtained derivative of (x_i, y_i) will be

$$\begin{bmatrix} dx_i'' \\ dy_i'' \end{bmatrix} = \begin{bmatrix} B_x & 0 & B_{xs} \\ B_y & 0 & B_{ys} \end{bmatrix} \begin{bmatrix} d\eta \\ ds_{1,i} \\ ds_{2,i} \end{bmatrix} = Bd\zeta , \tag{4.92}$$

where

$$
B = \begin{bmatrix} B_x & 0 & B_{xs} \\ B_y & 0 & B_{ys} \end{bmatrix} = \begin{bmatrix} \Delta x_{2i} & \Delta y_{2i} \\ a_{x_i} & a_{y_i} \end{bmatrix}^{-1} \begin{bmatrix} \beta_{i1} & 0 & s_{2,i} \\ \beta_{i2} & 0 & 0 \end{bmatrix};
$$

$$
\beta_{i1} = \begin{bmatrix} \Delta x_{2i}, \Delta y_{2i}, 0, 0, 0 \end{bmatrix};
$$

and

$$
\beta_{i2} = \begin{bmatrix} 0 \\ -A_{\alpha_i} \end{bmatrix} + A_{\beta_i} \begin{bmatrix} A_{\beta_1} \\ A_{\beta_2} \end{bmatrix}^{-1} \begin{bmatrix} a_{x_1} & a_{y_1} & 0 & 0 & A_{\alpha_1} \\ 0 & 0 & a_{x_2} & a_{y_2} & A_{\alpha_2} \end{bmatrix}.
$$

4.8.2 THE ε_σ ERROR MODEL

Positional error at any point between the explicit points Q_1 and Q_2 can be described by its variance based on Equations 4.91 and 4.92. Obtained from these two equations is the weighted average of differential equation, with weights equal to $1/\widehat{s}_{1,i}$ and $1/\widehat{s}_{2,i}$, where $\widehat{s}_{j,i}$ is the arc length between Q_i and Q_j for $j = 1$ or 2, respectively:

$$
\begin{bmatrix} dx_i \\ dy_i \end{bmatrix} = (1 - r_1) \begin{bmatrix} dx_i' \\ dy_i' \end{bmatrix} + r_1 \begin{bmatrix} dx_i'' \\ dy_i'' \end{bmatrix} = (A - r_1 A + r_1 B) \cdot d\zeta, \tag{4.93}
$$

where $r_1 = \widehat{s}_{1,i} / \widehat{s}_{1,2}$, and $\widehat{s}_{1,2}$ is the arc length between explicit points Q_1 and Q_2.

According to the error propagation law, the variance-covariance matrix of point k_i is expressed as

$$
D_i = \begin{bmatrix} \sigma_{x_i}^2 & \sigma_{x_i y_i} \\ \sigma_{y_i x_i} & \sigma_{y_i}^2 \end{bmatrix} = (A - r_1 A + r_1 B) \cdot D_\zeta \cdot (A - r_1 A + r_1 B)^T, \tag{4.94}
$$

where $\sigma_{x_i}^2$, $\sigma_{y_i}^2$, and $\sigma_{x_i y_i}$ are the variance of x_i, the variance of y_i and the covariance of x_i and y_i, respectively; D_ζ is the variance-covariance matrix of Q_1 and Q_2. The root mean square error (ε_σ) in the normal direction of a curve at implicit point Q_i is derived next in the following as an error indicator for this point.

Let θ denote the azimuth of this normal direction. Point Q_i in the normal direction with azimuth of θ is then expressed as

$$
u_\theta = \cos\theta x_i + \sin\theta y_i \tag{4.95}
$$

and its variance equals

$$
\sigma_\theta^2 = \cos^2\theta \cdot \sigma_{x_i}^2 + \sin^2\theta \cdot \sigma_{y_i}^2 + \sin 2\theta \cdot \sigma_{x_i y_i}. \tag{4.96}
$$

The standard deviation σ_θ of Q_i in this normal direction is called ε_σ and is illustrated in Figure 4.12. This is the distance between Q_i and intersection point G_θ of the error

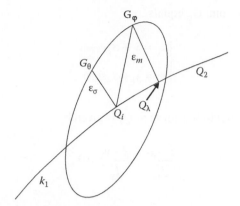

FIGURE 4.12   The ε_σ and ε_m error models for a curve.

ellipse of Q_i and its normal direction. The locus of ε_σ along the curve generates an error region around the curve. This error region is defined as the ε_σ band.

4.8.3   THE ε_m ERROR MODEL

As discussed in the above section, based on the error ellipse of the intermediate point on the curve, we have derived the standard deviation of the point in the normal direction of the curve. By analyzing the error ellipses of the infinite points on the curve, we find, however, that the error indicator ε_σ represented by the standard deviation in the normal direction of the curve may not be the maximum between the error ellipse of Q_i and the curve. Therefore, in this section, we investigate the maximum distance from the point on the error ellipse of an arbitrary point perpendicular to the curve. The aggregation of the maximum distances of all the intermediate points on the curve defines the maximum error region of the curve, based on the error ellipses of all the points. In Figure 4.12, for example, the distance ε_σ (between point Q_i and its error ellipse in its normal direction G_θ) is smaller than the distance ε_m (between the point G_φ and the point Q_λ) on the curve in the normal direction at Q_λ. Therefore, the maximum error model (ε_m) is further proposed to assess the positional error on a curve.

As illustrated in Figure 4.12, let $Q_\lambda = (x_\lambda, y_\lambda)$ denote the point on a curve that has the maximum distance to a point on the error ellipse (with the error ellipse's center at Q_i). The distance is in the normal direction to the curve toward to a point on the error ellipse. Let $G_\varphi = (x_\varphi, y_\varphi)$ denote the intersection point between the error ellipse and the normal direction line passing through the point Q_λ. The distance between Q_λ and G_φ is defined as an error indicator for point Q_i, the maximum error model, which can be derived below.

Let φ denote the azimuth of the straight-line segment of endpoints Q_i and G_φ. The variance of point Q_i in the direction with azimuth of φ can be represented as

$$\sigma_\varphi^2 = \cos^2\varphi \cdot \sigma_{x_i}^2 + \sin^2\varphi \cdot \sigma_{y_i}^2 + \sin 2\varphi \cdot \sigma_{x_i y_i} \tag{4.97}$$

and the coordinate of point G_φ equals

$$\begin{cases} x_\varphi = x_i + \sigma_\varphi \cos\varphi \\ y_\varphi = y_i + \sigma_\varphi \sin\varphi \end{cases}. \tag{4.98}$$

Furthermore, the normal to the curve at Q_λ is given by

$$\frac{x - x_\lambda}{f'_{x_\lambda}} - \frac{y - y_\lambda}{f'_{y_\lambda}} = 0, \tag{4.99}$$

where

$$f'_{x_\lambda} = \left(\partial f / \partial x\right)\big|_{(x_\lambda, y_\lambda)} \text{ and } f'_{y_\lambda} = \left(\partial f / \partial y\right)\big|_{(x_\lambda, y_\lambda)}.$$

Point G_φ passing through this normal must also satisfy Equation 4.100:

$$\left(x_i - x_\lambda + \sigma_\varphi \cos\varphi\right) f'_{y_\lambda} - \left(y_i - y_\lambda + \sigma_\varphi \sin\varphi\right) f'_{x_\lambda} = 0. \tag{4.100}$$

To maximize the distance $s_{\varphi,\lambda}$ between point G_φ and Q_λ, we have the nonlinear program:

$$\text{maximum } s_{\varphi,\lambda}^2 = \left(x_i - x_\lambda + \sigma_\varphi \cos\varphi\right)^2 + \left(y_i - y_\lambda + \sigma_\varphi \sin\varphi\right)^2 \tag{4.101}$$

subject to the constraints given in Equations 4.86, 4.97, and 4.100.

When addressing these constraints with first-order Taylor series expansions around φ^0, σ_φ^0, x_λ^0, and y_λ^0, the optimization problem given in Equation 4.101 becomes

$$\text{maximum } s_{\varphi,\lambda}^2 = \left(x_i - x_\lambda + \sigma_\varphi \cos\varphi\right)^2 + \left(y_i - y_\lambda + \sigma_\varphi \sin\varphi\right)^2 \tag{4.102}$$

$$\text{subject to } \alpha_{\varphi_1} \delta\varphi + \alpha_{\sigma_1} \delta\sigma_\varphi + w_1 = 0$$

$$\alpha_{\varphi_2} \delta\varphi + \alpha_{\sigma_2} \delta\sigma_\varphi + \alpha_{x\lambda_2} \delta x_\lambda + \alpha_{y\lambda_2} \delta y_\lambda + w_2 = 0$$

$$\alpha_{x\lambda_3} \delta x_\lambda + \alpha_{y\lambda_3} \delta y_\lambda + w_3 = 0,$$

where $\alpha_{\varphi_1} = \sigma_{x_i}^2 \sin2\varphi^0 - \sigma_{y_i}^2 \sin2\varphi^0 - 2\sigma_{x_i y_i} \cos2\varphi^0$;

$\alpha_{\sigma_1} = 2\sigma_\varphi^0$;

$w_1 = \left(\sigma_\varphi^0\right)^2 - \left(\sigma_{x_i}^2 \cos^2\varphi^0 + \sigma_{y_i}^2 \sin^2\varphi^0 + \sigma_{x_i y_i} \sin2\varphi^0\right)$;

$\alpha_{\varphi_2} = -\sin\varphi^0 \sigma_\varphi^0 \left(f'_{y_\lambda}\right)^0 - \cos\varphi^0 \sigma_\varphi^0 \left(f'_{x_\lambda}\right)^0$;

$$\alpha_{\sigma_2} = \cos\varphi^0 \left(f'_{y_\lambda}\right)^0 + \sin\varphi^0 \left(f'_{x_\lambda}\right)^0 ;$$

$$\alpha_{x\lambda_2} = -\left(f'_{y_\lambda}\right)^0 + \left(x_i - x_\lambda^0 + \sigma_\varphi^0 \cos\varphi^0\right)\left(f''_{y_\lambda}\right)^0 - \left(y_i - y_\lambda^0 + \sigma_\varphi^0 \sin\varphi^0\right)\left(f''_{x_\lambda}\right)^0 ;$$

$$\alpha_{y\lambda_2} = \left(f'_{x_\lambda}\right)^0 + \left(x_i - x_\lambda^0 + \sigma_\varphi^0 \cos\varphi^0\right)\left(f''_{y_\lambda}\right)^0 - \left(y_i - y_\lambda^0 + \sigma_\varphi^0 \sin\varphi^0\right)\left(f''_{x_\lambda}\right)^0 ;$$

$$w_2 = \left(x_i - x_\lambda^0 + \sigma_\varphi^0 \cos\varphi^0\right)\left(f'_{y_\lambda}\right)^0 - \left(y_i - y_\lambda^0 + \sigma_\varphi^0 \sin\varphi^0\right)\left(f'_{x_\lambda}\right)^0 ;$$

$$\alpha_{x\lambda_3} = \left(f'_{x_\lambda}\right)^0 ; \text{ and}$$

$$\alpha_{y\lambda_3} = \left(f'_{y_\lambda}\right)^0$$

$$w_3 = f^0 .$$

The Lagrangian theory associated with the nonlinear program over the equality constraints is then applied as

$$L = \left(x_i - x_\lambda + \sigma_\varphi \cos\varphi\right)^2 + \left(y_i - y_\lambda + \sigma_\varphi \sin\varphi\right)^2$$

$$+ 2v_1 \left(\alpha_{\varphi_1}\delta\varphi + \alpha_{\sigma_1}\delta\sigma_\varphi + w_1\right)$$

$$+ 2v_2 \left(\alpha_{\varphi_2}\delta\varphi + \alpha_{\sigma_2}\delta\sigma_\varphi + \alpha_{x\lambda_2}\delta x_\lambda + \alpha_{y\lambda_2}\delta y_\lambda + w_2\right)$$

$$+ 2v_3 \left(\alpha_{x\lambda_3}\delta x_\lambda + \alpha_{y\lambda_3}\delta y_\lambda + w_3\right) , \tag{4.103}$$

where v_1, v_2, and v_3 are the Lagrange multipliers for the constraints given in Equation 4.102.

Setting the derivative of L with respect to each of its variables to be zero, we get the following set of first-order optimality necessary conditions:

$$\frac{\partial L}{\partial \delta\varphi} = -2u_1\sigma_\varphi^0 \sin\varphi^0 + 2u_2\sigma_\varphi^0 \cos\varphi^0 + 2k_1\alpha_{\varphi_1} + 2k_2\alpha_{\varphi_2} = 0$$

$$\frac{\partial L}{\partial \delta\sigma_\varphi} = 2u_1 \cos\varphi^0 + 2u_2 \sin\varphi^0 + 2k_1\alpha_{\sigma_1} + 2k_2\alpha_{\sigma_2} = 0 \tag{4.104}$$

$$\frac{\partial L}{\partial \delta x_\lambda} = -2u_1 + 2k_2\alpha_{x\lambda_2} + 2k_3\alpha_{x\lambda_3} = 0$$

$$\frac{\partial L}{\partial \delta y_\lambda} = -2u_2 + 2k_2\alpha_{y\lambda_2} + 2k_3\alpha_{y\lambda_3} = 0 ,$$

where $u_1 = x_i - x_\lambda + \sigma_\varphi\cos\varphi$ and $u_2 = y_i - y_\lambda + \sigma_\varphi\sin\varphi$.

If the functions of variables u_1 and u_2 are expanded with first-order Taylor series with respect to the variables around the points u_1^0 and u_2^0, respectively, the following equation will be obtained:

$$\begin{cases} u_1 = u_1^0 + \cos\varphi^0 \, \delta\sigma_\varphi - \sigma_\varphi^0 \sin\varphi^0 \, \delta\varphi - \delta x_\lambda \\ u_2 = u_2^0 + \sin\varphi^0 \, \delta\sigma_\varphi + \sigma_\varphi^0 \cos\varphi^0 \, \delta\varphi - \delta y_\lambda \end{cases}. \tag{4.105}$$

Equation 4.105 is then substituted into Equation 4.104, and the result is grouped with the three constraints given in Equation 4.102. This yields the following set of equations:

$$\begin{cases} \left(\sigma_\varphi^0\right)^2 \delta\varphi + \sigma_\varphi^0 \, \sin\varphi^0 \delta x_\lambda - \sigma_\varphi^0 \, \cos\varphi^0 \delta y_\lambda + \alpha_{\varphi_1} v_1 + \alpha_{\varphi_2} v_2 \\ \quad + \left(-u_1^0 \sigma_\varphi^0 \, \sin\varphi^0 + u_2^0 \sigma_\varphi^0 \, \cos\varphi^0\right) = 0 \\ \delta\sigma_\varphi - \cos\varphi^0 \delta x_\lambda - \sin\varphi^0 \delta y_\lambda + \alpha_{\sigma_1} v_1 + \alpha_{\sigma_2} v_2 \\ \quad + \left(u_1^0 \, \cos\varphi^0 + u_2^0 \, \sin\varphi^0\right) = 0 \\ \sin\varphi^0 \sigma_\varphi^0 \delta\varphi - \cos\varphi^0 \delta\sigma_\varphi + \delta x_\lambda + \alpha_{x\lambda_2} v_2 + \alpha_{x\lambda_3} v_3 - u_1^0 = 0 \\ -\cos\varphi^0 \sigma_\varphi^0 \delta\varphi - \sin\varphi^0 \delta\sigma_\varphi + \delta y_\lambda + \alpha_{y\lambda_2} v_2 + \alpha_{y\lambda_3} v_3 - u_2^0 = 0 \\ \alpha_{\varphi_1} \delta\varphi + \alpha_{\sigma_1} \delta\sigma_\varphi + w_1 = 0 \\ \alpha_{\varphi_2} \delta\varphi + \alpha_{\sigma_2} \delta\sigma_\varphi + \alpha_{x\lambda_2} \delta x_\lambda + \alpha_{y\lambda_2} \delta y_\lambda + w_2 = 0 \\ \alpha_{x\lambda_3} \delta x_\lambda + \alpha_{y\lambda_3} \delta y_\lambda + w_3 = 0 \end{cases}. \tag{4.106}$$

Choosing the approximated initial values: $\sigma_\phi^0 = \sigma_\varepsilon$, $\phi^0 = \theta$, $x_\lambda^0 = x_i$, and $y_\lambda^0 = y_i$, the values of the parameters $\delta\varphi$, $\delta\sigma_\varphi$, δx_λ, and δy_λ can be obtained by solving Equation 4.106 with an iterative method. The maximum error $s_{\varphi,\lambda}$ for point Q_i and its azimuth φ are then determined. This maximum error is denoted as ε_m at Q_i in Figure 4.12. Therefore, the locus of ε_m around the curve is defined as the ε_m model for the curve.

4.8.4   AN ERROR MODEL FOR IRREGULAR CURVE: THIRD-ORDER SPLINE CURVE

The previous section has addressed the error model for a regular curve, while this current section further introduces an error model for an irregular curve, which is approximated by a third-order spline curve. Here, the ε_m error model is applied to describe positional error in the curve. For integers k_1 and k_2, a spline function of degree 3 with knots at x_{k1} and x_{k2} is

$$B(x) = \sum_{j=k_1-1}^{k_2+1} c_j \Omega_j(x), \tag{4.107}$$

where k_1 and k_2 are any two integers of the third-order spline, c_j is the interpolation coefficient of the third-order spline function, and Ω_j is the basic function of the third-order spline and represented by

$$\Omega_j(x) = \Omega(x-j)$$

$$= \begin{cases} -\left|x-j\right|^3\Big/6 + \left(x-j\right)^2 - 2\left|x-j\right| + 4/3 & 1 < \left|x-j\right| < 2 \\ \left|x-j\right|^3\Big/2 - \left(x-j\right)^2 + 2/3 & \left|x-j\right| \le 1 \\ 0 & \left|x-j\right| \ge 2 \end{cases} \tag{4.108}$$

Let $Q_{i(k1)} = (x_{i(k1)}, y_{i(k1)})$ be any point between points (x_{k1}, y_{k1}) and (x_{k2}, y_{k2}) on the spline curve, where $x_{k1} < x_{i(k1)} < x_{k2}$. The value of the spline function at this point is determined by the basic functions $x = x_{k1-1}, x_{k1}, x_{k2}$, and x_{k2+1}, and the interpolation coefficient of the spline function:

$$y_i = c_{k1-1}\Omega_{-1}(z_i) + c_{k1}\Omega_0(z_i) + c_{k1+1}\Omega_1(z_i) + c_{k1+2}\Omega_2(z_i) \tag{4.109}$$

$$= c_{k1-1}\left(1-z_i\right)^3\Big/6 + c_{k1}\left[\left(2-z_i\right)^3\Big/6 - 4\left(1-z_i\right)^3\Big/6\right] + c_{k_1+1}\left[\left(1+z_i\right)^3\Big/6 - 4z_i^3/6\right]$$

$$+ c_{k_1+2}\, z_i^3\Big/6\,,$$

where $z_i = (x_i - x_{k1})/(x_{k2} - x_{k1})$.

To find the maximum error at $Q_{i(k1)}$, Equation 4.109 is differentiated:

$$dy_i = a_{x_i}dx_i + a_{-1i}dc_{k_1-1} + a_{0i}dc_{k_1} + a_{1i}dc_{k_1+1} + a_{2i}dc_{k_1+2}\,, \tag{4.110}$$

where

$$a_{x_i} = -c_{k1-1}\left(1-z_i\right)^3\Big/2 - c_{k1}\left[\left(2-z_i\right)^2 - 4\left(1-z_i\right)^2\right]$$

$$+ c_{k1+1}\left[\left(1+z_i\right)^2 - 4z_i^2\right]\Big/2 + c_{k1+2}\, z_i^2/2$$

$$a_{-1i} = \left(1-z_i\right)^3\Big/6$$

$$a_{0i} = \left(2-z_i\right)^3\Big/6 - 4\left(1-z_i\right)^3\Big/6$$

$$a_{1i} = \left(1+z_i\right)^3\Big/6 - 4z_i^3\Big/6$$

$$a_{2i} = z_i^3\Big/6\,.$$

When $x_{i(k1)} = x_{k1}$ (or $x_{i(k1)} = x_{k2}$), we have $z_{k1} = 0$ (or $z_{k2} = 0$). In either case, Equation 4.110 becomes

$$\begin{cases} dy_{k1} = a_{x1}dx_{k1} + \left(dc_{k1-1} + 4dc_{k1} + dc_{k2+1} \right)/6 \\ dy_{k2} = a_{x2}dx_{k2} + \left(dc_{k1} + 4dc_{k1+1} + dc_{k1+2} \right)/6 \end{cases}, \qquad (4.111)$$

where

$$a_{x1} = -c_{k1-1}/2 + c_{k1+1}/2 \text{ and } a_{x2} = -c_{k1}/2 + c_{k1+2}/2.$$

Let $s_{1,i}$ denote the distance between Q_{k1} and $Q_{i(k)}$. We have

$$\begin{cases} s_{1,i}^2 = \left(x_i - x_{k1} \right)^2 + \left(y_i - y_{k1} \right)^2 \\ s_{1,i}ds_{1,i} = \Delta x_{1i}\left(dx_i - dx_{k1} \right) + \Delta y_{1i}\left(dy_i - dy_{k1} \right) \end{cases}. \qquad (4.112)$$

Based on Equations 4.110, 4.111, and 4.112, the following equation is obtained:

$$\begin{bmatrix} dx_i' \\ dy_i' \end{bmatrix} = \begin{bmatrix} -a_{x_i} & 1 \\ \Delta x_{1i} & \Delta y_{1i} \end{bmatrix}^{-1} \begin{bmatrix} A_x & A_c & 0 & 0 \\ A_{sx} & 0 & s_{1i} & 0 \end{bmatrix} \begin{bmatrix} dX \\ dc \\ ds_{1i} \\ ds_{2i} \end{bmatrix}, \qquad (4.113)$$

where $dX = [dx_{k1}, dy_{k1}, dx_{k2}, dy_{k2}]^T$

$$dc = \begin{bmatrix} dc_{k1,} & dc_{k1+1} \end{bmatrix}^T,$$

$$A_x = 6\begin{bmatrix} a_{-1i}, & a_{2i} \end{bmatrix} \begin{bmatrix} -a_{x1} & 1 & 0 & 0 \\ 0 & 0 & -a_{x2} & 1 \end{bmatrix},$$

$$A_c = \begin{bmatrix} a_{0i} - 4a_{-1i} - a_{2i}, & a_{1i} - a_{-1i} - 4a_{2i} \end{bmatrix},$$

$$A_{sx} = \begin{bmatrix} \Delta x_{1i}, & \Delta y_{1i}, & 0, & 0 \end{bmatrix}.$$

Similarly, when the distance $s_{2,i}$ between Q_{k2} and $Q_{i(k)}$ is considered, we have

$$\begin{bmatrix} dx_i'' \\ dy_i'' \end{bmatrix} = \begin{bmatrix} -a_{x_i} & 1 \\ \Delta x_{2i} & \Delta y_{2i} \end{bmatrix}^{-1} \begin{bmatrix} A_x & A_c & 0 & 0 \\ B_{sx} & 0 & 0 & s_{2i} \end{bmatrix} \begin{bmatrix} dX \\ dc \\ ds_{1i} \\ ds_{2i} \end{bmatrix}, \qquad (4.114)$$

where $B_{sx} = [0, 0, \Delta x_{2i}, \Delta y_{2i}]$.

According to the methods given in Sections 4.8.1 and 4.8.2, the variance of any point $Q_{i(k)}$ on the third-order spline curve and the root mean square error ε_σ in the normal direction at this point can be calculated, respectively. The ε_σ band for this spline curve is thus formed.

Furthermore, the ε_m band for the spline curve can be determined in a similar way based on the method introduced in Section 4.8.3. When the third-order spline curve is expressed in the form of parameter functions,

$$\begin{cases} x = x(z_i) \\ y = y(z_i) \end{cases} \qquad (4.115)$$

and the normal to the spline curve at point $G_\lambda = (x_\lambda, y_\lambda)$ is

$$\dot{x}_\lambda(x - x_\lambda) + \dot{y}_\lambda(y - y_\lambda) = 0, \qquad (4.116)$$

where

$$\dot{x}_\lambda = (dx/dz)\Big|_{z=z_\lambda} = x_{k2} - x_{k1}$$

and

$$\dot{y}_\lambda = (dx/dz)\Big|_{z=z_\lambda} = -c_{k1-1}(1-z_\lambda)^2/2 - c_{k1}\left[(2-z_\lambda)^2 - 4(1-z_\lambda)^2\right]/2$$
$$+ c_{k1+1}\left[(1+z_\lambda)^2 - 4z_\lambda^2\right]/2 + c_{k2+2}z_\lambda^2/2.$$

In order to calculate the maximum error at $Q_{i(k)}$ in the direction of φ, variables, φ, σ_φ, and z_λ are considered as unknown parameters. According to the approach that forms the conditional adjustment equations discussed above, the ε_m error band for the third-order spline curve can be generated.

4.8.5 CASE STUDY AND ANALYSIS

A simulation experiment was carried out to test the ε_σ and ε_m error models. Table 4.2 shows coordinates of ten points $(x_1, y_1), (x_2, y_2), \ldots, (x_{10}, y_{10})$, being explicit points on a curve. These ten points are subject to the boundary constraint in which the second order of the derivative of the first and last points are equal to zero, $\ddot{y}_1 = \ddot{y}_{10} = 0$, where $\ddot{y}_i = d^2 y_i/dx^2$ for $i = 1$ or 10.

TABLE 4.2
Original Data for Digitizing a Curve

i	x_i (cm)	y_i (cm)
1	1.0	244.0
2	2.0	221.0
3	3.0	208.0
4	4.0	208.0
5	5.0	211.5
6	6.0	216.0
7	7.0	219.0
8	8.0	221.0
9	9.0	221.5
10	10.0	220.0

For the third-order spline curve passing through these ten points, the interpolation coefficients on the interval [0, 1] are

$$\begin{cases} c_0 = 268.8 \\ c_1 = 244.0 \\ c_2 = 219.2 \\ c_3 = 205.2 \end{cases}.$$

By substituting the values of these interpolation coefficients into Equation 4.110, any point on the spline curve on [1, 2] can thus be calculated.

Suppose positional errors at points (x_1, y_1) and (x_2, y_2) have identical variances of 8.71 cm^2 in the x- and y-directions, and the errors in these two directions are linearly independent:

$$\sigma_{x_1}^2 = \sigma_{y_1}^2 = \sigma_{x_2}^2 = \sigma_{y_2}^2 = 8.71 \text{ cm}^2 \text{ and } \sigma_{x_1 y_1} = \sigma_{x_2 y_2} = 0.$$

Further, suppose that the variances of a distance measurement $(s_{1,i})$ and interpolation coefficients $(c_0, c_1, c_2 \text{ and } c_3)$ are 1 cm^2. That is,

$$\sigma_{s_{1,i}}^2 = \sigma_{c0}^2 = \sigma_{c1}^2 = \sigma_{c2}^2 = \sigma_{c3}^2 = 1 \text{ cm}^2.$$

The parameters to the error ellipse of these two points and the error indicators for an arbitrary implicit point Q_{r1} between (x_1, y_1) and (x_2, y_2) on the spline curve are then calculated.

The results are given in Table 4.3, where r_1 is the ratio of the length of the curve between (x_1, y_1) and Q_{r1} to the length of the curve between (x_1, y_1) and (x_2, y_2); A_1 and B_1 are the semimajor and semiminor axes of the error ellipse of Q_{r1}, respectively, and Φ_i is the primary axis direction of Q_{r1} on the curve.

The results in Table 4.3 show that the values of the two error indicators for each point Q_{r1} with different r_1 are similar. The reason is that the angle of the primary

TABLE 4.3
The ε_σ and ε_m Error Indicators for the Curve on Interval [1, 2]

r_1	$\sigma_{x_i}^2$	$\sigma_{y_i}^2$	$\sigma_{x_i y_i}^2$	A_1	B_1	$\ddot{O}_i$	ε_σ	ε_m
0.10	4.098	8.002	−0.158	2.830	2.023	92-19-12	2.0243	2.0244
0.20	1.696	6.544	−0.199	2.560	1.242	92-20-54	1.3021	1.3022
0.30	0.766	5.622	−0.202	2.373	0.870	92-22-27	0.8751	0.8751
0.40	0.388	5.061	−0.196	2.251	0.616	92-24-2	0.6229	0.6230
0.50	0.259	4.847	−0.194	2.203	0.501	92-24-47	0.5093	0.5093
0.60	0.294	5.042	−0.199	2.247	0.534	92-23-50	0.5423	0.5423
0.70	0.544	5.615	−0.208	2.371	0.732	92-20-56	0.7373	0.7375
0.80	1.391	6.642	−0.205	2.579	1.176	92-14-6	1.1794	1.1794
0.90	3.603	8.029	−0.146	2.834	1.897	91-53-8	1.8981	1.8983

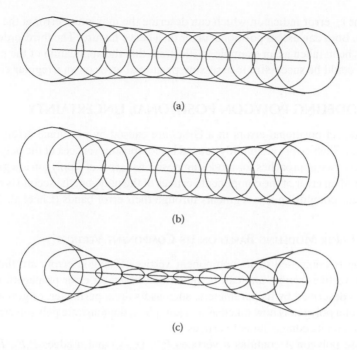

(a)

(b)

(c)

FIGURE 4.13 The error bands for (a) the straight line segment, (b) the circular curve, and (c) the spline curve.

axis direction of the error ellipse of Q_{r1} is approximate to 90°. Thus, little difference between ε_σ and ε_m occur.

The error band for the spline curve on the interval is next, compared with that for a straight-line segment of explicit points (x_1, y_1) and (x_2, y_2) and a circular curve passing through the two explicit points, under the same assumption of the error in the explicit points. This comparison aims to study the shape of the error band for different lines. Figure 4.13 shows the error bands for the straight-line segment in (a) the circular curve, in (b), and the spline curve in (c), respectively. The error band for the straight-line segment is significantly different from that of the spline curve. This is because the error model for the spline curve not only considers the error at the known points of the spline curve but also those in the distance measurement and the interpolation coefficients of the spline curve. Therefore, describing a curve with a series of straight-line segments in GIS may lead to a larger positional error, as compared to the use of curves.

In other words, the example shows that it is more precise to represent a curved object in GISci by a curve function than with a series of straight-line segments.

4.8.6 Note on Curve Error Models

The positional error on a curve feature has been described in Section 4.8. The curve feature error has been assessed by two error indicators: (a) the ε_σ error indicator is used to measure the any point error on a curve in the direction of the curve normal, and (b) the ε_m error indicator, which can describe the maximum error at the point to the curve. Both can assess positional error on a curve feature. The third-order spline curve has been taken as an example to illustrate that the application of the proposed approach could be successful in assessing positional error on an irregular curve.

4.9   MODELING POLYGON POSITIONAL UNCERTAINTY

Polygon object positional errors in a GISci are caused by positional errors in their boundaries. Such errors are mainly caused by composing vertices of the polygons. In this section, two approaches for modeling positional error of a polygon are given: the first is based on error of the component vertices of the polygon; the second is based on errors of the composing line segments through their error bands (Liu et al., 2005).

4.9.1   Error Modeling Based on Its Component Vertices

A polygon is composed of its component vertices. Polygon errors are thus determined by vertices errors, and can be quantified based on error propagation law. A polygon is described by its parameters, such as its area, perimeter, or gravity point. The area of a polygon is now taken as an example to demonstrate polygon error modeling based on its compositional vertices.

Suppose polygon A contains n vertices $P_i = (x_i, y_i)$ and n edges P_1P_2, P_2P_3, ..., $P_{n-1}P_n$, and P_nP_1. The covariance matrices of the individual vertices are expressed as

$$\Sigma_{P_iP_i} = \begin{bmatrix} \sigma^2_{x_i} & \sigma_{x_iy_i} \\ \sigma_{y_ix_i} & \sigma^2_{y_i} \end{bmatrix}. \tag{4.117}$$

The area S of the polygon is computed from

$$S = \frac{1}{2}\sum_{i=1}^{n}\left[x_i\left(y_{i+1} - y_{i-1}\right)\right] = \frac{1}{2}\sum_{i=1}^{n}\left[x_i\Delta y_{i-1,i+1}\right]. \tag{4.118}$$

The differential of the area is given as

$$dS = \frac{1}{2}\sum_{i=1}^{n}\left[\Delta y_{i-1,i+1}dx_i + \Delta x_{i-1,i+1}dy_i\right]. \tag{4.119}$$

This differential is used to compute the variance of the area of the polygon:

$$\sigma^2_{S1} = \frac{1}{4}\sum_{i=1}^{n}\left[\Delta y^2_{i-1,i+2}\sigma^2_{x_i} + \Delta x^2_{i-1,i+1}\sigma^2_{y_i} + 2\Delta x_{i-1,i+1}\Delta y_{i-1,i+1}\sigma_{x_iy_i}\right]. \tag{4.120}$$

When $\sigma_{xi}^2 = \sigma_{yi}^2 = \sigma_0^2$ and $\sigma_{xiyi} = 0$, the variance simplifies

$$\sigma_{S1}^2 = \frac{1}{4} \sum_{i=1}^{n} \left[\Delta y_{i-1,i+2}^2 + \Delta x_{i-1,i+1}^2 \right] \sigma_0^2 = \frac{1}{4} \sum_{i=1}^{n} l_{i-1,i+1}^2 \sigma_0^2 , \qquad (4.121)$$

where $l_{i-1,\,i+1}$ is the distance between vertices P_{i-1} and P_{i+1}. For an n-sided regular polygon, $l_{i-1,\,i+1}$ is identical for all $i = 1, 2, \ldots, n$. The standard deviation of its area equals

$$\sigma_{S1} = \sqrt{n} \sin \left[\frac{\pi}{2} - \frac{\pi}{n} \right] \cdot l \cdot \sigma_0 . \qquad (4.122)$$

4.9.2 ERROR MODELING BASED ON ITS COMPOSING LINE SEGMENTS

The area of a polygon is also a function of the lengths of its compositional line segments, according to Equation 4.118. This implies that the positional error for the area of the polygon is subject to a positional error in these line segments. The impact of the positional error in the line segment or error in the area of the polygon is elaborated in the following.

Let Δ_i denote the area of an error band for edge $P_i P_{i+1}$ of polygon A for $i = 1, 2, \ldots, n$. A relationship between the positional error Δ_S for the area of the polygon and the area Δ_i of the error band is given as

$$\Delta_S = \Delta_1 + \Delta_2 + \ldots + \Delta_n. \qquad (4.123)$$

If $\Delta_1, \Delta_2, \ldots, \Delta_n$ are independent, the standard deviation of the area of the polygon will be

$$\sigma_{S2} = \sqrt{\sigma_1^2 + \sigma_2^2 + \ldots + \sigma_n^2} , \qquad (4.124)$$

where σ_i is the standard deviation of the area of the error band for edge $P_i P_{i+1}$. The standard deviation σ_i is the half of the area of the standard error band derived in Section 4.8.2, that is, $\sigma_i = \Delta_i / 2$. For an n-sided regular polygon, the standard deviation of the area of the polygon is

$$\sigma_{S2} = \sqrt{\sum_{i=1}^{n} \sigma_i^2} = \frac{\sqrt{n} \Delta_S}{2n} = \frac{\Delta_S}{2\sqrt{n}} . \qquad (4.125)$$

Both Equations 4.124 and 4.125 demonstrate the following: the positional error for the area of the polygon increases if the positional error in the boundary line segments of the polygon increases, but decreases if the number of compositional vertices increases.

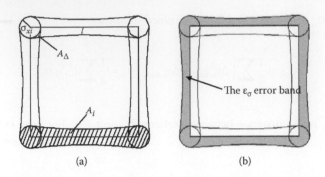

(a)   (b)

FIGURE 4.14   (a) The ε_σ error bands around the four edges and (b) that around square A (from Liu et al., 2005)

A polygon is drawn by linking line segments to form a closed area. Positional error for these line segments can be described based on the ε_σ error model. This error model can also be adopted to describe the positional error for the polygon.

Figure 4.14a gives a square A with four edges surrounded with the corresponding ε_σ error bands, whereas the error band for square A is shaded in gray in Figure 4.14b. If the standard deviation of the area of the error band for edge P_iP_{i+1} is equal for all i, the area of the error band for the square is

$$S = 4\left(A_i - \pi\sigma_{x_i}^2 - A_\Delta\right), \tag{4.126}$$

where

$$A_i = 2l \int_{r=0}^{1} \sigma_r\,dr + \pi\sigma_{x_i}^2,$$

$$\sigma_r = \sqrt{\left(1-r\right)^2\sigma_{x_i}^2 + r^2\sigma_{x_{i+1}}^2 + 2r\left(1-r\right)\sigma_{x_ix_{i+1}}},$$

and A_Δ is the area of the overlap region of the polygon and its error band.

4.9.3   CASE STUDY AND ANALYSIS

Two models for assessing positional error in a polygon have been introduced in the previous sections. In this section, the application of the two models is illustrated by case studies with computational results from real data sets.

Suppose the polygon perimeter is 100 times as large as the standard deviation of each polygon vertices. Table 4.4 and Figure 4.15 give the variance of the polygon area (σ_{S1}). Computations are based on error of vertices, the area Δ_S of the polygon error band, and the standard deviation of the polygon area (σ_{S2}). Computations are based on Equations 4.120 and 4.124.

TABLE 4.4

Computation Results of Different Error Indicators for a Polygon

Case	Polygon	Area of the Polygon	σ_{S_1}	Δ_S	σ_{S_2}
1	Regular quadrilateral	10000	141.42	586.6	142.15
2	Regular pentagon	23777	212.66	979.35	218.99
3	Regular hexagon	25981	212.13	982.30	200.51
4	Regular heptagon	27365	206.85	987.55	186.63
5	Regular octagon	28285	200.00	995.77	176.03
6	Regular enneagon	28926	192.84	1000.70	166.78
7	Regular decagon	29389	185.87	1007.28	159.27
8	Rectangle	20000	223.60	975.10	243.78
9	Trapezoid	15000	187.08	814.01	203.50
10	Pentagon	15000	183.71	849.88	190.04
11	Heptagon	35000	264.57	1462.77	276.44
12	Octagon	8750	129.90	811.69	143.49
13	Decagon	32500	244.95	1624.62	256.88

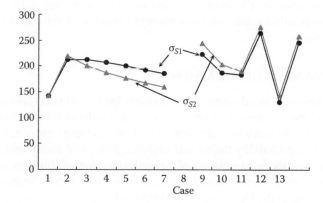

FIGURE 4.15 (See color figure following page 38.) Comparison of the error in the area of a polygon and the area of the error band for the polygon from case 1 to 13 (from Liu et al., 2005).

Table 4.4 shows that there is no significant difference between σ_{S1} and σ_{S2}. In fact, their difference is approximately 1/5 of the standard deviation of the vertices of the polygon. This result also shows that the area error of a polygon is caused by the positional error of the compositional vertices in the boundary. Therefore, an error analysis for the composing vertices of the polygon boundary is a feasible solution of the error modeling of the polygon area (or other parameters).

From Table 4.4, it can be see that the sum Δ_S of all areas of the ε_σ error bands for the edges of the polygon can be used to describe the polygon area error. This sum is larger than both σ_{S1} and σ_{S2}.

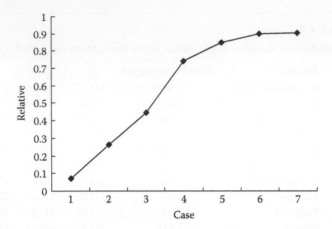

FIGURE 4.16 The ratio of the two error descriptors versus the polygon's area (from Liu et al., 2005).

The two error descriptors for the polygon area vary with the magnitude of the area itself. The relative value of the difference between these two error descriptors of the polygon area is computed and is given in Figure 4.16. This figure shows that the relative value increases when the number of the edges of a regular polygon increases.

4.10 SUMMARY AND COMMENT

This chapter has introduced methods for modeling positional uncertainties of spatial features within the framework of object-based data model in GISci. These methods cover point, polyline (with line segment as its basis element), curved, and polygon objects. Based on probability theory and statistics, positional uncertainties of spatial features have been modeled in terms of (a) confidence region, (b) error distribution, and (c) quantity of the errors, and have been presented in this chapter. Positional uncertainty in spatial data has been well addressed by researchers, as compared with other types of uncertainties such as temporal uncertainty. This is because the nature of positional uncertainty has been long discovered and therefore mathematical modeling solutions devised. In the next chapter, modeling another type of uncertainty—attribute uncertainty—is discussed.

REFERENCES

Bolstad, P. V., Gessler, P., and Lillesand, T. M., 1990. Positional error in manually digitized map data. *Int. J. Geog. Inform. Sci.* 4: 399–412.

Chapman, M. A., Alesheikh, A., and Karimi, H. A., 1997. Error modeling and management for data in GIS. In *Proceedings of the Coast GIS'97*. Aberdeen, Scotland.

Chrisman, N. R., 1982. A theory of cartographic error and its measurement in digital data base. In *Proceedings of Auto-Carto 5*, Bethesda, MD, American Congress on Surveying and Mapping, 159–168.

Dutton, G., 1992. Handling positional error in spatial databases. In *Proceedings of the 5th International Symposium on Spatial Data Handling*, Columbia, SC: International Geographic Union, 460–469.

Fan, A. M. and Guo, D. Z., 2001. The uncertainty band model of error entropy. *Acta Geodaetica et Cartographica Sinica* 30: 48–53 (in Chinese).

Goodchild, M. F. and Hunter, G. J., 1997. A simple positional accuracy measure for linear features. *Int. J. Geog. Inform. Sci.* 11: 299–306.

Grafarend, E. W. and Sanso, F., 1985. *Optimization and Design of Geodetic Networks*. Springer-Verlag, Berlin, New York.

Huang, Y. C. and Liu, W. B., 1997. Building the estimation model of digitizing error. *Photogram. Eng. Rem. Sens.* 63: 1203–1209.

Li, D. J., 2003. *A Study of Information Entropy-Based Models of Positional Uncertainty in Spatial Data*. Ph.D. dissertation. Wuhan University Press (in Chinese).

Liu, C. and Shi, D. J.. 2005. Relationship of uncertainty between polygon segment and line segment for spatial data in GIS, Geomatics and Informaton Science of Wuhan University, 30(1): 65–68.

Meng, X. L., Shi, W. Z., and Liu, D. J. 1998. Statistical tests of the distribution of errors in manually digitized cartographic lines. *Int. J. Geog. Inform. Sci.* 4: 52–58.

Perkal, J., 1966. *On the Length of Empirical Curves: Discussion Paper 10*, Ann Arbor (Michigan Inter-University Community of Mathematical Geographers).

Shi, W. Z., 1994. *Modeling Positional and Thematic Error in Integration of GIS and Remote Sensing*. ITC Publication, Enschede, Number 22, 147.

Shi, W. Z., 1997. Statistical modeling uncertainties of three-dimensional GIS features. *Cart. Geog. Inform. Sys.* 24: 21–26.

Shi, W. Z., 1998. A generic statistical approach for modeling error of geometric features in GIS. *Int. J. Geog. Inform. Sci.* 12: 131–143.

Shi, W. Z. and Liu, W. B., 2000. A stochastic process-based model for the positional error of line segments in GIS. *Int. J. Geog. Inform. Sci.* 14: 51–66.

Shi, W. Z., Tong, X. H., and Liu, D. J., 2000. An approach for modeling error of generic curve features in GIS. *Acta Geodaetica et Cartographica Sinica* 29: 52–58 (in Chinese).

Tong, X. H., Shi, W. Z. and Liu, D. J., 2000. Error distribution, error tests, and processing of digitized data in GIS. In *Accuracy 2000: Proceedings of the 4th International Symposium on Spatial Accuracy Assessment in Natural Resource and Environmental Sciences*, Amsterdam, edited by Heuvelink, G. B. M. and Lemmens, M. J. P. M., 642–646. Coronet Books, Inc.

5 Modeling Attribute Uncertainty

This chapter addresses attribute uncertainty modeling in GISci. Firstly, various concepts of attribute uncertainty and their sources are examined. Secondly, the methods for handling attribute uncertainties are introduced, including sample methods, error matrix, internal and external testing, the probability vector and parameters, the incidence of defect method, and sensitivity analysis.

5.1 INTRODUCTION

5.1.1 ATTRIBUTE AND ATTRIBUTE DATA

Attributes can be defined as the characteristics or nature of objects. In GISci, an attribute is regarded as a property inherent in a spatial entity. For example, the ownership of a land parcel is an attribute of the land parcel. Characteristics, variables, and values of spatial entities in GISci are described by attribute data.

The natural or real world can be represented in either of two ways: continuous or discrete representation. According to their continuous or discrete nature, attribute data can also be classified as categorical attribute data or continuous attribute data. For instance, land cover type is a categorical attribute and "forest" is a categorical attribute value, whereas population is an example of continuous attribute and "6.9 million people in Hong Kong" is an example of a population attribute value.

In GISci, attribute data may be classified either as categorical (discrete) and a continuous representation, or a qualitative and quantitative attribute representation. The classification, "categorical" and "continuous" for attribute data is used in this book. A categorical variable can only be represented by a finite number of values from a defined data set, such as a set composed of all members of land cover classes. A continuous variable, on the other hand, can take on any number, usually within a defined domain. Categorical attribute data may or may not have a ranking. For example, environment conditions can be considered to fall into four classes, with Class 1 having the highest environment condition quality ranking and class 4 the lowest. In this situation, the ranking represents the categorical value. In another example, the numbers 1 to 4 may stand for four categories of land cover: water, vegetation, soil, and forest. In this case, the numbers do not imply ranking. An example of a continuous attribute variable may be the (average) altitude of a land parcel. The domain range for the variable, for example, can be defined, as (–10,000, 10,000) in meters. The altitude value can be any real number within this range.

5.1.2 Attribute Uncertainty

Attribute uncertainty is defined by the closeness of an attribute value to its true value in the real world, where objects are usually complex. For example, materials of a spatial object may not be evenly distributed within the boundary of the object. Furthermore, boundaries of spatial objects are often not crisp.

The nature and quantity of an attribute uncertainty is determined by a series of factors. These may include, for example, (a) the attribute characteristics of the spatial objects themselves, (b) the process of recognizing the attributes, (c) the methods and technologies used for measuring or acquiring the attribute values, and (d) mathematical characteristics of spatial analysis and modeling approaches applied on the attribute data.

The nature of attribute uncertainty can be inaccuracy, randomization, or fuzziness. After a spatial data capturing process, attribute uncertainty of a spatial entity can be, for example, a random variable in a space or time domain. Attribute uncertainties can also be related to the scale of spatial data, resolution, sampling size and scheme, as well as other factors.

Attribute uncertainty may heavily affect the quality of spatial decision making. In many GIS applications, the attribute accuracy requirements are even higher than those for positional accuracy. For example, in urban planning applications, the accuracy requirement of the class type of a land parcel (attribute accuracy) is even higher than that of boundary positional accuracy.

The two kinds of attribute data—categorical and continuous, defined above— require corresponding, and thus different, uncertainty modeling methods and accuracy indicators. Continuous attribute data error can be modeled in a similar way to that of positional error. For example, error propagation law may be applied. This law is commonly used for modeling positional error propagation, and can also be used to estimate attribute uncertainty propagation for continuous attribute data.

5.1.3 Sources of Attribute Uncertainty

Attribute uncertainty arises from the following four sources:

a. The uncertain nature of objects in the natural world. The material within a spatial object may not be homogenous, but rather heterogeneous. Therefore, the spectral characteristics of such objects are not pure but rather complex. Another example is the fuzzy boundary of a spatial object, such as the boundary of a soil unit. This is a transition zone, rather than a crisp boundary.

b. Potential errors or uncertainty introduced by the methods and technologies applied for measuring or acquiring attribute values. Satellite imaging technology is often applied to capture remotely sensed images from the Earth's surface. The attribute values of spatial objects (such as the land cover types) obtained, based on the spectral information of the satellite images, possess uncertainty, due to the limited spectrum information in the images.

c. Spatial object attribute cognition uncertainties arising from limitations. Methods whereby attributes of spatial objects can be understood are either through automatic classification or human interpretation. For example, the automatic remote sensing classification of the satellite image is a commonly used automatic attribute cognition method. In such a method, the uncertainty in the final attribute cognizance may be related to the quality of the satellite image (e.g., regarding its spatial resolution spectral resolution, and temporal resolution), the methods applied for classification (e.g., maximum likelihood classification or minimum distance classification), and local knowledge about the area to be recognized.

d. Depending on the mathematical characteristics of spatial analysis or modeling applied to the attribute data, various uncertainties may accumulate and propagate in the attribute data. For example, the classified remote sensing images may be further integrated with vector map data by an overlay spatial analysis. Uncertainty in the classified remote sensing image then propagates. The quantity and nature of the propagated uncertainty depends upon the nature of the spatial analysis, such as Boolean or algebraic spatial analysis. Attribute uncertainty propagation models for logical operation and arithmetic operation can be established according to fuzzy set theory and probability theory, respectively.

In summary, attribute uncertainties are generated, propagated, and accumulated through uncertainty inherent in the natural world, cognition, modeling, measurement, and spatial analysis. These uncertainties may finally affect decision making applications. Attribute uncertainty sources and accumulation are illustrated in Figure 5.1.

Two commonly used models for spatial object modeling exist in GISci: object-based and field-based models. These models provide two different ways of abstracting the features of the real world. Both models induce uncertainties (attribute, positional, and other uncertainties) in representing spatial entities from the real world in a GISci. Mathematical function is the third category method for spatial object modeling in GISci (as described in Section 2.4 of Chapter 2).

In an object-based model, a group of elementary geometries such as points, lines, and polygons are deployed to represent spatial entities in terms of positional, attribute, and topological relationships. Pure mathematical points, lines, and polygons, however, do not exist in the real or natural world.

In reality, the GIS representation of the spatial elements of points, lines, and polygons are an approximation of pure mathematical points, lines, and polygons. The difference between the pure mathematical representation of the elements (points, lines, and polygons) and the approximated representation of these elements in GIS is a fundamental source of uncertainty of the object-based model. This source may govern attribute, positional, and relational uncertainties of the spatial entities in GISci.

Object-based models are suitable for presenting positional uncertainties of discrete objects. The limitation of the current GIS, designed according to the object-based model, is that positional error indicators are treated as an additional attribute in the data model, rather than as uncertainty-based spatial elements (points, lines, and polygons with uncertainty) in the object-based model. The eventual solution to this issue is the development of an uncertainty-based object model.

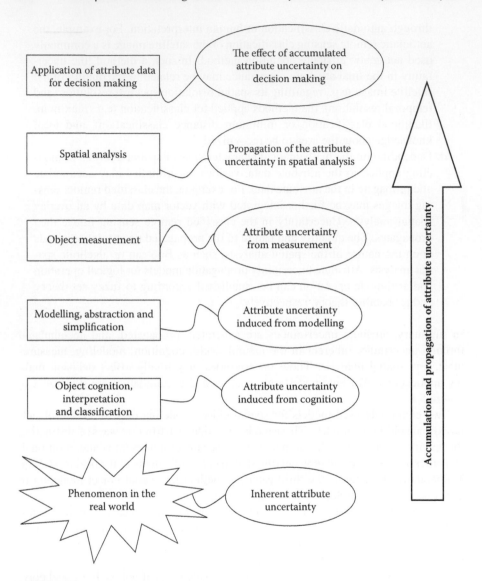

FIGURE 5.1   Source of attribute uncertainty.

It may be easy for object-based models to explicitly represent spatial entities and corresponding positional uncertainties, but it is not as easy to represent the spatial distribution of those uncertainties in such a model, based on the generation of heterogeneous material of spatial objects and gradual changes of the object boundary.

Difficulties arise in providing a comprehensive description of attribute uncertainty for an object-based model. Alternatively, field-based models are suitable for representing attribute uncertainties and the spatial distribution can be well described, both as regards spatial distribution of uncertainty, due to heterogeneous materials of spatial objects and also uncertainty in a gradually changing boundary.

5.1.4 Relationship Between Attribute and Positional Uncertainties

Although differences exist between attribute and positional uncertainties, both types of uncertainty are closely related. A spatial entity in GISci is integrated, as represented by its components: spatial element, attribute element, temporal element, scale, and spatial relations. Therefore, uncertainties in these elements are integrated and applied to a spatial object. For example, it might be difficult to separate positional and attribute uncertainties for a particular geographic object with a fuzzy boundary.

In an object-based model, an object (such as a building) is represented by its basic elements, such as points, lines, or polygons and associated attributes (e.g., address and ownership of the building). Therefore, uncertainties of the basic elements and those associated attributes may be applied to the spatial objects simultaneously. This is also applicable to the object description in a field-based model.

Boundary uncertainty of an area object is a typical example of integration, simultaneously affected by positional and attribute uncertainties. The combined effects of such positional and attribute uncertainties of an object imply that its boundary is no longer a geometric line, but a transitional region of a certain width.

For example, positional uncertainty existing in a transitional region may cause attribute uncertainty. An object from a remotely sensed image classification with attribute uncertainty in that transitional region may be caused by the positional error of the image. The central part of an area object is the most certain, and this degree of certainty gradually reduces from that area toward the object boundary. Hence, it is essential to define the object area by the transitional region with a particular width, instead of by a crisp line. The "S-band" model (Shi, 1994) is proposed for modeling the integrated effects of positional and attributes uncertainties in transitional regions. The model is designed around the same mathematical bases using probability theory in modeling both positional and attribute uncertainties. This model is described in detail in Chapter 6.

5.1.5 Uncertainty in Classification

Modeling uncertainty or accuracy of classification from satellite images is one of the most typical examples of modeling attribute uncertainty of categorical data. Classification accuracy assessment for the classification satellite images is a complex procedure and has received considerable attention in the field of remote sensing. The difficulties in assessing classification accuracy arise because of the significant effect of such factors as the number of classes, the variety of shape and size of individual areas, test point selection methods, and classes that can be confused, one with the other (Aronoff, 1985).

Classification is a procedure to decide whether an object (pixel) o belongs to class A. Therefore, the uncertainty of a classification may include three aspects: (a) definition of a class, (b) measurement of an object, and (c) the decision regarding whether the object o belongs to class A.

Uncertainty may exist in class definition. Class A may not be well defined, and one may not be able to identify the class clearly as A or another class as B. Soil classifications, for example, are typically fuzzy. Measuring an object o is the second source of uncertainty.

In making the decision $o \in A$, accuracy confidence may vary within the area of a polygon. For example, at the center of a polygon, there is strong confidence that the class is A, but less confidence of the class definition at the boundary of the polygon. Burrough and Heuvelink (1992) defined a transition zone to describe this uncertainty. An area delineated by a polygon may be heterogeneous, for example, a forest zone where the area may be 80% class A and 20% class B. However, the indicators 20% and 80% do not give any clues as to the spatial distribution of uncertainty. The concept of a probability vector (introduced in the latter part of this chapter), however, can give a description of spatial distribution of uncertainty on a per-pixel basis.

5.2 METHODS FOR MODELING ATTRIBUTE UNCERTAINTIES

Several methods for describing attribute uncertainties in spatial data are introduced in this section. They include sampling methods, error matrix, probability vector, and corresponding error parameters, rate of defect method, and sensitivity analysis. The theoretical bases supporting these and other attribute uncertainty modeling methods include statistics, probability theory, and fuzzy set.

5.2.1 SAMPLING METHODS

Sampling is an important method for assessing uncertainty in attribute data, especially categorical attribute data. Three statistical measures, dealing with categorical attribute uncertainty are selected from current literature. All are based on an experimental approach: (1) the percentage correctly measured per class or for the whole set (Rosenfield, 1986); (2) the percentage range correctly measured per class or for the whole set at a given confidence level (Aronoff, 1985; Hord and Brooner, 1976); and (3) the percentage of correctly measured per class or for the whole set (Greenland et al., 1985; Rosenfield and Fitzpatrick-Lins, 1986). The attribute data can be such as an area-class map or a classified remotely sensed image.

These methods are suitable, in a variety of ways, for categorical attribute data assessment. The emphasis in this section is on remote sensing classification assessment. Categorical attribute data accuracy can be assessed by evaluating a sample of the classification result. The classes determined by classification are defined as reference data. By comparing the measured data with the reference data, an error matrix can be formed. This matrix describes the quality of the classification as a whole, and that of individual classes. Based on this matrix, further analysis of classification accuracy can be carried out. The error matrix concept is introduced in Section 5.2.3 of this chapter.

The sampled data act as an input to an error matrix or other advanced statistical analyses in the sampling-based attribute uncertainty assessment methods. In this process, the selection of minimum significant samples is of importance. Two basic factors affect the quality of the sample: size and scheme.

Sampling size is important in the assessment of the accuracy of classified remotely sensed image data. Collecting ground samples is expensive; therefore, the number of sample points has to be kept to a minimum. However, as indicated by the work of such as van Genderen and Lock (1977); Rosenfield (1982); Congalton (1988); Fukunaga and Hayes (1989), it is important for the sample size to be large enough to enable the conducted analysis to be statistically valid.

The other important factor, as stated above, is the sampling scheme used. Selection of the proper sampling scheme is critical in enabling the generation of an assessment result that is representative of the whole classified image. A poor sampling scheme may result in significant biases, which could then result in over- or under-estimated accuracy. Five commonly used sampling schemes are simple random sampling (Cochran, 1977), clustered sampling (Kish, 1965), stratified random sampling (Barrett and Nutt, 1979), systematic sampling (Barrett and Nutt, 1979), and stratified systematic unaligned sampling (Berry and Baker, 1968).

Congalton (1988) conducted sampling simulations on diverse areas and concluded that in all cases, simple random and stratified random sampling provided satisfactory results. Depending on the spatial auto-correlation of the area in question, other sampling schemes may also be appropriate. Once the sample size is selected and the sample scheme chosen, sampling can be practically implemented.

5.2.2 INTERNAL AND EXTERNAL TESTING

Three approaches can be applied to determine the attribute data quality statistically: deduction, internal testing, and external testing (Kennedy-Smith, 1986). The usual method for deducing attribute quality is to use survey values that have their attribute quality derived by either external or internal testing.

Internal testing is carried out by comparing several independent repeated measurements, with the average being considered the "truth." The result of internal testing is a precision measure in quality control.

External testing is carried out by comparing the survey result with the truth or what can be treated as the truth. The result of external testing is a measure of accuracy. External testing represents the solution which is most satisfying to the "customer" (Kennedy-Smith, 1986), but it fails to separate source and procedure errors during data manipulation. The result encompasses the effects of all errors.

When determining attribute data quality by using external testing, it is first necessary to select a certain number of check points. Random sampling is advocated for the location of checkpoints (Hay, 1979; van Genderen and Lock, 1977). Stratified random sampling is recommended to ensure that each class has a certain number of sampling points. The procedure, according to Hord and Brooner (1976), for determining attribute data quality by using external testing is shown by the following example. The aim is to determine the range of true map accuracy.

1. Define a confidence level (e.g., 99.7%), and look up the Z_α associated with such a confidence level in the normal distribution table; we have $Z_\alpha = 3.0$.
2. Determine the number (N) of sample (e.g., $N = 200$).

3. Calculate percentage (P) of correctly classified checkpoints (e.g., $P = 89\%$).
4. Determine the test accuracy range x using the inequality (Drummond, 1991)

$$(-Z_\alpha^2 - N)x^2 + (Z_\alpha^2 + 2NP)x - NP^2 > 0,$$

which gives $0.81 < x < 0.95$ when $N = 200$ and $Z_\alpha = 3.0$.

Thus, it can be stated that the true classification accuracy is correct, in the range of 81% to 95% with a 99.7% confidence level for a sample accuracy of 89% from 200 checkpoints. It is obvious that the width of the determined classification accuracy will be increased by reducing the number of checkpoints or raising the confidence level.

The user's quality indicator may only be supported by a single unambiguous accuracy statistic, which is the lower limit of the true accuracy range for a high confidence level (e.g., 99.7%). In the example, the single unambiguous accuracy statistic is 81%. The disadvantage of the measure is that the whole classification result may be rejected, even though for a certain single class, the result is acceptable.

For some applications, classification results should have a minimum percentage of correctly classified pixels. In this situation, hypothesis testing is most appropriate. The hypothesis testing of a predetermined accuracy is a method generally applied in quality control. Acceptance sampling is important for quality control. The minimum sampling size can be determined when risks and predetermined accuracy are defined. A comprehensive discussion of statistical quality control is presented by Grant and Leavenworth (1988).

5.2.3 ERROR MATRIX METHOD

An error matrix, also called a confusion matrix, is a square array of numbers that expresses the unit number assigned to a particular category, compared to the "truth." The truth is either from a ground survey or from other reference data. An error matrix can be used to assess uncertainty of categorical attribute data based on a set of samples. For example, this method can be applied to assess such as the accuracy of a classified satellite image, or thematic maps.

The error matrix and related concepts are introduced through the example illustrated in Table 5.1. In the example, the matrix is used for assessing the accuracy of

TABLE 5.1

An Example of Error Matrix

Reference Data/ Classification Result	A	B	C	D	Sum of Row
A	65	4	22	24	115
B	6	81	5	8	100
C	0	11	85	19	115
D	4	7	3	90	104
Sum of column	75	103	115	141	434

remote sensing image classification. The columns in the table list the reference data (truth), and the rows list the classification results generated from the remotely sensed images. An error matrix can indicate the accuracy of the overall image classification and also accuracy of the classification for each category class. Both commission and omission errors, or producer accuracy and user accuracy, *kappa* coefficients can also be derived from the error matrix.

The value of the following classification accuracy indicators can be calculated from the error matrix in Table 5.1.

Overall accuracy = $(65 + 81 + 85 + 90)/434 = 321/434 = 74\%$
Producer accuracy:
    Class A: $65/75 = 87\%$
    Class B: $81/103 = 79\%$
    Class C: $85/115 = 74\%$
    Class D: $90/141 = 64\%$
User accuracy:
    Class A: $65/115 = 57\%$
    Class B: $81/100 = 81\%$
    Class C: $85/115 = 74\%$
    Class D: $90/104 = 87\%$

As well as obtaining the overall accuracy from an error matrix, it is also possible to obtain the producer and user accuracies for each of the classes. From Table 5.1, it can be seen that the overall accuracy of the classification, obtained from the error matrix, is 74%. At the same time, a detailed accuracy description (accuracy for each classified class) is also derived, such as for Class A, Class B, Class C, and Class D.

In terms of mathematical description, it is assumed that n surveyed samples are distributed in k^2 Euclidean space. Each sample belongs to one category in k-classified attributes. Equally, each of these n samples possesses the corresponding classification reference. The contrasted results of the samples are given in Table 5.2, where n_{ij} represents the obtained results that belonged, originally, to the ith class are classed as the jth class ($i = 1, 2, \cdots, k, j = 1, 2, \cdots, k$).

TABLE 5.2
Mathematical Expression of an Error Matrix

Reference Data (j)/ Classification Result (i)	1	2	k	Sum of Row n_{i+}
1	n_{11}	n_{12}	n_{1k}	n_{1+}
2	n_{21}	n_{22}	n_{2k}	n_{2+}
k	n_{k1}	n_{k2}	n_{kk}	n_{k+}
Sum of columns n_{+j}	n_{+1}	n_{+2}	n_{+k}	n

The total sampled data in the ith class of the classifications:

$$n_{i+} = \sum_{j=1}^{k} n_{ij} .$$ (5.1)

Based on reference data, the total sampled data that belongs to the jth class is

$$n_{+j} = \sum_{i=1}^{k} n_{ij} .$$ (5.2)

The overall error for the contrast of the classified data and reference data is

$$\text{Overall accuracy} = \frac{\sum_{i=1}^{k} n_{ii}}{n} .$$ (5.3)

The producer accuracy is

$$\text{Producer accuracy}_j = \frac{n_{jj}}{n_{+j}} .$$ (5.4)

Equally, the user accuracy is

$$\text{User accuracy}_i = \frac{n_{ii}}{n_{i+}} .$$ (5.5)

In additional to the above overall accuracy—that is, producer accuracy and user accuracy—values of other uncertainty indicators can also be derived. The two most commonly used measures of attribute accuracy are binormal probabilities and kappa coefficients of agreement. Binormal probabilities are based on the "percent correct" and therefore do not count for errors of commission or omission. The kappa coefficient provides an alternative in the form of a difference measurement between the observed agreement of two data sets and the agreement, contributed by chance.

The KHAT statistics is widely used in the assessment for classification accuracy of remote sensing images. It is defined as

$$\hat{K} = \frac{N \sum_{i=1}^{r} x_{ii} - \sum_{i=1}^{r} x_{i+} x_{+i}}{N^2 - \sum_{i=1}^{r} x_{i+} x_{+i}} ,$$ (5.6)

where + in the subscription represents the sum of a row or a column, r represents the classified number of classes, while N represents the total number of samples. Therefore, the calculation of kappa coefficients can be implemented in an error matrix.

A kappa coefficient of 0.70 can be interpreted as a 70% better classification than that which would be expected by random assignment of classes. The advantages of kappa are that errors of commission and omission are included and a conditional kappa coefficient may represent accuracy for individual categories (Campbell, 1987). Rosenfield (1986) and Chrisman (1984) suggested that kappa should be a standard measure form for attribute classification accuracy as a whole and proposed a conditional kappa coefficient for individual classes.

A common model to describe the uncertainty of a classified remote sensing image is the error matrix. As stated above, two key factors affect the results of accuracy assessment for a given classification result: sample size and sample scheme. Based on an error matrix, a number of error indicators can be derived, such as user accuracy, producer accuracy, kappa coefficient, and others. These indicators can be used to describe the overall classification accuracy and classification accuracy for a specific class. Uncertainty of each pixel and spatial distribution of a classified image, however, cannot be provided by the error matrix method. Therefore, a probability vector and corresponding parameters have been further developed (Shi, 1994) and are introduced in Section 5.2.4.

5.2.4 PROBABILITY VECTOR AND THE FOUR PARAMETERS

5.2.4.1 Maximum Likelihood Classification

The concept of probability vector and four parameters derived from the probability vector are introduced in this section. Both the probability vector and the four parameters can be used to describe attribute uncertainty of a classified image from a maximum likelihood classification. The probability vector can also be used to describe the spatial distribution of that uncertainty.

In a maximum likelihood classification, conditional probability $P(C_i|Z_T)$ provides a probability that a designated pixel (Z_T) belongs to C_i class. The classification is processed, according to the following rules:

$$Z_T \in C_i \qquad P\left(C_i|Z_T\right) > P\left(C_j|Z_T\right) \text{ for all } j \neq i. \tag{5.7}$$

That is, that if $P(C_i|Z_T)$ is at its maximum, the pixel will then be categorized as one member in class C_i. This is a special case of a Bayesian classification.

In the Formula 5.7, $P(C_i|Z_T)$ is usually an unknown, but under the condition with enough trained data set and the priori knowledge of land cover (i.e., the occupied percentage of each land cover), the known value can be estimated. Bayes' rule can be expressed as the conditional probabilities $P(C_i|Z_T)$ and $P(Z_T|C_i)$ as follows:

$$P(C_i|Z_T) = P(Z_T|C_i) \cdot P(C_i) / P(Z_T), \tag{5.8}$$

where $P(C_i|Z_T)$ is the probability of Z_T value appeared in class C_i, and $P(C_i)$ indicates the probability of class C_i appeared in the whole image. It is possible to calculate the value by estimating the contrast between the area occupied by class C_i and the total area.

5.2.4.2 Probability Vector

From Formula 5.7, it can be seen that it is not possible to assign a pixel to a certain class with 100% certainty. It is, thus, a decision-making process with uncertainty. The probability vector, generated during the maximum likelihood classification, is used to describe the uncertainty in the classification.

According to Formula 5.8, we have the following vector:

$$\left[P\left(C_1 \middle| Z_T\right), P\left(C_2 \middle| Z_T\right), \cdots, P\left(C_n \middle| Z_T\right) \right]^T$$

(5.9)

$$PV(Z_T) = \left[P\left(C_{l1} \middle| Z_T\right), P\left(C_{l2} \middle| Z_T\right), \cdots, P\left(C_{lk} \middle| Z_T\right) \right]^T.$$

The zero elements in the vector are removed and sorted by descending order. This leads to the probability vector with the following form:

$$PV(Z_T) = \left[P\left(C_{l1} \middle| Z_T\right), P\left(C_{l2} \middle| Z_T\right), \cdots, P\left(C_{lk} \middle| Z_T\right) \right]^T.$$

(5.10)

Here, k is the number of elements in the Vector 5.9, and also $P(C_{li}|Z_T) \geq P(C_{lj}|Z_T)$ for $i < j$. Vector 5.10 is defined as the probability vector of pixel Z_T generated in the maximum likelihood classification. In a maximum likelihood classification, each pixel is classified as the class, to which the first element in the vector belongs.

5.2.4.3 The Four Parameters

According to the probability vector defined by Equation 5.10, Shi (1994) further defined four parameters to describe the uncertainties of classified pixels—absolute error, relative error, mixture level of pixels, and completeness of evidence.

1. Absolute accuracy (U_A)

   The absolute accuracy (U_A) is defined as

   $$U_A(Z_T) = P\left(C_{l1} \middle| Z_T\right) \middle/ \left[1 - P\left(C_{l1} \middle| Z_T\right) \right],$$

   (5.11)

   where the range of U_A is $(0, +\infty)$. The larger the value of U_A, the lower the uncertainty that Z_T is classified as class C_i. For examples, when $P(C_{l1}|Z_T)$ is 90%, $U_A(Z_T)$ is 9.0 according to Equation 5.11. If $P(C_{l1}|Z_T)$ is 20%, $U_A(Z_T)$ is 0.25. From these two examples, it easy to see that there is a smaller uncertainty in the classification for a pixel with higher absolute accuracy $U_A(Z_T)$ value, and vice versa.

2. Relative accuracy (U_R)

The second parameter is the relative accuracy (U_R), which is defined to describe the probable misclassification between two classes C_{li} and C_{lj} for a pixel. The parameter is defined as:

$$U_R(Z_T, i, j) = \left| P\left(C_{li}\middle|Z_T\right) - P\left(C_{lj}\middle|Z_T\right) \right|. \tag{5.12}$$

The value of the parameter $U_R(Z_T, i, j)$ is within the range [0, 1]. The larger the value of the parameter, the easier the discrimination between the classes C_{li} and C_{lj} for the pixel Z_T. That is to say, there is less chance for the pixel Z_T to be misclassified between classes C_{li} and C_{lj}. For example, according to Formula (5.12), if $P(C_{l1}|Z_T)$ is 80% and $P(C_{l2}|Z_T)$ is 10%, then $U_R(Z_T, 1, 2)$ is 0.70. This means the pixel Z_T can be confidently classified as class C_{l1}, rather than class C_{l2}. In another example, $P(C_{l1}|Z_T) = 33\%$ and $P(C_{l2}|Z_T) = 30\%$, then $U_R(Z_T, 1, 2) = 0.03$. In this situation, it is not very sure if the pixel Z_T should be classified as a member of class C_{l1} or C_{l2}. Misclassification may happen in such circumstances.

3. Mixture level (M)

The third parameter measures the extent to which a pixel belongs to a mixed pixel, that is, a pixel consists of components from various different classes. This parameter is defined as:

$$M = \sum_{i=1}^{n} \sum_{j=i+1}^{n} | P(x \in C_i) - P(x \in C_j) |, \text{ for all } i \neq j. \tag{5.13}$$

Here, $P(x \in C_i)$ is the probability of a pixel belonged to C_i class, and $P(x \in C_j)$ is the probability of a pixel belonged to class C_j.

In cases of similar measured values of all elements in the probabilistic vector for a pixel, a comparatively smaller value of M is obtained. This indicates that the chance that pixel is a mixed pixel is relatively higher.

4. Incompleteness (Θ)

The fourth parameter describes the incompleteness of evidence. The parameter is represented by the symbol Θ and is defined as the following for the pixel Z_T:

$$\sum_{i=1}^{k} P\left(C_{li}\middle|Z_T\right) + \Theta = 1. \tag{5.14}$$

The existence of Θ indicates that we may not be able to ensure that a pixel is classified to any of the classes during a classification. This pixel is then categorized as "unclassified" category. The value of parameter Θ is within the range [0, 1]. The larger the value of Θ, the higher the chance that the pixel is classed as an "unclassified" category.

The above four parameters describe attribute uncertainties of a classification from four different aspects. One advantage of the probabilistic vector is that it can express the spatial distribution of the attribute uncertainty of the classified remote sensing images. However, it may generate large data volume. The proposed four parameters can be applied, and data volume cost reduced in describing the attribute uncertainties.

5.2.5 THE INCIDENCE OF DEFECTS METHOD

The incidence of defects method is proposed for evaluating spatial data quality regarding attribute uncertainty in GISci (Shi et al., 2002). The defect rate is defined as the ratio between the number of spatial objects with defects and the total number of spatial objects in the unit. The unit can be a map sheet or a layer of spatial objects, such as the total number of buildings in the building layer of the whole area. Although the incidence of defects method is proposed for modeling attribute uncertainty, it can also be used to model other types of uncertainties, such as incompleteness and logic inconsistency.

The stratified sampling technique is adopted for sampling, and the defect rate is computed for assessing the attribute quality. This is due to the considerations that most of the existing spatial data are conceptually and physically modeled in layers. Attribute uncertainty of one layer may be different from that of the others. For example, attribute accuracy for a "building" layer may be different from that of a "vegetation" layer. The stratified sampling strategy is thus recommended.

Depending on the uncertainty level, defects are classified into three classes: minor defect (minor_def), moderate defect (moderate_def), and serious defect (serious_def). Defects are counted by number. For example, there may be one serious defect and 20 minor defects for a map sheet in terms of attribute uncertainty. In the following, the mathematical description of the incidence of defects method is given.

The population of a spatial data set to be inspected is N. If one data unit (such as a layer of spatial data) is treated as an inspected unit, the overall population is the sum of the populations of each data unit. Sampling size, the overall sampled data volume for the examination, is n. If a single data unit is considered as an inspected unit during the examination, the sampling size is the sum of sampling sizes of all the data units.

Let the population of attribute data set X be N, sampling (without the replacement) is applied to the data unit. There are m times sampling. The sample size for each sampling time is n. The numbers of defects obtained from the sampled inspection are y_i ($i = 1, 2, \cdots, m$), with the following equation

$$y_i = W_{SD\_a} \cdot \text{minor\_def} + W_{SD\_b} \cdot \text{moderate\_def} + W_{SD\_c} \cdot \text{serious\_def}, \quad (5.15)$$

where W_{SD_a} stands for the weight of a minor defect, W_{SD_b} represents the weight of a moderate defect, and W_{SD_c} is the weight of serious defect. The weights are determined based on the producer's understanding of the data set, and also the user's data quality requirements. For example, the weights can be set as follows: $W_{SD\_a} = 0.2$, $W_{SD\_b} = 0.3$, $W_{SD\_c} = 0.5$.

As the sample size for each sampling time is n, for the ith sampling, we have:

$$
y_{ij} = \begin{cases} 1, & j^{th} \text{ unit is a defect} \\ 0, & j^{th} \text{ unit is not a defect} \end{cases}, \tag{5.16}
$$

where i stands for the ith sampling, and j stands for the jth sampled unit in the ith sampling. Then, the defect number y_i is:

$$
y_i = \sum_{j=1}^{n} y_{ij}. \tag{5.17}
$$

The sample size for each sampling time is n and the defect number for each sampling inspection is y_i, their ratio is the estimated rate of defect for each time of sampling, with the following representation:

$$
\hat{u}_i = \frac{y_i}{n}. \tag{5.18}
$$

N is the population of X and the total defect number is Y; $\hat{u}_i$ is then an estimation of the total incidence of defect $u = Y/N$.

If the population N of X is large enough, there are m times sampling, based on statistics, it is common to agree that the defect number y_{ij} obeys the Poisson distribution. In addition,

$$
E(y_{ij}) = \bar{y}_i = \frac{1}{n} \sum_{j=1}^{n} y_{ij},
$$

where y_{ij} represents the defect number of each sampling unit and the sample size for each time of sampling is n. Therefore, $\hat{u}_i$ is estimated by:

$$
\hat{u}_i = \bar{y}_i = \hat{\lambda} = \frac{1}{n} \sum_{j=1}^{n} y_{ij}, \tag{5.19}
$$

where n is the sample size for each time of sampling.

If in the ith time of sampling, there is a sample size n, and the detected defect number is y_i, then y_i obeys the Poisson distribution. Similarly, $E(y_i) = n\lambda$ and $V(y_i) = n\lambda$, and then the mathematical expectation and variance of the rate of defect are

$$E(\hat{u}_i) = E\left(\frac{y_i}{n}\right) = \lambda = u , \qquad (5.20)$$

$$V(\hat{u}_i) = V\left(\frac{y_i}{n}\right) = \frac{\lambda}{n} = \frac{u}{n} . \qquad (5.21)$$

The expectation of the rate of defect is the measurement of the defect number in the sample data, whereas its variance is the measurement of the scattering extent of the defect among samples of the attribute data. As the defect number obeys the Poisson distribution, the mathematical expectation and variance of the defect rate can thus be used as the quality indicators of attribute data.

A stratified sampling scheme, rather than random sampling, is adopted due to the consideration that the quality of attribute data among the different layers may be different. In the attribute quality assessment, the rate of defect for each of the layers is first estimated. The overall defect rate is then obtained via the weighted average of all the layers.

Let the rate of defect of kth layer ($k = 1, 2, ..., h$) be U_k, the weighted average of the overall rate of defect is:

$$U_w = \sum_{k=1}^{h} W_{layer(k)} U_k = W_{layer(1)} U_1 + W_{layer(2)} U_2 + \cdots + W_{layer(h)} U_h , \qquad (5.22)$$

where $W_{layer(k)}$ ($k = 1, 2, ..., h$) is the weight for each layer, and

$$\sum_{k=1}^{h} W_{layer(k)} = 1 .$$

Similarly, we have

$$\hat{u}_w = \sum_{k=1}^{h} W_{layer(k)} \hat{u}_k , \qquad (5.23)$$

where $\hat{u}_k$ represents the estimation of the rate of defect for each sampling data k ($k = 1, 2, ..., h$) layer, and $\hat{u}_w$ is the estimation of the rate of defect setoff the population.

Computing each layer variance of the weighted average variance forms, an estimation of variance of the incidence of defects for the population is given as

$$V(\hat{u}_w) = \sum_{k=1}^{h} W_{layer(k)}^2 V(\hat{u}_k) . \qquad (5.24)$$

In the following, an example using the incidence of defects method for assessing attribute spatial data quality is illustrated.

Example

Two data sets of urban land use (*A* and *B*) are evaluated by applying the rate of defect method. There are four classes in the land use data set: fishery, cultivated land, residence, and forestry. Table 5.3 lists the attributes of the land use data to be assessed. A stratified sampling is applied to both data sets. Table 5.4 illustrates sampling and evaluation results based on the incidence of defect method.

The weights of the four land use types are given based on their importance and scale. The defects are classified as minor defect, moderate defect, and serious defect. The quality of the attribute data is estimate based on the two indicators: (a) incidence of defect and (b) variance. The rates of each of the three types of defects for both data sets A and B are illustrated in Figure 5.2. Table 5.5 shows the estimated variance of these rates of defect, which indicate the reliability of the estimated rate of defect.

It is observed from Figure 5.2 and Table 5.5, that (a) although the rates for moderate and serious defects in data set *A* are lower than those in data set *B*, because *A*'s rate of minor defect is much higher than that in *B*, the sum defect rate for data set *A* is higher. That is to say that the overall quality of data set *A* is lower than the quality of *B*. The bar chart in Figure 5.2 not only provides the rate for each type of defect, but also the defect sum. Therefore, the overall quality of a data set and also the detailed information about data quality indicated by different levels of defects can be gained. (b) The absolute value of the incidence of defect indicates the quality of the data set. In the example, the absolute value for both data sets *A* and *B* are very small. This means the overall quality of both data sets is high. In practical applications, an acceptable quality level can be set. If the defect rate is lower than the quality level, the overall data can be accepted in terms of its attribute data quality.

The incidence of defects method for attribute data quality assessment possesses a sound theoretical basis and is also easy to implement. The incidence of effect method covers more types of uncertainties in measuring attribute quality. This is different from the error matrix and kappa coefficient, which mainly concentrates on the classification error.

Whereas the acquisition of the value of the defect rate depends on sampled data, whether sampled data can extensively reflect the population is obviously determined by the sampling size and scheme applied. The use of the stratified sampling scheme, however, considers the fact that spatial data in GIS may have a different accuracy level from one layer to another. Therefore, the stratified sampling method is a highly appropriate method to complement the incidence of defects method.

5.2.6 Modeling Error Propagation for Continuous Attribute Data

Error propagation methods for continuous attribute data are now considered. Both positional data and continuous attribute data are based on continuous random variables. Therefore, the methods for handling error propagation for continuous attribute data are similar to the methods used for modeling positional uncertainty of spatial data. Error propagation law and Monte Carlo simulation methods used for modeling

TABLE 5.3

Attribute Contents of the Land Use Data

Class of Land Use	No. of Layer	Area	Authority	Administrative Region	Boundary	Productivity	Contract Period	Vegetated Attribute	Attribute-Recorded Length
Fishery	1	✓	✓	✓	✓	✓	✓		7
Cultivated land	2	✓	✓	✓	✓	✓			6
Residence	3	✓	✓	✓	✓				5
Forestry	4	✓	✓	✓	✓	✓	✓	✓	8

TABLE 5.4

Inspected Results by the Incidence of Defect Method

Data Set	No. of Layer k	Weight of Layer $W_{layer(k)}$	Sample Size of Each Layer n_k	Data-Recorded Length r_k	Minor Defect $y_{kj}^{(1)}$	Moderate Defect $y_{kj}^{(2)}$	Serious Defect $y_{kj}^{(3)}$	Sum of Defects y_{kj}
A	1	0.30	100	7	10	3	1	24
	2	0.20	150	6	15	2	0	21
	3	0.25	200	5	20	1	1	28
	4	0.25	150	8	15	2	1	26
B	1	0.30	200	7	16	5	2	41
	2	0.20	150	6	10	3	0	19
	3	0.25	100	5	5	2	1	16
	4	0.25	200	8	10	3	2	29

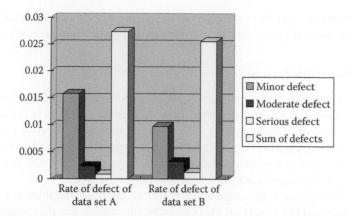

FIGURE 5.2 (See color figure following page 38.) A bar chart of attribute quality of data sets *A* and *B* with three types of defect: minor moderate and serious defects. (From Shi et al., 2005.)

TABLE 5.5
Variance of Incidence of Defect of Data Sets *A* and *B*

Data Set	Variance of Defect	Value of Minor Defect $y_{kj}^{(1)}$	Value of Moderate Defect $y_{kj}^{(2)}$	Value of Serious Defect $y_{kj}^{(3)}$	Sum of Defects y_{kj}
A	Variance of defect $D(\hat{u})$	1.74E-05	2.93E-06	1.04E-06	3.14E-05
B	Variance of defect $D(\hat{u})$	1.09E-05	3.8E-06	1.5E-06	2.98E-05

positional uncertainties in spatial data can also be applied to the analysis of error propagation of continuous attribute data. The only difference is that the random variables need to be changed from coordinate measurements (for positional data) to attribute measurement values (for attribute data) of a location. The transformation for positional data is defined by the following function:

$$y_k = g_k(x_1, x_2, \ldots, x_n),$$

the positional variables y_k, x_1, x_2, and x_n are now changed into random variables with attribute values at a location s ($s \in R^2$), that is,

$$y_k(s) = g_k(x_1(s), x_2(s), \ldots, x_n(s)),$$

where $x_1(s)$, $x_2(s)$, $\ldots$, $x_n(s)$ are random variables of attribute values with known stochastic characteristics at location s. $y_k(s)$ is a functional variable at location s. $y_k(s)$ is related to $x_1(s)$, $x_2(s)$, $\ldots$, $x_n(s)$ through the function g_k. The problem is now to derive the stochastic characteristics of $y_k(s)$ based on the known statistic characteristics of $x_1(s)$, $x_2(s)$, $\ldots$, $x_n(s)$. The behavior of the random variable $y_k(s)$ can then be derived.

The techniques used for positional error propagation, for modeling error propagation of continuous attribute data are achieved, based on this transformation.

5.2.7 SENSITIVITY ANALYSIS

A sensitivity analysis can be applied for studying propagation of attribute uncertainty in a spatial analysis. An error propagation analysis for spatial data in GISci usually assumes that errors of the input data are known and the process of error propagation and errors in the output are then analyzed after the spatial analysis. If errors of input data are unknown, a sensitivity analysis can be applied to study the relationships between input and output errors in a function transformation.

In a sensitivity analysis, simulated theoretical interfere variables are added to the input data and the effects on the outputs by simulations are then analyzed. The interfere variables can be used to represent either human error or errors in input data sets themselves. A sensitivity analysis may demand large computer resources—both computation time and storage for large data volume. One limitation of a sensitivity analysis is that it cannot give the mathematical expression of attribute uncertainty propagation, analytically.

Lodwick et al. (1988) reviewed methods and applications of sensitivity analysis for attribute uncertainties analysis. Five indicators, including attribute uncertainty, positional uncertainty, and area errors, were defined for the analysis. Two algorithms were proposed for determining the confidence level of the outputs by the use of those five indicators. Bonin (1998) studied the propagation of attribute uncertainties in vector-based GIS based on sensitivity analysis. The attribute uncertainties of roads were taken as an example to demonstrate the usage of the method. The probability model for the estimation of the attribute noise in vector-based GIS, was also proposed (Bonin, 2000). The sensitivity analysis was also applied to raster-based GIS. It was found that both continuous and discrete variables are not presented with isotropy, and that semantics may affect different level outputs (Fisher, 1991; McMaster, 1996).

5.3 SUMMARY AND COMMENT

The focus of this chapter has been on methods for modeling attribute uncertainty, and these methods covered sampling techniques, the error matrix, probability vector and four parameters, internal and external test testing, the incidence of defect method, and the sensitivity analysis. Among the methods, sampling techniques form a basis for both the error matrix and the incidence of defect method. The error matrix can indicate attribute uncertainty for each class, in addition to overall classification. The incidence of defect model can provide information about different levels of defects for a data set with attribute uncertainty. However, a probability vector can be used to describe spatial distribution of attribute uncertainty of a maximum likelihood classified image. The proposed four parameters, based on the probability vector, describe classification uncertainties in four different aspects: absolute error, relative error, level of mixture, and incompleteness.

Modeling attribute uncertainties for spatial data and spatial analysis is an ongoing research initiative, with the following issues needing to be further investigated: (a) inherent attribute uncertainties in the natural world, such as examples of the same material, but with different spectral performance; (b) modeling the intermediate boundary of spatial objects caused by integrated attribute and positional uncertainties; (c) modeling semantic uncertainty in spatial query and spatial analysis; (d) attribute uncertainty propagation in spatial analysis and its effect to the spatial analysis; and (e) the influences of logic inconsistencies among objects toward attribute uncertainty.

REFERENCES

Aronoff, S., 1985. The minimum accuracy value as an index of classification accuracy. *Photogram. Eng. Rem. Sens.* 51(1), 1687–1694.

Barrett, J. P. and Nutt, M. E., 1979. *Survey Sampling in the Environmental Sciences: A Computer Approach.* Compress, Inc., Wentworth, NH.

Berry, B. J. L. and Baker, A. M., 1968. In Berry, B. J. L. and Marble, D. F., eds., *Geographic Sampling. Spatial Analysis: A Reader. Statistical Geography.* Prentice Hall, Englewood Cliffs, NJ. 91–100.

Bonin, O., 1998. Attribute uncertainty propagation in vector geographic information systems: Sensitivity analysis. In Kelly, K., ed., *Proceedings of the Tenth International Conference on Scientific and Statistical Database Management.* Capri, Italy: IEEE Computer Society: 254–259.

Bonin, O., 2000. New advances in error simulation in vector geographical databases. In Accuracy 200. In Heuvelink, G. B. M. and Lemmens, M. J. P. M., eds., *Proceedings of the 4th International Symposium on Spatial Accuracy Assessment in Natural Resources and Environmental Sciences.* University of Amsterdam, The Netherlands. 59–65.

Burrough, P. A. and Heuvelink, G., 1992. The sensitivity of boolean and continuous (fuzzy) logical modelling to uncertainty data. *Proceedings of EGIS '92.* pp. 1032–1039.

Campbell, J. B., 1987. *Introduction to Remote Sensing.* The Guilford Press, New York–London.

Chrisman, N. R., 1984. The role of quality information in the long-term functioning of a geographic information system. *Cartographica.* Vol. 21, Nos. 2 and 3, summer/autumn.

Cochran, W. G., 1977. *Sampling Techniques.* John Wiley & Sons, New York.

Congalton, R. G., 1988. A comparison of sampling schemes used in generating error matrices for assessing the accuracy of maps generated from remotely sensed data. *Photogram. Eng. Rem. Sens.* 54(5), 593–600.

Drummond, J. E., 1991. Determining and Processing Quality Parameters in Geographic Information Systems, Ph.D. thesis, University of Newcastle upon Tyne, UK.

Fisher, P. F., 1991. Modeling soil map-unit inclusions by Monte Carlo simulation. *Int. J. Geog. Inform. Syst.* 5(2): 193–208.

Fukunaga, K. and Hayes, R. R., 1989. Effects of sample size in classifier design. *IEEE Transactions on Pattern Analysis and Machine Intelligence.* PAMI-11(8). 873–885.

Grant, M. E. and Leavenworth, R. S., 1988. *Statistical Quality Control.* McGraw-Hill International Editions, New York.

Greenland, A., Socher, R. M., and Thompson, M. R., 1985. Statistical evaluation of accuracy for digital cartographic database. *Proceedings of Auto-Carto 7.* ASP-ACSM, Washington DC, 1985.

Hay, A. M., 1979. Sampling designs to test land-use map accuracy. *Photogram. Eng. Rem. Sens.* 45(4): 529–533.

Hord, R. M. and Brooner, W., 1976. Land use map accuracy criteria. *Photogram. Eng. Rem. Sens.* 42(5): 671–677.

Kennedy-Smith, G. M., 1986. Data quality—a management philosophy. *Proceedings of Auto-Carto.* London, 1986.

Kish, L., 1965. *Survey Sampling.* John Wiley & Sons, New York.

Lodwick, W. A., Munson W., and Svoboda, L., 1988. Sensitivity Analysis in Geographic Information Systems, Part 1: Suitability Analysis. Research Paper RP8861, University of Colorado–Denver.

McMaster, S., 1996. Assessing the impact of data quality on forest management decisions using geographical sensitivity analysis. In *GISDATA'96* Summer Institute.

Rosenfield, G. H., 1982. Sample design for estimating change in land use and land cover. *Photogram. Eng. Rem. Sens.* 48(5): 793–801.

Rosenfield, G. H., 1986. Analysis of thematic map classification error matrices. *Photogram. Eng. Rem. Sens.* 52(5): 681–686.

Rosenfield, G. H. and FitzPatrick-Lins, K., 1986. A coefficient of agreement as a measure of thematic classification accuracy. *Photogram. Eng. Rem. Sens.* 52(2): 223–227.

Shi, W. Z., 1994. *Modeling Positional and Attribute Uncertainties in Integration of Remote Sensing and Geographic Information Systems.* Enschede, The Netherlands: ITC Publication. 147 pages.

Shi, W. Z., Liu, C., and Liu, D. J., 2002. Accuracy assessment for attribute data in GIS based on simple random sampling. *Geomatics and Information Science of Wuhan University.* 27(5): 445–450.

van Genderen, J. L. and Lock, B. F., 1977. Testing land use map accuracy. *Photogram. Eng. Rem. Sens.* 43(9)L: 1135–1137.

6 Modeling Integrated Positional and Attribute Uncertainty

Modeling integrated positional and attribute uncertainty by the proposed "S-band" model is presented in this chapter, and corresponding solutions are given. The methods of modeling positional uncertainty have been introduced in Chapter 4, and methods of modeling attribute uncertainty have been described in Chapter 5. The proposed model is to integrate the two types of uncertainties presented in these two chapters. When multiple data sources are integrated for a spatial analysis, uncertainties in the corresponding data sources are combined. The objective of modeling integrated positional and attribute uncertainty is to estimate the level of such combined uncertainty.

6.1 INTRODUCTION

Although differences exist between attribute and positional uncertainties, both types of uncertainty, in many cases, are closely related. A spatial entity in a GIS is comprehensively represented by its elements in spatial, attribute, temporal, scale, and spatial relationship domains. In fact, uncertainties of these elements of a spatial object in the domains will affect the spatial object as a whole; therefore, integrated uncertainty, which is composed of multiple sources of uncertainties, needs to be investigated. Integrated positional and attribute uncertainty is one example of integrated uncertainty.

In an object-based GIS model, an object (such as a building) is represented by its basic geometrical elements: points, lines, or polygons and their associated attributes (such as address and ownership of a building). Therefore, any uncertainties in the basic geometric elements and associated attributes have simultaneous effects on the identification of the spatial object. This is the same for objects in a field-based model, affected simultaneously by positional and attribute uncertainties.

Typically, boundaries of an object are affected by both positional and attribute uncertainties in an integrated way. Therefore, the object boundary should not be represented as a crisp line but as a transitional region of a certain width. A positional error, existing in such a region may also result in object attribute uncertainty. This uncertainty of the attribute value, the category of the land cover class that is identified from a remotely sensed image, may be caused by positional error of the image or attribute error in the classification. The central part of the area object is more certain, and the uncertainty within the object gradually increases from the center toward the boundary.

Integrated modeling of positional and attribute uncertainty is a key issue in multiple sources-based spatial analyses, given that some data may contain positional uncertainty, and some may possess attribute uncertainty. When these data are combined for a spatial analysis, the uncertainties in the corresponding data sets are also combined and even propagated. Hence, it is essential to develop adequate and pertinent estimation methods for these combined uncertainties.

The need for a combination of positional and attribute uncertainty estimation methods is illustrated by the following example. In a land resources inventory, the land area within a given administrative boundary, such as a county, needs to be found. Using remote sensing techniques, a satellite image is classified and a land cover thematic map is then obtained. The administrative boundaries from a paper map, such as a boundary map of the county, are digitized and stored in a GIS. These two layers (the land cover thematic map and the administrative boundaries map) are then overlaid and the total area for each of the land cover types within that county is then calculated. This procedure for a land resource inventory by integrating remote sensing and GIS data is illustrated in Figure 6.1.

Normally, both the classified image and the digitized map are assumed error-free. In reality, however, both image and map layers possess uncertainties. For instance, the satellite image is classified using the maximum likelihood classification method. If a pixel is categorized as "forest" with only 70% certainty, there is a 70% probability that the pixel is forest and a 30% probability that the pixel is not forest. Thus, the classification result contains attribute uncertainty. However, the digitized administrative boundary in GIS has positional error. If a point is located close to a boundary line segment, it may not be sure that this point is really within the boundary. This uncertainty arises from the random error in the digitized points composing the polygon boundary, for example, resulting from errors during data

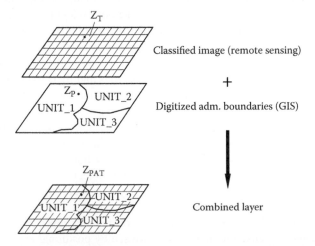

FIGURE 6.1 The procedure of land resources inventory by integrating classified map from satellite image and vector boundary map from GIS.

capture. This is positional uncertainty. The probability that the point belongs to the county polygon can then be determined by modeling the positional uncertainty of area objects.

When a classified raster image is overlapped with a vector administrative boundary map, the corresponding attribute and positional uncertainties that exist in both data layers are combined. A theory for modeling such an integrated uncertainty problem is thus required. The "S-band" model and corresponding solutions presented in this chapter are designed to solve this problem.

6.2   PROBLEM DEFINITION

The problem of integrating positional and attribute uncertainties is defined as follows. Suppose that a maximum likelihood classifier is used to classify a remotely sensed image. For a given pixel, Z_T, the probability is that the particular class C_{IF} the pixel belongs to equals

$$P[Z_T \in C_i / Z_T(x)],$$

where $Z_T(x) = (b_1, b_2, ..., b_n)^T$ is a vector in N-dimensional spectral domain, and C_i is one of overall classes.

For a polygon object in a GIS, the probability that the boundary point Z_P belongs to O_i equals

$$P[Z_P \in O_i / Z_P(x)].$$

That is, for a given point $Z_P(x)$ the probability of belonging to the polygon object O_i can be computed. This probability is actually a positional uncertainty indicator. $Z_p(x) \in R^2$ is a point in a two-dimensional Euclidean space, while O_i is an area object, for instance, as in the example given above, a county.

The problem can now be defined as: What is the probability of a point Z_{PAT} simultaneously belonging to both C_i and O_i after the combination of two layers? That is:

$$P\{[Z_T \in C_i / Z_T(x)] \wedge [Z_P \in O_i / Z_P(x)]\} = ?,$$

where $Z_{PAT} \in R^{n+2}$ is shorthand for $Z_{PAT} = (b_1, b_2, ..., b_n, x, y)^T$. This is the theoretical research problem to be solved for modeling uncertainty, generated by combing attribute and positional uncertainties.

For convenience, the following symbols are used to describe this problem.

H_T: $Z_T \in C_i$
H_P: $Z_P \in O_i$
$H_{P \wedge T}$: $[Z_T \in C_i / Z_T(x)] \wedge [Z_P \in O_i / Z_P(x)].$

In order to model the integration of positional and attribute uncertainties, the S-band model is proposed and is described, below, in Section 6.3. Two corresponding implementation solutions are provided in Sections 6.4 and 6.5.

6.3   THE S-BAND MODEL

The S-band model (Shi, 1994) has been designed to model integration of positional and attribute uncertainties, with emphasis on the uncertainties in the area object transition regions. Compared with commonly used approaches where positional and attribute uncertainties are separated, the philosophy behind the S-band model is that the uncertainty within the transition boundary region of an object is related to (a) positional uncertainty, (b) attribute uncertainty, and (c) correlation between (a) and (b). The S-band model is given by the following formula:

$$U_{PAT}(x) = F\left(U_P(x), U_T(x), \rho\left(U_P(x), U_T(x)\right)\right), \tag{6.1}$$

where $x \in R^2$ is a lattice or point coordinate vector located in a two-dimensional Euclidean space; x varies over an index set $D \in$ R2; $U_P(x)$ is the positional uncertainty at x; $U_T(x)$ is the attribute uncertainty at x; and $\rho(U_P(x), U_T(x))$ is the correlation between $U_P(x)$ and $U_T(x)$ at x. Finally, $U_{PAT}(x)$ is the integrated positional and attribute (PAT) uncertainty at x.

The name of the S-band model is based on the shape of the boundaries of an object with integrated positional and attribute uncertainty—in other words, an irregular s-shaped band. The model indicates that the uncertainty at a location x is related to its positional and attribute uncertainties, and the correlation between them. $U_P(x)$ in the above formula is a quantitative value, whereas $U_T(x)$ can be either a quantitative or qualitative value, and $\rho(U_P(x), U_T(x))$ can be one of them. It is therefore, difficult to define a concrete form of F for all the possible cases of integrating positional and attribute uncertainties for this model. A specific form of F can be given only when a specific application problem is well defined.

6.4   PROBABILITY THEORY-BASED SOLUTION

When H_T and H_P are coindependent events, the product rule of the probability theory to solve the above defined problem, can be as shown as follows:

$$P\left[H_{P \wedge T}\right] = P\left\{\left[Z_T \in C_i / Z_T(x)\right] \wedge \left[Z_P \in O_i / Z_P(x)\right]\right\} \tag{6.2}$$

$$= P\left[H_T / Z_T(x)\right] P\left[H_P / Z_P(x)\right].$$

It is essential to assume that the attribute and positional uncertainties are mutually independent. If this criterion is not satisfied, the general form of production rule, requiring the use of conditional probabilities, is to be applied. Because the product rule is a general probability reasoning method, a large sample size is required. This situation increases the cost. Hence, more practical solutions that require a smaller sample size need to be developed.

The following cases illustrate reasoning applications. Some evidence to support the hypothesis is positive and some negative. For example, when determining the land cover type for a parcel using multiple data sources, it can be concluded from remote sensing data that the parcel is forest, while from other data sources, it can be concluded that the parcel is not forest. In such a case, an uncertainty expression that covers the range of both the positive and negative beliefs—for example, [−1, 1]—is needed. But, the probability value range is [0, 1] and cannot be applied to solve this type of problem. The certainty factor model, which can overcome this constraint, is thus introduced.

6.5 CERTAINTY FACTOR-BASED SOLUTION

For solving the integration of positional and attribute uncertainties, the certainty factor model with probabilistic interpretation, developed by Heckerman (1986) and based on the original definition of the certainty factor, is adopted. The certainty factor model was used in MYCIN, an expert system used in medical diagnosis (Shortliffe and Buchanan, 1975) for uncertainty-based reasoning. In the following, the certainty factor model and its probabilistic interpretation are described. The adoption of the model for dealing with the problem indicated at the beginning of this section is then described.

6.5.1 THE MYCIN CERTAINTY FACTOR

The MYCIN certainty factor model is a reasoning method for managing uncertainty. The basic goal is to provide a method for handling uncertainty that avoids the requirement of large amounts of data and the intractability of computation in general probabilistic reasoning.

MYCIN's knowledge is stored as rules in the form: IF {evidence} THEN {hypothesis}. In medical science, relationships between evidence and hypothesis are often uncertain. The "certainty factors" are used to accommodate these nondeterministic relationships. A certainty factor, which represents the change of belief in a hypothesis based on given evidence, is attached to each rule. The certainty factor range is between −1 and 1. Positive numbers correspond to an increase in belief in a hypothesis, whereas negative numbers correspond to a decrease in belief. It is important to note that certainty factors do not correspond to measures of absolute belief.

The following notation represents a rule:

$$E \xrightarrow{\;CF(H,E)\;} H \;,$$

where H is a hypothesis, E is evidence relating to the hypothesis, and $CF(H, E)$ is the certainty factor attached to the rule.

There are at least two views on probability: the objective and the subjective. In the objective interpretation, a probability is regarded as a multitested average value, and in the other, the so-called subjective interpretation, a probability is considered as a measure of reliability. The certainty factor is defined according to the subjective interpretation of probability (Shortliffe and Buchanan, 1975).

The certainty factor $CF\,(H, F)$ expresses the reliability change in hypothesis H based on the evidence E. The definition of $CF\,(H, E)$ is

$$CF(H,E)=\begin{bmatrix} \dfrac{P(H/E)-P(H)}{1-P(H)}, & P(H/E)>P(H) \\[4mm] \dfrac{P(H/E)-P(H)}{P(H)}, & P(H/E)<P(H) \end{bmatrix},\qquad (6.3)$$

where $P(H)$ is the prior probability of the hypothesis H, while $P(H/E)$ is the posterior probability of H given E. This definition indicates that certainty factors are indicators of reliability change.

In MYCIN, it is possible that several items of evidence have a bearing on the same hypothesis. It is also possible for a hypothesis to serve as evidence for another hypothesis. The result is a network of rules. The network structure is often called an inference network (Duda et al., 1976). A major component of the certainty factor model is a prescription for propagating uncertainty through an inference network. The propagation of uncertainty through a complex inference network is accomplished by the repeated application of combination schemes for elementary networks.

In a parallel combination, two or more items of evidence simultaneously support the same hypothesis. If evidence E_1 and E_2, simultaneously support the hypothesis H, the combination of two certainty factors $CF(H, E_1)$ and $CF(H, E_2)$ then produces one certainty factor $CF(H, E_1E_2)$. The combined certainty factor for the hypothetical rule: IF {E1 AND E2} THEN {H}, is shown in Figure 6.2.

Parallel combinative function of certainty factors is defined as follows (Heckerman, 1986):

$$Z=\begin{cases} x+y-xy, & x,y\geq 0 \\[3mm] \dfrac{x+y}{1-\min\left(|x|,|y|\right)}, & x\cdot y<0, \\[3mm] x+y+xy, & x,y<0 \end{cases}\qquad (6.4)$$

where $x = CF(H, E_1)$, $y = CF(H, E_2)$, and $z = CF(H, E_1E_2)$.

The sequential combination can be defined as a hypothesis that serves as evidence of another hypothetical rule. For example, in the upper part of Figure 6.3, the combination

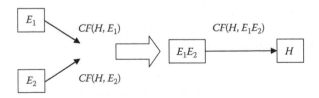

FIGURE 6.2   Parallel combination of certainty factor.

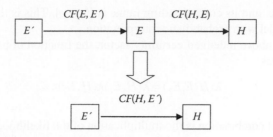

FIGURE 6.3 Sequential combination of certainty factor.

of two certainty factors $CF(E, E')$ and $CF(H, E)$ in two hypothetical rules produces a new hypothetical certainty factor $CF(H, E')$, shown in the lower part of Figure 6.3.

The function of the sequential certainty factor combination is defined as

$$z = \begin{cases} wx & w \geq 0 \\ -wy & w < 0 \end{cases}, \qquad (6.5)$$

where $w = CF(E, E')$, $x = CF(H, E)$, $y = CF(H, \neg E)$, and $z = CF(H, E')$. Based on the two certainty factor combination models represented by (6.3) and (6.4), the parallel and sequential combinations in an inference network to obtain a reasoning result for the whole inference network can be repeatedly computed. The details about certainty factors can be found in Heckerman (1986).

6.5.2 PROBABILISTIC INTERPRETATION OF CERTAINTY FACTORS

Two restrictions exist in the original definition of certainty factors as shown in (6.3): firstly, the definition is inconsistent with the quantity-combinative function used in MYCIN, and the other restriction is that the original certainty factor model is not well related to the probability theory (Heckerman, 1986). Therefore, Heckerman reformulated and redefined it as follows:

$$CF_1(H,E) = \begin{cases} [\lambda(H,E)-1]/\lambda(H,E), & \lambda \geq 1 \\ \lambda(H,E)-1, & \lambda < 1 \end{cases}, \qquad (6.6)$$

where the likelihood ratio λ is defined as

$$\lambda(H,E) = P(E/H)/P(E/\neg H), \qquad (6.7)$$

where $P(E/H)$ is the probability established by E under the condition of a given hypothesis H, and $P(E/\neg H)$ is the conditional probability occasioned by E under the condition of a given $\neg H$ as non-H. The likelihood ratio λ is actually an indicator for reliability change, and its corresponding range is $[0, +\infty]$. This is the so-called certainty factor model with probabilistic interpretation.

Based on the above redefined certainty factor, the function of parallel combination is redefined as

$$\lambda(H, E_1 E_2) = \lambda(H, E_1)\lambda(H, E_2) . \qquad (6.8)$$

Thus, the parallel combination is the multiplication of the likelihood ratio.

Based on the newly defined certainty factor, the sequential combination is redefined as the following function:

$$\lambda(H, E') = \frac{\lambda(E, E')\lambda(H, E)(1 - \lambda(H, \bar{E})) + \lambda(H, \bar{E})(\lambda(H, E) - 1)}{\lambda(E, E')(1 - \lambda(H, \bar{E})) + (\lambda(H, E) - 1)} . \qquad (6.9)$$

A detailed description of the above two combination functions can be found in Heckerman (1986). After the parallel and sequential combinations, with the use of the Equation 6.6, the likelihood ratio can be transformed into a certainty factor. The certainty factor value, after combination, is thus obtained.

6.5.3 THE CERTAINTY FACTOR-BASED SOLUTION

The above section has introduced two certainty factor models: the original definition of certainty factor and the probabilistic interpretation-possessed certainty factor. Here, the latter one is adopted. This is owing to the considerations that both positional and attribute uncertainties are modeled based on the probability theory.

The certainty factor model with the probabilistic interpretation is applied to solve the problem for the combination of positional and attribute uncertainties: What is the value of $P[H_{P \wedge T}]$ under the condition with given $P[H_P / Z_P(x)]$ and $P[H_T / Z_T(x)]$? This problem is defined in Section 6.2.

In the above situations, evidence (E) is a coordinate in spectral and Euclidean space, represented with $Z_T(x)$ and $Z_P(x)$, respectively.

6.5.4 DERIVATION OF THE FORMULAS

Since:

$$\left[Z_T \in C_i / Z_T(x)\right] \wedge \left[Z_P \in O_i / Z_P(x)\right] \subseteq \left[Z_P \in O_i / Z_P(x)\right],$$

$$\left[Z_T \in C_i / Z_T(x)\right] \wedge \left[Z_P \in O_i / Z_P(x)\right] \subseteq \left[Z_T \in C_i / Z_T(x)\right],$$

the definition of the likelihood ratio (λ) from the above mentioned condition can then be:

$$\lambda\left(H_{P \wedge T}, Z_P(x)Z_T(x)\right) \leq \lambda\left(H_P, Z_P(x)\right) \qquad (6.10)$$

and

$$\lambda\left(H_{P\wedge T}, Z_P(x)Z_T(x)\right) \le \lambda\left(H_T, Z_T(x)\right). \tag{6.11}$$

Thus, combining the Formula 6.6, for the parallel combination problem, we have:

$$\lambda\left(H_{P\wedge T}, Z_P(x)Z_T(x)\right) \le \lambda\left(H_{P\wedge T}, Z_P(x)\right)\lambda\left(H_{P\wedge T}, Z_T(x)\right).$$

According to the Formulas 6.10 and 6.11, we have:

$$\lambda\left(H_{P\wedge T}, Z_P(x)Z_T(x)\right) \le \lambda\left(H_P, Z_P(x)\right)\lambda\left(H_T, Z_T(x)\right) \tag{6.12}$$

$$T = \lambda\left(H_P, Z_P(x)\right)\lambda\left(H_T, Z_T(x)\right), \tag{6.13}$$

where T is a threshold value of the upper bound of $\lambda(H_{P\wedge T}, Z_P(x)Z_T(x))$, that is, the less or equal combined likelihood value to this value, or:

$$\lambda(H_{P\wedge T}, Z_P(x)Z_T(x)) \le T.$$

By applying Bayes' formula and Formula 6.7, the likelihood value can be obtained as the following:

$$\lambda(H_T, Z_T(x)) = \frac{P[H_T/Z_T(x)]}{P[\neg H_T/Z_T(x)]} \cdot \frac{P(\neg H_T)}{P(H_T)}. \tag{6.14}$$

In a classified image, the attribute conditional probability $P[H_T/Z_T(x)]$ for each pixel can be obtained from maximum likelihood classification, whereas the $P(\neg H_T, Z_T(x))$ can be computed by $1 - P[H_T/Z_T(x)]$. The priori probabilities $P(H_T)$ and $P(\neg H_T)$ can be calculated by using commonly used techniques in classification. Applying (6.14) and the above computed values, $\lambda(H_T, Z_T(x))$ can thus be solved.

The likelihood ratio for the positional uncertainty can be obtained from the following formula:

$$\lambda(H_P, Z_P(x)) = \frac{P[H_P/Z_P(x)]}{P[\neg H_P/Z_P(x)]} \cdot \frac{P(\neg H_P)}{P(H_P)}, \tag{6.15}$$

where $P[H_P, Z_P(x)]$ and $P[\neg H_P, Z_P(x)]$ can be derived from the positional uncertainty modeling method for area objects. The $P(H_P)$ can be estimated by using the object area and the total area, and $P(\neg H_P)$ is calculated by $1 - P(H_P)$. Put in the above values in formula 6.15, $\lambda(H_P, Z_P(x))$ can finally be solved.

6.5.5 The Computational Procedure

Based on the above principle, the following computational procedure for combining positional and attribute uncertainty is proposed.

Step 1: To calculate $P[\lnot H_P, Z_P(x)]$ and $P(\lnot H_T, Z_T(x))$, according to $P[H_P/Z_P(x)]$ and $P[H_T/Z_T(x)]$

$$P(\lnot H_P, Z_P(x)) = 1 - P[H_P/Z_P(x)],$$

$$P(\lnot H_T, Z_T(x)) = 1 - P[H_T/Z_T(x)].$$

$P(H_P)$ and $P(H_T)$ are determined from the priori knowledge about the map or image.

Step 2: To calculate $\lambda(H_T, Z_T(x))$ and $\lambda(H_P, Z_P(x))$, based on Formulas 6.14 and 6.15.

Step 3: To calculate the threshold value (T) of $\lambda(H_{P \land T}, Z_P(x)Z_T(x))$ according to the formula (6.13).

Step 4: To calculate the T-corresponded certainty factor by the Formula 6.6. The value of the certainty factor forms the upper bound of the integrated positional and attribute uncertainty. In other words, the combined uncertainty value, that is, the certainty factor value is less or equal to the T-corresponded certainty factor value $CF(T)$.

A threshold value for certainty factor values can be obtained. The value is the largest certainty factor value for the hypothesis $H_{P \land T}$ under the condition of a given $(Z_P(x) \cdot Z_T(x))$. This is, in fact, the indicator of combined positional and attribute uncertainty, based on the certainty factor model with probabilistic interpretation.

The above procedure can be practically implemented in an integrated GIS and remote sensing environment for inference in an inference network. According to the nature of the combination problems, either parallel or sequential combination can be chosen. A certainty factor for each pixel can finally be obtained. Each factor is an indicator of integrated positional and attribute uncertainty and gives an upper bound (by the threshold value) for each pixel; the uncertainty of the pixel is less than this value.

6.6 AN EXAMPLE OF MODELING INTEGRATED UNCERTAINTY

In making an inventory of the land cover in a county, the boundary of the county is digitized from a paper map and the land cover is classified from a remote sensing image. In so doing, it is necessary to know the land cover area for each of the classes and it is also necessary to know the uncertainties and the spatial distribution of the inventory data.

In a situation where there is no requirement for uncertainty information, this problem could be solved by a commonly used overlay operation in GIS. The S-band model is applied for modeling uncertainty and spatial data distribution. In such a case, positional and attribute uncertainties are modeled first and the integrated uncertainty then determined.

6.6.1 Modeling Positional Uncertainty

The product rule according to Formula 6.2 is first applied. The model holds true on the assumption of perfect georeferencing—that there is no positional error in the classified remote sensing image. This may not always be true in the real world, but it is an assumption in this study. The positional uncertainty of the assumed county boundary is shown in Figure 6.4a. Figure 6.4b shows the positional uncertainty inside the boundary of the county. The darker the gray value, the larger the probability and the smaller the uncertainty. The statistical results are given in Table 6.1.

Table 6.1 indicates that, among all the total 1638 pixels, 989 pixels are within the probability range 0.91 to 1, inferring that this range contains 60% of the total pixels with the probabilities ranging from 0.91 to 1 that they belong to area objects.

(a)   (b)

(c)   (d)

FIGURE 6.4   Positional uncertainty, attribute uncertainty and their integration with an area object.

TABLE 6.1
Statistical Results of the Positional Uncertainties Inside the County Boundary

Probability interval	0~0.1	0.11~0.2	0.21~0.3	0.31~0.4	0.41~0.5	0.51~0.6	0.61~0.7	0.71~0.8	0.81~0.9	0.91~1	Sum
Number of pixels	0	0	0	0	20	107	154	170	198	989	1638

6.6.2 MODELING ATTRIBUTE UNCERTAINTY

A classified image is used to illustrate the spatial distribution of various classes. The total number of pixels per class is given in Table 6.2. These results however do not indicate uncertainty in the classification. Often, an error matrix is used to give an assessment of the classification accuracy, but such an assessment is very much dependent on the sampling method used to obtain the reference data and, does not necessarily provide the spatial uncertainty distribution of the area of interest.

The attribute uncertainty distribution based on the probability vectors of the selected region is shown in Figure 6.4d, displaying the maximum probability of each pixel. The darker the pixel, the higher the probability that the pixel belongs to the assigned class, and the smaller the classification uncertainty.

To compute the frequency, the distribution probability values are divided into ten intervals, 0–0.10, 0.11–0.20, ..., 0.91–1. Statistical results are listed in Table 6.3. This table enables the acquisition of knowledge of the total area of each class and also provides a confidence level rating regarding the classification. This is an obvious to users of this data set. For example, 56 of the total 117 pixels classified as "urban" have only a 0.21 to 0.30 probability. Thus, the classification result of "urban" is very uncertain. On the other hand, 138 of the total 222 pixels classified as "grass" are in

TABLE 6.2
Statistic Result by the Maximum
Likelihood Classification

Class_number	Class Name	Area (Number of Pixels)
1	Urban	117
2	Grass	222
3	Water	19
4	Forest	1280
	Total	**1638**

TABLE 6.3
Classification Results in Various Probability Intervals

Probability interval	0–0.10	0.11–0.2	0.21–0.3	0.31–0.4	0.41–0.5	0.51–0.6	0.61–0.7	0.71–0.8	0.81–0.9	0.91–0.1	Sum
Urban	0	0	56	13	3	9	7	5	11	13	117
Grass	0	0	18	17	4	15	7	11	12	138	222
Water	0	0	12	0	0	0	0	0	0	7	19
Forest	0	0	394	57	65	48	65	73	126	452	1280
										Total	**1638**

the probability range of 0.91–1; thus, this classification of "grass" can be assumed to be relatively certain.

6.6.3 Modeling Integrated Positional and Attribute Uncertainty

Given a boundary layer with positional uncertainty and a classified remote sensing image with attribute uncertainty, an overlay operation is applied. In this overlay process, positional and attribute uncertainties are combined. The combination is modeled using Formula 6.2. The combination result is shown in Figure 6.4c. Comparing (c) with (d) in Figure 6.4, it can be seen that the grey values around the boundary region have become lighter; the uncertainty in this region increases after the combination. But there is no change within the interior region, indicating that positional uncertainty has nil regional effect. The statistics for uncertainty after combining positional and attribute data are given in Table 6.4.

Comparing Tables 6.3 and 6.4, it is noticed that after the combination of the classified remote sensing image layer and the boundary layer, more pixels possess lower probability values. For example, within the interval 0.91–1, the number of "urban" pixels is reduced from 13 to 4, for "grass" from 138 to 57, for "water" from 7 to 0, and for "forest" from 452 to 240. However, the number of pixels with higher uncertainty has increased. For example, in the interval 0.11–0.20, the number of "urban" pixels has increased from 0 to 24; the number of "grass" pixels from 0 to 9; the number of "water" pixels from 0 to 4; and the number of "forest" pixels from 0 to 80. The trend of these changes reflects the fact of the increase of the integrated uncertainty, after the positional and attribute uncertainties are combined.

6.6.4 Visualization of Uncertainties

Figure 6.4 presents the uncertainty distribution represented by gray levels, a commonly used method to visualize uncertainty. However, because of the limitations of the human eye in distinguishing grey values (normally less than 64 levels), the gray-level coding may not work very well in some cases. In Figure 6.4, for example, in (c), only a few pixels can be seen with corresponding locations. In (d) the pixels

---

TABLE 6.4
Statistic Results of Integrated Positional and Attribute Uncertainty After Overlay

Probability interval	0–0.1	0.11–0.2	0.21–0.3	0.31–0.4	0.41–0.5	0.51–0.6	0.61–0.7	0.71–0.8	0.81–0.9	0.91–0.1	Sum
Urban	0	24	42	7	7	15	12	4	2	4	117
Grass	0	9	19	13	18	26	25	29	29	57	222
Water	0	4	8	0	2	4	1	0	0	0	19
Forest	0	80	344	62	95	93	109	133	124	240	1280

Total 1638

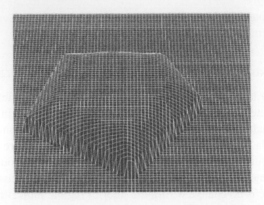

FIGURE 6.5   Positional uncertainty of an area object.

have different gray values, but their difference cannot be realized clearly. Therefore, a method which uses three-dimensional, color, and gray-value techniques for visualizing uncertainties is presented as follows.

If the demand is for the visualization of one component, such as the positional uncertainty itself, then a three-dimensional visualization method can be applied directly. Figure 6.5 which shows the positional uncertainty of an area object is one such example.

In Figure 6.5, the plane coordinates represent the location of a pixel; the third dimension (height) represents the uncertainty value of a pixel. The higher the height value, the lower the uncertainty. From this figure, it can be clearly seen that uncertainty increases from small to large when a point moves from inside to outside the area object. Height is constant for the points within the interior area of the object, meaning that the uncertainty values are the same in that area.

If the requirement is the representation of uncertainty and the corresponding attribute value, a three-dimensional plus color representation can be used. For example, Figure 6.6 represents the classification result of an image and the probability by which a pixel is assigned to a class. The color represents the class type: red represents "urban," green represents "forest," yellow represents "grass," and blue represents "water." The height represents the probability value. The higher the height value, the lower the uncertainty.

Similarly, in Figure 6.7 the uncertainty after the integration of positional and attribute uncertainties is visualized. Here, the height indicates the probability that a pixel belongs to a class (e.g., "urban") and lies within the area (county A). By comparing Figures 6.6 and 6.7, it can be clearly seen how uncertainties differ before and after the overlay operation, combining positional and attribute uncertainties. The difference is more clearly illustrated here than in (c) and (d) of Figure 6.4. From Figure 6.6, not only the class type can be seen but also the pixel class probability. For example, the two pixels located near the center of the area are classified as "water" (in blue) and have a very low probability value, that is, the confidence that these pixels are indeed "water" is low.

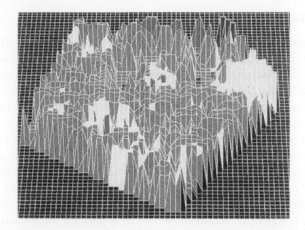

FIGURE 6.6 **(See color figure following page 38.)** Attributes and uncertainty of a classified image. (From Shi, W. Z. 1994. *Modeling Positional and Thematic Uncertainties in Integration of Remote Sensing and Geographic Information Systems*. Enschede, The Netherlands.)

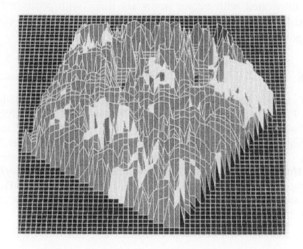

FIGURE 6.7 **(See color figure following page 38.)** Integrated positional and attribute uncertainty.

The visualization result can be used directly by both data provider and data user. For the data provider, those areas with low probability values should be checked more closely as these are the areas where misclassification is likely to occur. To reduce the risk of using wrong information, however, the data users can give greater weight to those areas with higher probability values and lower weight to those with smaller probability values in their data set choice.

6.7   SUMMARY AND COMMENT

This chapter addresses the issue of the integration of positional and attribute uncertainties, which is essential in multiple source data integration, such as integration of field-based classified images from remote sensing and object-based vector data from GIS. The S-band model has been developed to described the integrated positional and attribute uncertainty, with a focus on modeling uncertainty for transitional region of area objects.

The S-band model is a generic approach for handling integrated positional and attribute uncertainty. The model elements are as follows: positional uncertainty, attribute uncertainty, and the mutual correlation between them. Two solutions have been provided for the S-band model: the probability theory-based solution and the certainty factor model-based solution. With the assumption that two data layers are independent, a product-rule-based model can be applied for the S-band model. With the certainty factor model-based solution, the value range of uncertainty expressions is extended from [0, 1] to [−1, 1]. This is particularly important for a reasoning, which includes uncertainty indicators covering both positive and negative ranges.

The application of the S-band model has been demonstrated by an example for land resource inventory, in which a classified remote sensing image with attribute uncertainty is integrated with vector boundary maps with positional uncertainty. The results of the example include both statistic results and graphic visualization of the integrated positional and attribute uncertainty.

REFERENCES

Duda, R., Hart, P., and Nilsson, N., 1976. Subjective Bayesian method for rule-based inference system. In *Proceedings of Natural Computer Conference* Vol. 45, AFIPS: 1075–1082.

Heckerman, D., 1986. Probabilistic interpretation of MYCIN's certainty factors. In *Uncertainty in Artificial Intelligence*, ed. Kanal, L. N., Lemmer, J. F., 167–196. Elsevier, North Holland. New York.

Shi, W. Z., 1994. *Modeling Positional and Thematic Uncertainties in Integration of Remote Sensing and Geographic Information Systems*. Enschede, The Netherlands: ITC Publication.

Shortliffe, E. H. and Buchanan, B. G., 1975. A model of inexact reasoning in medicine. *Math. Biosci.* 23: 351–379.

Section III

Modeling Uncertainties
in Spatial Model

7 Modeling Uncertain Topological Relationships

Uncertainty modeling in spatial data has been introduced in Part II, and now, in Part III (Chapters 7 and 8) uncertainty modeling in a spatial model, is further explored and described. Modeling uncertain topological relations is addressed in this chapter. First, the concepts of topological relationships between objects in GISci are introduced, and this is followed by a description of modeling these relationships. Finally, based on the concept of quasi coincidence and quasi difference in fuzzy topology, an uncertain topological relationship model is proposed.

7.1 AN OVERVIEW OF TOPOLOGICAL RELATIONSHIP MODELS

Most GISci are traditionally designed on the basis that the represented spatial objects, such as rivers, roads, trees, and buildings, are error free. As has been indicated in Chapter 2, this assumption may not always be true, owing either to vagueness or fuzziness in the definition or identification of spatial objects or randomness during the measuring process. For example, it is impossible to describe a fuzzy boundary between urban and rural areas using crisp lines, without uncertainty, as has been the practice in traditional GIS representations. It has become obvious that there is a need to enhance the currently used GIS functions by introducing a means to describe single objects. Such descriptions would include any uncertainty present and, also, enable the description of fuzzy topological relationships between such uncertain spatial objects. On this basis, it could be concluded that descriptions of objects with crisp boundaries using the classical set theory (Apostol, 1974), is possibly no longer suitable (Wang, Hall, and Subaryono, 1990). However, it is likely that fuzzy set theory could be used to describe fuzzy uncertain objects in GISci. In addition, fuzzy topology provides a theoretical basis for modeling fuzzy topological relationships between spatial objects.

Topological relationships are defined as invariant under geometric distortion, and form a subset of spatial relationships. The representation of topological relationships between spatial objects is one of the major issues for spatial data modeling in GISci. The application of topological relationships makes it possible to accelerate spatial queries related to spatial relationships and also facilitate topologic consistency checking in GIS.

7.1.1 The Four-Intersection Model

Based on point-set topology, a spatial object A can be divided into two parts: its interior and its boundary, denoted by A^o and ∂A, respectively. The topological relationship between objects can be described by the four intersections of interiors and boundaries, denoted by

$$\begin{pmatrix} \partial A \cap \partial B & \partial A \cap B^o \\ A^o \cap \partial B & A^o \cap B^o \end{pmatrix}.$$

By considering the invariance property, empty and nonempty, of each intersection, 16 possible relations exist between objects in two-dimensional space (Egenhofer and Franzosa, 1991).

7.1.2 The Nine-Intersection Model

If only the topological spatial relations between polygonal areas in the plane are considered, there are nine relations between two spatial regions. Based on point-set topology and the embedding of set A into R^2, the set A in R^2 can be divided into three parts: the interior (A^o), the boundary (∂A), and the exterior (A^c). This extends the four-intersection model to a new intersection model by considering the intersection of interior (A^o), boundary (∂A) and exterior (A^c) with interior (B^o), and boundary (∂B) and exterior (B^c). The interaction of these six parts, provide a total of nine intersection combinations,

$$\begin{pmatrix} A^o \cap B^o & \partial A \cap B^o & A^c \cap B^o \\ A^o \cap \partial B & \partial A \cap \partial B & A^c \cap \partial B \\ A^o \cap B^c & \partial A \cap B^c & A^c \cap B^c \end{pmatrix}.$$

This can be used to represent the topological relationships between spatial objects.

By considering the invariance property, empty and nonempty, of each component, there are 512 possible relations between spatial objects. From these 512 cases, there are only eight relations: "disjoint," "contains," "inside," "equal," "meet," "covers," "covered by," and "overlap" that can be realized, if the objects are spatial regions in R^2 (Egenhofer 1993).

7.1.3 Algebraic Model

When dealing with indeterminate boundaries, Clementini and Di Felice (1996) defined a region with a broad boundary by using two simple regions, denoted by ΔA. More precisely, the broad boundary is a simple connect subset of R^2 with a hole. Based on the empty and nonempty invariance, Clementini and Di Felice's algebraic model provides a total of 44 relations between two spatial regions with a broad boundary.

7.1.4 THE "EGG-YOLK" MODEL

In dealing with nonexact spatial objects and the vagueness/fuzziness spatial objects, Cohn and Gotts (1996) suggest using two concentric subregions, indicating the degree of "membership" in a vague/fuzzy region, in which "egg" represents the precise part and "yolk" represents the vague/fuzzy part. Based on region connection calculus theory (Randell et al., 1992), eight basic relations can then be defined. They are disconnected, externally connected, partially overlapping, tangential proper part, nontangential proper part, equal, proper part inverse, and tangential proper part inverse.

The egg-yolk model is an extension of region connection calculus theory into the vague/fuzzy region. There are 46 relations that can be identified to represent conservatively defined limits on the possible "complete crispings" or precise versions of a vague region.

Compared with 44 relations in the algebraic model (Clementini and Di Felice 1996), there are 46 relations in the egg-yolk model. Two different set relations, due to the empty and nonempty nine-intersection model (Clementini and Di Felice, 1996), cannot distinguish between 30th and 31st, or between 37th and 38th (Cohn and Gotts, 1996).

7.1.5 A NINE-INTERSECTION MODEL BETWEEN SIMPLE FUZZY REGIONS

To investigate the topological relationship between fuzzy regions, Tang and Kainz (2001) decompose a fuzzy set A into several topological parts:

  i. The core, $A^{\oplus}$, which is the fuzzy interior part with value equal to one,
 ii. The c-boundary, $\partial^c A$, the fuzzy subset of ∂A with $\partial A(x) = \overline{A}(x)$,
iii. B-closure, $A^{\perp}$, the fuzzy subset of $\overline{A}$ with $\overline{A}(x) > \partial A(x)$, and
 iv. The outer, $A^{=}$, the fuzzy complement of A with value equal to one.

A nine-intersection matrix is thus formalized and a 4 by 4 intersection matrix, based on the different topological parts of two fuzzy sets. For the nine-intersection matrix,

$$\begin{pmatrix} A^{\oplus} \wedge B^{\oplus} & A^{\oplus} \wedge \ell B & A^{\oplus} \wedge B^{=} \\ \ell A \wedge B^{\oplus} & \ell A \wedge \ell B & \ell A \wedge B^{=} \\ A^{=} \wedge B^{\oplus} & A^{=} \wedge \ell B & A^{=} \wedge B^{=} \end{pmatrix},$$

the core ($A^{\oplus}$), fringe (ℓA), and outer ($A^{=}$) are adopted to formalize, where the fringe is the union of c-boundary and b-closure. There are in total of 44 relations between two simply fuzzy regions. For the 4 by 4 intersection matrix,

$$\begin{pmatrix} A^{\oplus} \wedge B^{\oplus} & A^{\oplus} \wedge B^{\perp} & A^{\oplus} \wedge \partial^c B & A^{\oplus} \wedge B^{=} \\ A^{\perp} \wedge B^{\oplus} & A^{\perp} \wedge B^{\perp} & A^{\perp} \wedge \partial^c B & A^{\perp} \wedge B^{=} \\ \partial^c A \wedge B^{\oplus} & \partial^c A \wedge B^{\perp} & \partial^c A \wedge \partial^c B & \partial^c A \wedge B^{=} \\ A^{=} \wedge B^{\oplus} & A^{=} \wedge B^{\perp} & A^{=} \wedge \partial^c B & A^{=} \wedge B^{=} \end{pmatrix},$$

the core ($A^{\oplus}$), c-boundary ($\partial^c A$), b-closure ($A^{\perp}$), and outer ($A^{=}$) are adopted to formalize the fuzzy topological relationships between spatial objects.

7.2 MODELING TOPOLOGICAL RELATIONSHIPS BETWEEN OBJECTS

This section presents a formal representation and calculation of the topological relations between uncertain spatial objects. Uncertain geographic objects fall into three types: (1) a geometrically uncertain core area surrounded by an uncertain boundary; (2) a geometrically uncertain core area surrounded by a certain boundary; and (3) a geometrically certain core area surrounded by an uncertain boundary. Here, we present solutions for case (3) in this section. However, geographical objects in the cases (1) and (2) may exist in the natural or real world.

7.2.1 Representation of Spatial Objects

Spatial object types can be identified in N-dimensional space. For example, four types of spatial objects, such as point, line, surface, and body, can be identified in a three-dimensional space.

7.2.1.1 N-Cell and N-Cell Complex

An N-cell is a bordered subset of an N-manifold and thus homeomorphic to an N-manifold with (N-1)-manifold boundary. Depending on the dimension of the space, the cells of dimension 0, 1, 2, and 3 are defined as follows:

0-cell $\equiv$ point
1-cell $\equiv$ line segment
2-cell $\equiv$ area
3-cell $\equiv$ sphere

The geometric representation of these cells is shown in Figure 7.1.

An N-cell complex represents the composition of all N-dimensional and lower-dimensional cells and satisfies the following conditions:

- Different elements of the N-cell complex have disjoint interiors.
- For each cell in the complex, the boundary is a union of elements of the complex.
- If two cells intersect, they also do so in a cell of the complex.

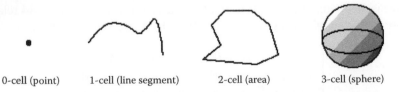

| 0-cell (point) | 1-cell (line segment) | 2-cell (area) | 3-cell (sphere) |

FIGURE 7.1   Cells of dimension 0, 1, 2, and 3.

7.2.1.2 The Definition of a Spatial Object

In mathematics, an N-dimensional space is a topological space, the dimension of which is N. Any certain N-dimensional spatial object with precise boundaries K can be represented as an N-cell complex. N-dimensional space can be subdivided into three components: the interior $K°$, the boundary ∂K, and the exterior K^-. Let $C(p, r)$ denote an N-dimensional disk with radius r and centered at p, see Equation 7.1:

$$C(p,r) = \{q \in \Gamma | d(p,q) < r\}. \tag{7.1}$$

The interior, boundary, and exterior of K can be defined as follows:

$$K^0 = \{p \in K : \exists \varepsilon > 0 : C(p,\varepsilon) \subset K\},$$

$$\partial K = \{p \in K : \forall r > 0 : C(p,r) \cap K \neq \varnothing \wedge C(p,r) \cap \Gamma - K \neq \varnothing\},$$

$$K^- = \{p \in \Gamma - K\},$$

where Γ represents the n-dimensional space 3^n.

By invoking methods to characterize certainty, imprecision, and randomness of spatial objects, a unified framework for representing spatial objects has been developed (Shi and Guo, 1999) and described as follows: Let Γ denote the N-dimensional space 3^n with the usual metric d, and suppose K is an N-dimensional spatial object, $\bar{x}$ is an arbitrary point in K, represented as $\bar{x} = (x_1, x_2, \ldots, x_n)$, where $(x_1, x_2, \ldots, x_n)$ is the spatial coordinate of $\bar{x}$. We can decompose K into three regions: (1) the core RK, (2) the crust BK, and (3) the exterior EK. The core region RK is that part of spatial object K that is a homogeneous N-dimensional manifold in Γ. It can be subdivided into two components: the interior $RK°$ and the boundary ∂RK. The crust region BK is the uncertain part of spatial object K. It can be the indeterminate boundary of an imprecise object or the confidence region of a random object. The exterior region EK of spatial object K is the set $\Gamma - K$. The relationship between the three regions can be concisely represented as follows:

$$K = RK \cup BK,$$

$$RK \cup BK \cup EK = \Gamma,$$

$$RK \cap BK \cap EK = \varnothing.$$

According to fuzzy statistics, the result of statistical and probabilistic models can be translated into a membership function in a fuzzy set. Fuzzy set theory is prominent as an appropriate means of modeling imprecision. Thus, an uncertain spatial object can be represented as a fuzzy object K. The fuzzy membership of K is denoted by Equation 7.2 below:

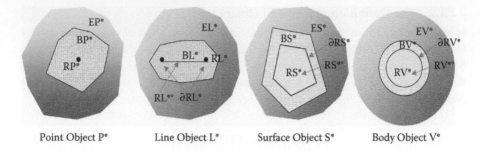

Point Object P*   Line Object L*   Surface Object S*   Body Object V*

FIGURE 7.2 The core, crust, and exterior of uncertain spatial objects.

$$\mu_K(\bar{x}) = \begin{cases} 1, & \bar{x} \in RK \\ \mu_K, & \bar{x} \in BK, \\ 0, & \bar{x} \in EK \end{cases} \qquad (7.2)$$

where $0 < \mu_k < 1$, and $\bar{x}$ is an arbitrary point in n-dimensional space.

For example, in a three-dimensional space, the core, crust, and exterior regions of a point object P*, a line object L*, a surface object S*, and a body object V* are represented as in Figure 7.2.

7.2.2 TOPOLOGICAL RELATIONSHIPS BETWEEN SPATIAL OBJECTS

Topological relations become more complex under uncertain situations, because uncertain spatial objects K consist of the three components which are the core RK, crust BK, and exterior EK. Meanwhile, the relationship between EK and the other components can be determined by the relationship between RK and BK due to the existing relationship RK∪BK∪EK = Γ, that is, EK = Γ − RK∪BK. The topological relationship between any two objects K_1 and K_2 can be concisely represented as one of the 13 basic relations below:

1. The *disjoint* relation

$$< K_1, disjoint, K_2 > \Leftrightarrow (K_1 \cap K_2 = \varnothing).$$

2. The *meet* relation

$$<K1, meet, K_2> \Leftrightarrow$$
$$(RK_1 \cap RK_2 = \varnothing) \wedge (BK_1 \cap BK_2 \neq \varnothing) \wedge (RK_1 \cap BK_2 = \varnothing) \wedge (BK_1 \cap RK_2 = \varnothing).$$

3. The *nearly meet* relation

$$<K_1, nearly\ meet, K_2> \Leftrightarrow$$
$$(RK_1 \cap RK_2 = \varnothing) \wedge (BK_1 \cap BK_2 \neq \varnothing) \wedge ((RK_1 \cap BK_2) \cup (BK_1 \cap RK_2) \neq \varnothing).$$

4. The *nearly overlap* relation

$\langle K_1$, *nearly overlap*, $K_2 \rangle \Leftrightarrow$
$(RK_1 \cap RK_2 = \varnothing) \wedge (BK_1 \cap BK_2 \neq \varnothing) \wedge ((RK_1 \cap BK_2) \neq \varnothing) \wedge (BK_1 \cap RK_2) \neq \varnothing).$

5. The *overlap* relation

$\langle K_1$, *overlap*, $K_2 \rangle \Leftrightarrow (RK_1 \cap RK_2 \neq \varnothing) \wedge (RK_1 \not\subset K_2) \wedge (RK_2 \not\subset K_1).$

6. The *nearly covered by* relation

$\langle K_1$, *nearly covered by*, $K_2 \rangle \Leftrightarrow (RK_1 \subseteq K_2).$

7. The *nearly covered* relation

$\langle K_1$, *nearly covered*, $K_2 \rangle \Leftrightarrow (RK_2 \subseteq K_1).$

8. The *covered by* relation

$\langle K_1$, *covered by*, $K_2 \rangle \Leftrightarrow (RK_1 \subseteq RK_2) \wedge (K_1 \subseteq K_2).$

9. The *covers* relation

$\langle K_1$, *covers*, $K_2 \rangle \Leftrightarrow (RK_2 \subseteq RK_1) \wedge (K_2 \subseteq K_1).$

10. The *inside* relation

$\langle K_1$, *inside*, $K_2 \rangle \Leftrightarrow (K_1 \subseteq RK_2).$

11. The *contains* relation

$\langle K_1$, *contains*, $K_2 \rangle \Leftrightarrow (K_2 \subseteq RK_1).$

12. The *nearly equal* relation

$\langle K_1$, *nearly equal*, $K_2 \rangle \Leftrightarrow (RK_1 \subseteq K_2) \wedge (RK_2 \subseteq K_1).$

13. The *equal* relation

$\langle K_1$, *equal*, $K_2 \rangle \Leftrightarrow (K_1 \subseteq K_2) \wedge (K_2 \subseteq K_1).$

7.2.3 A Decision Algorithm for Topological Relationship

Suppose the fuzzy memberships of spatial objects K_1 and K_2 are μ_{k1} and μ_{k2}, denoted by Equations 7.3 and 7.4:

$$\mu_{K_1}(\overline{x}) = \begin{cases} 1, & \overline{x} \in RK_1 \\ \mu_{K_1}, & \overline{x} \in BK_1, \\ 0, & \overline{x} \in EK_1 \end{cases} \quad (7.3)$$

$$\mu_{K_2}(\overline{x}) = \begin{cases} 1, & \overline{x} \in RK_2 \\ \mu_{K_2}, & \overline{x} \in BK_2, \\ 0, & \overline{x} \in EK_2 \end{cases} \quad (7.4)$$

where $0 < \mu_{k1} < 1$, $0 < \mu_{k2} < 1$, and $\overline{x}$ is any point in N-dimensional space.

The topological relationship between two spatial objects K_1 and K_2 can be determined by the fuzzy memberships of Equation 7.5

$$\mu(K_1, K_2) = \bigvee_{\overline{x}_i} \left[\mu_{K_1}(\overline{x}_i) \wedge \mu_{K_2}(\overline{x}_i) \right], \quad (7.5)$$

where the signs $\wedge$ and $\vee$ represent the minimum and maximum operations in fuzzy set theory.

The value of $\mu(K_1, K_2)$ describes the magnitude of the common region of intersection. In general, the larger the value, the larger the common region of intersection. The topological relationship is determined by the value of the intersection.

A formal representation of the topological relations between spatial objects with uncertainty has been given in this section. A spatial object has been geometrically defined according to the specified topological relations and with the corresponding mathematical expressions. As a result, these topological relations can be quantified. Such quantification is essential for practical implementation, in a GIS platform, of the proposed method.

7.3   MODELING UNCERTAIN TOPOLOGICAL RELATIONSHIPS

A fuzzy topology for modeling uncertain topological relationship between spatial objects is introduced in this section. Fuzzy topology is a generalization of ordinary topology (Chang 1968; Liu and Luo 1997), in which the concept of set is generalized from two values, {0, 1}, in the ordinary topology to the values of a continuous interval [0, 1] in the fuzzy topology. This generalization can be helped to understand fuzzy relations between objects in GISci and applied to describe the fuzzy relations numerically (Egenhofer and Franzosa, 1991; Cohn and Gotts, 1996; Clementini and Di Felice, 1996; Shi and Guo, 1999; Winter, 2000; Tang et al., 2005; Shi and Liu, 2004; Liu and Shi, 2006; Dilo et al., 2007).

The existing models are designed for describing the topological relations between regions. In fact, the topological relations are sometimes not merely a matter of region-to-region. Other cases also exist, such as that of fuzzy line-to-region relations.

Quasi coincidence and quasi difference are two ways of partitioning fuzzy sets via the sum and difference between their membership functions. The sum or difference can be used to describe the level of an effect of one fuzzy object on another. In this section, these two techniques are used for modeling topological relations between fuzzy objects in GISci, the details of which are provided in Shi and Liu (2004).

7.3.1 Fuzzy Definition of GIS Elements

Point, line, and region (polygon) are basic elements in GISci. They are defined based on fuzzy sets (Shi and Liu, 2004). The definition of a fuzzy point is adopted from *Fuzzy Topology* (Liu and Luo, 1997) and the simple fuzzy line and fuzzy region are newly defined below.

7.3.1.1 Fuzzy Point

Definition 7.3.1 (fuzzy point): An I-fuzzy point on X is an I-fuzzy subset $x_a \in I^X$, defined as:

$$x_a(y) = \begin{cases} a & \text{if } y = x \\ 0 & \text{otherwise} \end{cases}.$$

7.3.1.2 Simple Fuzzy Line

Definition 7.3.2 (simple fuzzy line): I-fuzzy subset $\ell \in I^X$ is called a simple I-fuzzy line if supp(ℓ) is a line segment in X (or R^2).

7.3.1.3 Fuzzy Region

Definition 7.3.3 (fuzzy region in X (or R^2)): A fuzzy set A in X (or R^2) is called a fuzzy region if supp(A) has a non-empty interior in the sense of background topology.

7.3.2 Quasi Coincidence

Quasi coincidence is a form of partitioning fuzzy sets via the sum of their membership functions. As the membership function of a fuzzy set is abstract, it can be applied to interpret complex situations, such as the spread of SARS, or the incidence of immunity. The sum of the membership values can be used to interpret the closeness of the relationship, and thus can be widely used in a GISci. Quasi coincidence is supported by the following definitions.

Definition 7.3.4 (quasi coincidence): Let A and B be two fuzzy sets in I^X. We say that A is quasi coincident with B (write $A_{\hat{q}}B$) at x if $A(x) + B(x) > 1$. Denote $A \wedge B = \{x \in X: A(x) + B(x) > 1\}$ the quasi coincident set.

The quasi coincident set is an ordinary set and is simply a collection of all x in X with the properties of $A(x) + B(x) > 1$. Thus, the quasi coincident fuzzy sets can be defined as

$$\min(A \wedge B) = \begin{cases} A(x) \wedge B(x) & \text{if } A(x) + B(x) > 1 \\ 0 & \text{otherwise} \end{cases}.$$

Proposition 7.3.6: $\min(A \wedge B) \leq A \wedge B$.

Proposition 7.3.7: Let (I^X, δ) and (I^Y, μ) be two I-fts', $A, B \in I^X$, $C, D \in I^Y$, $f : X \to Y$ be an ordinary mapping. Then

(i) $A_{\hat{q}} f^{\leftarrow}(C) \Leftrightarrow f^{\rightarrow}(A)_{\hat{q}} C$.

(ii) $A_{\hat{q}} B \Rightarrow f^{\rightarrow}(A)_{\hat{q}} f^{\rightarrow}(B)$.

(iii) $f^{\leftarrow}(C)_{\hat{q}} f^{\leftarrow}(D) \Rightarrow C_{\hat{q}} D$.

Remarks: Proposition 7.3.7 gives several invariants of quasi coincidence under mapping. The properties in the proposition are, of course, invariant under homeomorphism. If f is a homeomorphic map, we have the proposition below.

Definition 7.3.8: Define the fuzzy set $A_{\frac{1}{2}}$ by

$$A_{\frac{1}{2}}(x) = \begin{cases} A(x) & \text{if } A(x) > \dfrac{1}{2} \\ 0 & \text{otherwise} \end{cases}.$$

Proposition 7.3.9: Let $A \in I^X$ and $f^{\rightarrow} : (I^X, \delta) \to (I^Y, \mu)$ be an I-fuzzy homeomorphism, then

$$f^{\rightarrow}\left(A_{\frac{1}{2}}\right) = f^{\rightarrow}(A)_{\frac{1}{2}};$$

i.e., the set $A_{\frac{1}{2}}$ is invariant under homeomorphism.

Proposition 7.3.10: The quasi coincident, $A \wedge B$, set is divided into two parts: one is $A(x) > \frac{1}{2}$ and $B(x) > \frac{1}{2}$ denoted by $\min(A \wedge B)^{(A \& B > 0.5)}$ and the other is $A(x) \leq \frac{1}{2}$ or $B(x) \leq \frac{1}{2}$ denoted by $\min(A \wedge B)^{(A \text{ or } B \leq 0.5)}$.

Remarks:

i. Since the set $A \wedge B$ is invariant under homeomorphisms, it is a homeomorphic invariant topological component (or topological components). Thus, this direction can be used to partition the set $A \wedge B$ into several homeomorphic invariant parts, and the unchanged properties in a transformation in GISci are guaranteed.

ii. Many studies have focused on determining whether the empty and non-empty of homeomorphic components can present fuzzy topological relationships between two spatial objects. Thus, it can be seen that the homeomorphic invariant is important. In addition to a study of the empty and non-empty properties, other potentially useful topological properties of these components are also investigated.

Proposition 7.3.11: For $A(x) > \frac{1}{2}$ and $B(x) > \frac{1}{2}$, we have $A \wedge B^c \subset A \wedge B$ and $A^c \wedge B \subset A \wedge B$.

This proposition tells us that the fuzzy sets $A(x) > \frac{1}{2}$ and $B(x) > \frac{1}{2}$ can be further divided into three parts: $A \wedge B^c$, $A^c \wedge B$ and $\{x \in X: A(x) > \frac{1}{2}$ and $B(x) > \frac{1}{2}$ and $A(x) = B(x)\}$, and denoted by $\min(A \wedge B^c)_{(A+B>1)}^{(A \& B > 0.5)}$, $\min(A^c \wedge B)_{(A+B>1)}^{(A \& B > 0.5)}$, and $\{A = B\}$, respectively.

Proposition 7.3.12: $A \wedge B^c$, $A^c \wedge B$ and $\{A = B\}$ are invariant under homeomorphic fuzzy mapping.

So far, the fuzzy set $A \wedge B$ has been decomposed into several homeomorphic invariant parts: $A \wedge B_{(A+B\leq 1)}$, $\min(A \wedge B)_{(A+B>1)}^{(A \text{ or } B \leq 0.5)}$, $\min(A' \wedge B)_{(A+B>1)}^{(A \& B > 0.5)}$, $\{A = B\}$ and $\min(A \wedge B')_{(A+B>1)}^{(A \& B > 0.5)}$. Figure 7.3 shows the logic diagram of the decomposition of $A \wedge B$.

Figure 7.4a illustrates the change of topological relations with the concept of a quasi coincidence, $A \wedge B$, in R^2. The right column shows the top-down view of two fuzzy sets, whereas the left column shows the membership value of two fuzzy sets. The region enclosed by a dashed line on the left column and right column are the quasi coincidence in several different cases.

Figure 7.4b illustrates several topological relationships with the concept of quasi coincidence, $A \wedge B^c$, in R^2. The right column shows the top-down view of two fuzzy sets while the left column shows the membership value view of two fuzzy sets. The region enclosed by a dashed line on the left column and right column are the quasi coincidence in several different cases.

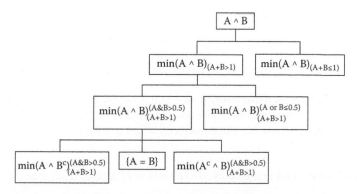

FIGURE 7.3 The logic diagram of the decomposition of A $\wedge$ B. (From Shi, W. Z. and Liu, K. F. 2004. Modeling fuzzy topological relations between uncertain objects in GIS, *Photogram. Eng. Remote Sens.* 70(8): 921–929. With permission.)

Disjoint, then A ∧ Bᶜ is equal supp(A):	
Non-empty A ∧ Bᶜ cases:	

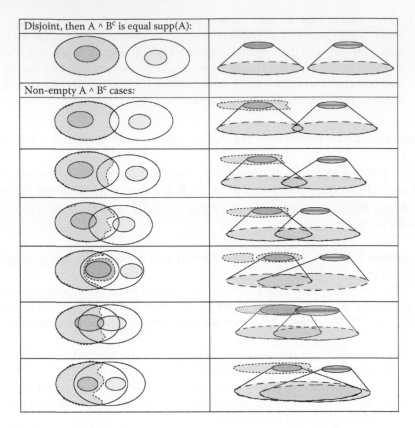

FIGURE 7.4(a)   (See color figure following page 38.) Quasi coincident fuzzy topological relations between two objects in GISci. (From Shi, W. Z. and Liu, K. F. 2004. Modeling fuzzy topological relations between uncertain objects in GIS, *Photogram. Eng. Remote Sens.* 70(8): 921–929. With permission.)

7.3.3   QUASI DIFFERENCE

Quasi difference is a form of partitioning fuzzy sets by means of the difference in their membership functions. The difference can be used to compare the fuzzy value of two fuzzy sets and also to analyze which part has a higher value than the other.

Definition 7.3.13 (quasi difference): Let A and B be two fuzzy sets in I^x. Define the quasi difference of A and B, denoted by $A\backslash\backslash B$, as $A\backslash\backslash B = \vee \{x_\lambda \in M(\downarrow A): B(x) = 0\} \vee\vee \{x_\lambda \in M(\downarrow A): \lambda \text{ not} \geq B(x) = 0\}$. The definitions of $\downarrow A$ and $M(\downarrow A)$ refer to those of Liu and Luo (1997).

Definition 7.3.14: Let A and B be two fuzzy sets in I^x. Define $A\backslash\backslash B_o = \vee \{x_\lambda \in M(\downarrow A): B(x) = 0\}$.

Lemma 7.3.15: Let $A \in I^x$ and $f^\rightarrow: (I^x, \delta) \rightarrow (I^y, \mu)$ be an I-fuzzy homeomorphism, and $f(x) = y$; then $f^\rightarrow (A) \rightarrow (y) = 0$ if and only if $A(x) = 0$.

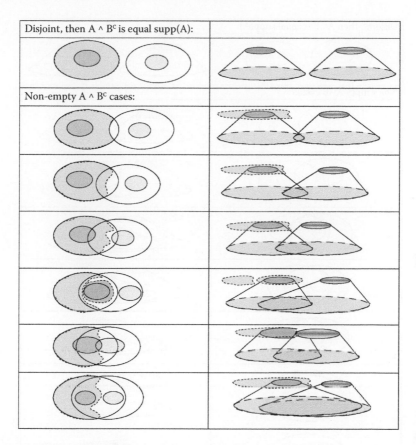

| Disjoint, then A ∧ Bᶜ is equal supp(A): | |
| Non-empty A ∧ Bᶜ cases: | |

FIGURE 7.4(b)   (See color figure following page 38.) Quasi coincident fuzzy topological relations between two objects in GISci. (From Shi, W. Z. and Liu, K. F. 2004. Modeling fuzzy topological relations between uncertain objects in GIS, *Photogram. Eng. Remote Sens.* 70(8): 921–929. With permission.)

Proposition 7.3.16: Let $A \in I^x$ and $f^{\rightarrow}$: $(I^x, \delta) \rightarrow (I^y, \mu)$ be an I-fuzzy homeomorphism, then $f^{\rightarrow}(A\backslash\backslash B_o) = f^{\rightarrow}(A)\backslash\backslash f^{\rightarrow}(B)_o$; i.e., the set $A\backslash\backslash B_o = \vee \{x_\lambda \in M(\downarrow A): B(x) = 0\}$ is invariant under homeomorphic mappings.

By using the above results, the fuzzy set $A \vee B$ can be decomposed into several homeomorphic invariant parts, $A\backslash\backslash B_o$, and $B\backslash\backslash A_o$. On $A \wedge B$, we have $(A(x) > \frac{1}{2}$ and $B(x) > \frac{1}{2})$, and $(A(x) \leq \frac{1}{2}$ or $B(x) \leq \frac{1}{2})$. On $A(x) > \frac{1}{2}$ and $B(x) > \frac{1}{2}$, we have three more parts, which are $\{A = B\}$, $\min(A \wedge B^c)^{(A \& B > 0.5)}_{(A+B > 1)}$, and $\min(A^c \wedge B)^{(A \& B > 0.5)}_{(A+B > 1)}$.

Combining previous works and the above results, the fuzzy set $A \vee B$ has been decomposed into several homeomorphic invariant parts, $A\backslash\backslash B_o$, $B\backslash\backslash A_o$, $A \wedge B_{(A+B \leq 1)}$, $\min(A \wedge B)^{(A \text{ or } B \leq 0.5)}_{(A+B > 1)}$, $\min(A^c \wedge B)^{(A \& B > 0.5)}_{(A+B > 1)}$, $\{A = B\}$ and $\min(A \wedge B^c)^{(A \& B > 0.5)}_{(A+B > 1)}$. Figure 7.5 illustrates topological relations using quasi difference, $A\backslash\backslash B$, in R^2. The right column is the top-down view of two fuzzy sets, whereas the left column illustrates the membership value view of two fuzzy sets. The region enclosed by a

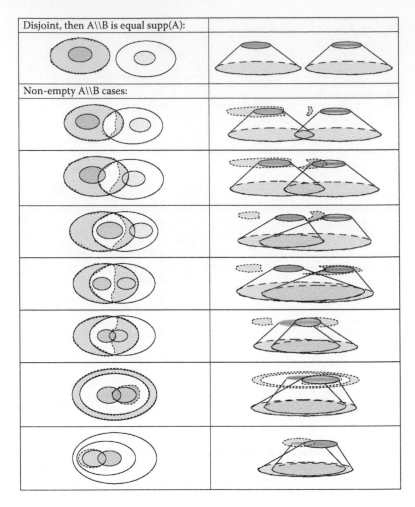

FIGURE 7.5   **(See color figure following page 38.)** Different cases of quasi different fuzzy topological relations between two objects in GIS. (From Shi, W. Z. and Liu, K. F. 2004, Modeling fuzzy topological relations between uncertain objects in GIS, *Photogram. Eng. Remote Sens.* 70(8): 921–929. With permission.)

dashed line on the left column and right column are the quasi difference in various cases.

7.3.4   TOPOLOGICAL RELATIONS BETWEEN TWO FUZZY SETS IN R^2

Definition 7.3.17 (topological relations): The topological relations between two fuzzy sets A and B are the topological properties of all the homeomorphic invariant topological components of A and B.

The topological relationships between fuzzy sets A and B are classified by using the results in Sections 7.3.2 and 7.3.3. The target topological components of A and B

will be $A\backslash\backslash B_o$, $B\backslash\backslash A_o$, $A \wedge B_{(A+B\le 1)}$, $\min(A \wedge B)_{(A+B>1)}^{(A\,or\,B\le 0.5)}$, $\min(A \wedge B^c)_{(A+B>1)}^{(A\&B>0.5)}$, $\min(A^c \wedge B)_{(A+B>1)}^{(A\&B>0.5)}$ and $\{A = B\}$.

The components $A \wedge B_{(A+B\le 1)}$, $\min(A \wedge B)_{(A+B>1)}^{(A\,or\,B\le 0.5)}$, $\min(A \wedge B^c)_{(A+B>1)}^{(A\&B>0.5)}$, $\min(A^c \wedge B)_{(A+B>1)}^{(A\&B>0.5)}$ and $\{A = B\}$ can be used to classify the depth of the relationship between two fuzzy sets. The components $A\backslash\backslash B_o$ and $B\backslash\backslash A_o$ can be used to indicate the depth of the independent part of each fuzzy set. The topological relationship between the fuzzy sets A and B can thus, be described by the topological properties (empty and nonempty, subspace properties, connectivity, compactness, etc.) of the seven-tuple topological components:

$$(A\backslash\backslash B_o, B\backslash\backslash A_o, A \wedge B_{(A+B\le 1)}, \min(A \wedge B)_{(A+B>1)}^{(A\,or\,B\le 0.5)}, \min(A \wedge B^c)_{(A+B>1)}^{(A\&B>0.5)},$$
$$\min(A^c \wedge B)_{(A+B>1)}^{(A\&B>0.5)}, A = B).$$

The structure of the seven topological components is illustrated in Figure 7.6. By providing the formulae of two fuzzy sets, their relationship by means of the above seven-tupled topological components can be calculated.

The following propositions describe several elementary properties of the above seven-tupled topological components.

Proposition 7.3.18: If $A\backslash\backslash B_o$ is empty, then supp$(B) \subset$ supp(A). By symmetry, we also have, if $B\backslash\backslash A_o$, then supp$(A) \subset$ supp(B).

Proposition 7.3.19: $A \wedge B_{(A+B\le 1)}$, $\min(A \wedge B)_{(A+B>1)}^{(A\,or\,B\le 0.5)}$, $\min(A \wedge B^c)_{(A+B>1)}^{(A\&B>0.5)}$, $\min(A^c \wedge B)_{(A+B>1)}^{(A\&B>0.5)}$ and $A = B$ are all empty, if and only if $A \wedge B$ is empty; hence, A and B are disjointed.

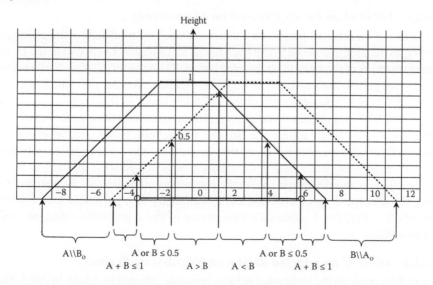

FIGURE 7.6 An illustration of the structure of the decomposition of $A \vee B$ (From Shi, W. Z. and Liu, K. F. 2004. Modeling fuzzy topological relations between uncertain objects in GIS, *Photogram. Eng. Remote Sens.* 70(8): 921–929. With permission.)

Proposition 7.3.20: One of $A \wedge B_{(A+B \leq 1)}$, $\min(A \wedge B)_{(A+B>1)}^{(A \text{ or } B \leq 0.5)}$, $\min(A \wedge B^c)_{(A+B>1)}^{(A \& B>0.5)}$, $\min(A^c \wedge B)_{(A+B>1)}^{(A \& B>0.5)}$ or $A = B$ is nonempty, if and only if $A \wedge B$ is nonempty; hence, A and B are not disjointed.

7.4   AN APPLICATION EXAMPLE

7.4.1   BACKGROUND

Quasi coincidence was used to study the spatial distribution of severe acute respiratory syndrome (SARS) within a community. In this example, the effect of an infected region on its neighbor regions was investigated by using fuzzy topology.

SARS is a serious disease that threatens the life and health of people. This disease killed nearly 300 people in Hong Kong within a few months in 2003. One problem needing to be solved was the identification of the spatial distribution patterns of the disease, within a community.

The following data sets were used for the study: a basic map of the study area, and within the study area, the daily infection rate within each building, the daily death rate, the daily rate of newly infected people from each building, the daily ICU patient occupancy rate from the study area. (For each building in the study area, the number of people infected each day included the number of newly infected and the unrecovered cases.) Some required data were not available and were simulated for this case study.

7.4.2   FUZZY TOPOLOGY FOR SARS ANALYSIS

7.4.2.1   Effect of an Infected Person on a Community

The effect of an infected person on a community can be modeled by the uncertain topological relationship between a fuzzy line and fuzzy region. Let $A(x[t])$ be the membership function of the trace of an infected person. The value of $A(x[t]) = 0$ means that the infected person is not affected by any other diseases, whereas the value $A(x[t]) = 1$ means that there is a dense concentration of SARS viruses in that body.

Let $B(x[t])$ be a region of a community representing the condition of certain people or the situation of point x at time t. $B(x[t])$ can, for example, be the population density function. When $B(x(t)) = 0$, this means that the location is in a good condition, that is, it is safe. In contrast, when $B(x[t]) = 1$, the location is in a poor condition.

The effect of an infected person on the community, can therefore be analyzed by using the concept of quasi coincidence; a community in a particular location $x(t)$, with $A(x[t]) + B(x[t]) > 1$ indicates a high chance of that community being infected, and vice versa.

7.4.2.2   Effect of an Infected Region on Neighboring Regions

Figure 7.7 represents the residential regions seriously affected by SARS in 2003. The two residential areas are (a) Amoy Gardens, with one of the first serious outbreaks of SARS; and (b) Lower Ngau Tau Kok Estate, a neighboring region with a high potential of being imminently affected.

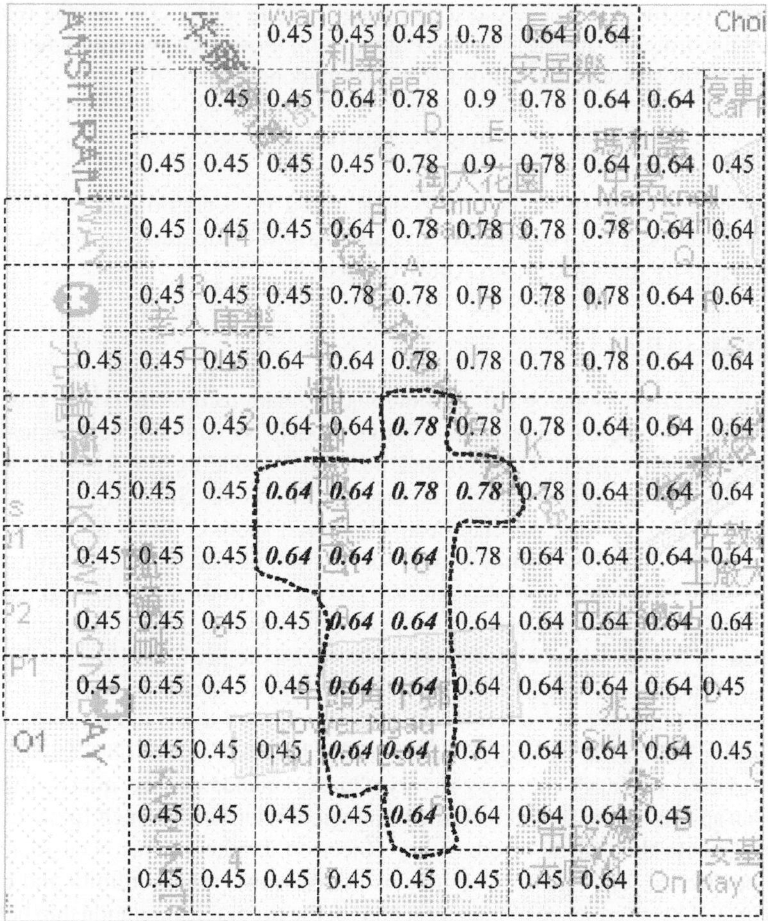

FIGURE 7.7 The effect of Amoy Gardens on Lower Ngau Tau Kok Estate can be computed by the concept of quasi coincidence. The region enclosed by a closed dashed line represents a potentially highly infectious region. (From Shi, W. Z. and Liu, K. F. 2004. Modeling fuzzy topological relations between uncertain objects in GIS, *Photogram. Eng. Remote Sens.* 70(8): 921–929. With permission.)

The objective of this study is to model the potential of SARS spread from the original region (Amoy Garden) into its neighbor region (Lower Ngau Tau Kok Estate). Within each grid, there are two numbers: the upper one ($A[x(t)]$), which stands for the density of the SARS virus from Amoy Garden, and the lower one ($B[x(t)]$), which represents the density of the SARS virus in Lower Ngau Tau Kok Estate.

From Figure 7.7, it is seen that Block E was the most seriously infected building (with very high $A(x[t])$ value of 0.9), and that the trend of infection declined steadily; the farther the distance from Block E, the lower the $A(x[t])$ value, decreasing from 0.78 to 0.65, 0.45, etc. In contrast, the membership function ($B[x(t)]$) for Lower Ngau Tau Kok Estate decreased from the center of the estate to its neighbor, that is, from 0.50, to 0.45, 0.25, etc.

By applying the concept of quasi coincidence (some quasi coincidence relationships are illustrated in Figure 7.4a and Figure 7.4b), those areas with a higher risk of infection are those areas that fulfill the following condition:

$$A(x[t]) + B(x[t]) > 1$$

The regions fulfilling this condition are indicated by a closed dashed line. This means that the people living within this area have a comparatively higher risk of being infected by the SARS from Amoy Gardens.

7.5 SUMMARY

Modeling uncertain topological spatial object relationships in GISci has been described in this chapter. Firstly, the existing models for depicting topological relations in GISci have been reviewed and analyzed, including the four-intersection model, the nine-intersection model, and the egg-yolk model. Secondly, a formal representation of the topological relationships between spatial objects has been given. Based on N-cell complex, N-dimensional space is subdivided into three components: the interior, the boundary, and the exterior. According to their topological relations and the corresponding mathematical expressions, the topological relationship can be quantified. A formal framework for representing topological relationships between spatial objects has also been presented. This framework can be applied to solve the topological and geometric problems using the formal logic and algebraic methods. Thirdly, uncertain topological relationships between spatial objects have been modeled based on fuzzy topology theory. Quasi coincidence and quasi difference have been applied to (a) distinguish the topological relations between fuzzy objects, and (b) indicate the effects of one fuzzy object on another. It has been found that (a) the number of topological relationships between two sets is infinite and can be approximated by a sequence of matrices, and (b) the topological relationship between two objects in GIS are shape-dependent.

REFERENCES

Apostol, T. M., 1974. *Mathematical Analysis.* Addison-Wesley, Reading, MA.
Chang, C. L., 1968, Fuzzy topological spaces. *J. Math. Anal. Appl.,* 24: 182–190.
Clementini, E. and Felice, D. P., 1996. An algebraic model for spatial objects with indeterminate boundaries. In *Geographic Objects with Indeterminate Boundaries,* ed. Burrough, P. A. and Frank, A. U. London: Taylor & Francis, pp. 155–169.
Cohn, A. G. and Gotts, N. M., 1996. The "Egg-Yolk" representation of regions with indeterminate boundaries. In *Geographic Objects with Indeterminate Boundaries,* ed. Burrough, P. A. and Frank, A. U., 171–187. London: Taylor & Francis.
Egenhofer, M. and Franzosa, R., 1991. Point-set topological spatial relations. *Int. J. Geog. Inform. Syst.,* 5(2): 161–174.
Egenhofer, M., 1993. A model for detailed binary topological relations. *Geomatica,* 47(3) and (4): 261–273.
Liu, Y. M. and Luo, M. K., 1997. *Fuzzy Topology.* Singapore: World Scientific.

Liu, K. F. and Shi, W. Z., 2006. Computation of fuzzy topological relations of spatial objects based on induced fuzzy topology, *Int. J. Geog. Inform. Syst.* 20(8): 857–883.

Randell, D. A., Cui, Z., and Cohn, A. G., 1992. A spatial logic based on regions and "connection." In *Proceedings of KR92*, 1992.

Shi, W. Z. and Guo, W., 1999. Topological relations between uncertainty spatial objects in three-dimensional space. *Proceedings of International Symposium on Spatial Data Quality*, July, Hong Kong: 487–495.

Shi, W. Z. and Liu, K. F., 2004. Modeling fuzzy topological relations between uncertain objects in GIS. *Photogramm. Eng. Remote Sens.* 70(8): 921–929.

Tang, X. M. and Kainz, W., 2001. Analysis of topological relations between fuzzy regions in general fuzzy topological space, *The SDH conference 02'*, Ottawa, Canada.

Tang, X. M., Kainz, W., and Fang, Y., 2005. Reasoning about changes of land covers with fuzzy settings. *Int. J. Rem. Sens.* 26(14): 3025–3046.

Wang, F., Hall, G. B., and Subaryono, 1990. Fuzzy information representation and processing in conventional GIS software: database design and application. *Int. J. Geog. Inform. Syst.* 4(3): 261–283.

Winter, S., 2000. Uncertain topological relations between imprecise regions. *Int. J. Geog. Inform. Sci.* 14(5): 411–430.

8 Modeling Uncertainty in Digital Elevation Models

Specific focus of this chapter is on uncertainty in a digital elevation model. Surface modeling approaches are introduced in Section 8.2, followed by a brief description of a digital elevation model (DEM) error sources. Error estimation for two types of DEM models is given: (a) for a triangular irregular network surface model (Section 8.4) and (b) for a regular grid DEM, generated from a nonlinear interpolation (Section 8.5).

8.1 INTRODUCTION

A digital elevation model (DEM) is a digital representation of a continuous surface by an array of z values, referenced to a common datum. Digital elevation models are commonly used to represent the terrain relief of the Earth surface. DEM has been widely used in many areas, such as mapping, engineering, geology, construction, road design, and agriculture.

A DEM is an approximation of the surface of the natural world. A difference exists between the surface elevations in that the natural world and the surface represented by the model. This difference is referred to as a DEM error. Such errors can be represented by various indicators, such as a DEM point error, or average surface errors. The errors are usually measured in terms of the root mean square. Such errors affect DEM applications, such as in determining topographic features (for example, slope gradient or slope aspect). Therefore, modeling uncertainties in a DEM model is not only a theoretical issue, but also an issue related to successful DEM applications.

8.2 SURFACE MODELING

A DEM can be generated by interpolating various sampling data sets of the real-world surface. These data sets may include, for example, rectangular grids, square grids, triangles, and randomly distributed points. Many interpolation algorithms have been documented in the literature, such as linear interpolation or nonlinear interpolation, and fitting methods. These interpolation algorithms are briefly introduced, as follows, with the aim of forming a basis for analyzing DEM uncertainty.

8.2.1 COMMON INTERPOLATION METHODS

Polynomial interpolation has been widely applied in generating a DEM. This interpolation method derives a function $z = f(x, y)$ from n known points $P(x_i, y_i, z_i)$ for $i = 1, 2, ..., n$, in a three-dimensional space. Such a function delineates a surface containing all known points $P(x_i, y_i, z_i)$ and provides a value of the elevation z at any location $P(x, y)$ on the surface. For example, a bivariate or patchwise interpolation model (Kidner, 2003) is expressed as

$$z = \sum_{k=0}^{m} \sum_{l=0}^{n} \alpha_{kl} x^k y^l . \tag{8.1}$$

It is an $(m + 1)(n + 1)$ term polynomial in which α_{kl} are the coefficients of the polynomial. The z value at point P is interpolated, after the coefficients are resolved, by putting the x and y values into Equation 8.1.

The coefficients of the polynomial can be estimated based on the three-dimensional coordinates of the known points. For $i = 1, 2, ..., n$, each known point is $P(x_i, y_i, z_i)$ substituted into Equation 5.1 and yields an equation containing $(m + 1)$ $(n + 1)$ unknown coefficients α_{kl}:

$$z_i = \sum_{k=0}^{m} \sum_{l=0}^{n} \alpha_{kl} x_i^k y_i^l . \tag{8.2}$$

A system of equations can be formed with a number of points with known x and y coordinates and z value. When the number of the equations is larger than the number of unknown coefficients, the least square adjustment can be applied to solve the coefficients.

Common polynomial interpolation methods include the level plane, the linear plane, bilinear interpolation, biquadratic interpolation, bicubic interpolation, piecewise cubic polynomial interpolation, and biquintic interpolation. The level plane has the form

$$z = a_{00}. \tag{8.3}$$

This interpolation is based on the understanding that any continuous function can be approximated by a step function. This is the simplest interpolation approach, and estimates point elevation either by using the point elevation that is the closest DEM vertex to that point, or by using the average elevation of the four surrounding DEM vertices of that point. The level plane generates a discontinuous representation of a terrain (a DEM surface) by level surface sets of different elevations. The level plane in a two-dimensional space, for example, is a stepwise line (see Figure 8.1). This stepwise line is an approximate representation of the original line (the smooth curve line in the figure). The intersection points of the stepwise line and the original

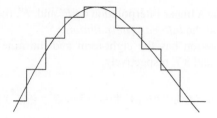

FIGURE 8.1 An example of the level plane.

line are the locations of known points, used for deriving the interpolation function of the level plane.

Linear interpolation is defined as

$$z = a_{00} + a_{10}x + a_{01}y. \tag{8.4}$$

This three-term polynomial is formed by fitting a triangle (plane) to the three closest DEM vertices.

Adding one additional "xy" term into the linear interpolation equation yields

$$z = a_{00} + a_{10}x + a_{01}y + a_{11}xy, \tag{8.5}$$

which is the bilinear interpolation. Bilinear interpolation generates a DEM surface using linear interpolation along each edge of a grid cell, composed of the closest four DEM vertices of each point inside the grid (Figure 8.2).

Given the elevations at points A, B, C, and D (the heights of A', B', C', and D') in Figure 8.2, the elevation at point O is derived as follows. Firstly, the elevation at point E (a point on the line joining A and D) is obtained by performing a linear interpolation on points A' and D'. Secondly, in a similar manner, the elevation at point F (a point on the line joining B and C) is computed from a linear interpolation on points

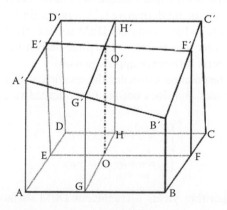

FIGURE 8.2 An example of the bilinear interpolation.

B' and C'. Finally, using a linear interpolation on E' and, F', the elevation at point O is yielded. This is called the *bilinear interpolation*.

Biquadratic interpolation has the eight-term and the nine-term expressions as given in Equations 8.6 and 8.7, respectively:

$$z = a_{00} + a_{10}x + a_{01}y + a_{20}x^2 + a_{11}xy + a_{02}y^2 + a_{21}x^2y + a_{12}xy^2 \tag{8.6}$$

and

$$z = a_{00} + a_{10}x + a_{01}y + a_{20}x^2 + a_{11}xy + a_{02}y^2 + a_{21}x^2y + a_{12}xy^2 + a_{22}x^2y^2. \tag{8.7}$$

In both expressions, the polynomial coefficients are derived from the four closest DEM vertices and the four midpoints of the individual grid edges enclosed by the four vertices.

Bilinear and bicubic interpolations are the most widely used interpolation algorithms in DEM generation. Bicubic interpolation has two forms: the 12-term and the 16-term bicubic interpolations. Their expressions are

$$z = a_{00} + a_{10}x + a_{01}y + a_{20}x^2 + a_{11}xy + a_{02}y^2 + a_{30}x^3$$
$$+ a_{21}x^2y + a_{12}xy^2 + a_{03}y^3 + a_{31}x^3y + a_{13}xy^3 \tag{8.8}$$

and

$$z = a_{00} + a_{10}x + a_{01}y + a_{20}x^2 + a_{11}xy + a_{02}y^2 + a_{30}x^3$$
$$+ a_{21}x^2y + a_{12}xy^2 + a_{03}y^3 + a_{31}x^3y + a_{22}x^2y^2$$
$$+ a_{13}xy^3 + a_{32}x^3y^2 + a_{23}x^2y^3 + a_{33}x^3y^3. \tag{8.9}$$

The coefficients of the 12-term polynomial are derived from the elevations at the four grid cell vertices, together with the first two derivatives (z_x' and z_y'). These derivatives express the surface slope in the x and y directions. For the coefficients of the sixteen-term polynomial, one additional derivative (z_{xy}'), representing the slope in both x and y directions, is required. To have a smooth DEM surface, the biquadratic interpolation is the minimum-order interpolation. The three derivatives, for instance, can be estimated based on central difference approximations.

Piecewise cubic interpolation estimates the coefficient of the polynomial based directly on DEM vertices and does not require a derivative estimation. It has the form:

$$\begin{cases} z_x = a + bx + cx^2 + dx^3 \\ z_y = a + by + cy^2 + dy^3 \end{cases}. \tag{8.10}$$

This interpolation algorithm yields a continuous DEM surface, but the generated DEM edge may not be smooth.

To guarantee that the first derivative across grid cell boundaries is smooth, the lower-order interpolation method is applied, such as the biquintic interpolation. The 36-term biquintic interpolation is expressed as

$$z = a_{00} + a_{10}x + a_{01}y + a_{11}xy + a_{20}x^2 + a_{21}x^2y + a_{02}y^2 + a_{12}xy^2 + a_{22}x^2y^2 + a_{30}x^3 \quad (8.11)$$

$$+ a_{31}x^3y + a_{32}x^3y^2 + a_{03}y^3 + a_{13}xy^3 + a_{23}x^2y^3 + a_{33}x^3y^3 + a_{40}x^4 + a_{41}x^4y + a_{42}x^4y^2$$

$$+ a_{43}x^4y^3 + a_{04}xy^4 + a_{14}xy^4 + a_{24}x^2y^4 + a_{34}x^3y^4 + a_{44}x^4y^4 + a_{50}x^5 + a_{51}x^5y + a_{52}x^5y^2$$

$$+ a_{53}x^5y^3 + a_{54}x^5y^4 + a_{05}y^5 + a_{15}xy^5 + a_{25}x^2y^5 + a_{35}x^3y^5 + a_{45}x^4y^5 + a_{55}x^5y^5.$$

This method is relatively complicated. In addition to the four DEM vertices, the first, second, third, and fourth derivatives are used for estimating the coefficients of the polynomial in Equation 8.11.

To give an interpolation approach that can generate a highly accurate DEM for any type of terrain surfaces could present some difficulties. Theoretically, a lower-order polynomial is more applicable to a less complex terrain, whereas a complex terrain can be well approximated by a higher-order polynomial. A higher-order polynomial, however, may raise computation load, and lead to the instability of the coefficients, resulting in an unrealistic surface. First-, second-, and third-order polynomials are used, mostly, in practical applications.

8.2.2 REGULAR GRID DIGITAL ELEVATION MODEL

A regular grid digital elevation model is a digital representation of a continuous surface by a regular array of z values, referenced to a common datum (Figure 8.3).

The grid size determines the DEM resolution. In practical applications, this is determined on the basis of the level of the surface details to be extracted. For example, in a hydrological application, detailed hydrologic features can be extracted from a DEM with a 10m resolution. It is difficult however to extract these features from a 30m resolution DEM. In general, the higher the DEM resolution, the more detailed the extracted features. However, a high resolution DEM means a higher data volume level and computation load.

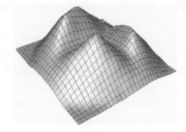

FIGURE 8.3 An example of the regular grid digital elevation model.

One weakness of the grid representation is its discontinuous representation of the terrain. Feature lines (such as ridge, gully, and others) can then not be well represented. Similarly, the grid cannot give a satisfactory representation of man-made objects (such as roads or buildings), either. It can provide an effective representation of smooth natural terrains.

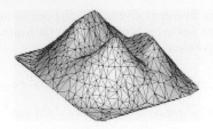

FIGURE 8.4   An example of the TIN digital elevation model.

8.2.3 A Triangulated Irregular Network Model

A triangulated irregular network (TIN) is a digital terrain model, constructed from irregularly spaced three-dimensional discrete points. These irregular discrete points form a set of adjacent and conterminous triangles. A triangulated irregular network is then generated, based on a triangulation algorithm, such as Delaunay triangulation, where the vertices of the triangles are the original discrete points (Figure 8.4).

The feature points (such as peaks, ridges, canyons, or passes) and feature lines (such as a ridge or break lines) can be represented as the vertices and edges of the triangles, respectively. Therefore, a TIN can have a better representation of terrain than the regular grid DEM.

Another advantage of a TIN is that its resolution can easily be varied depending on the terrain complexity level in different locations. One part of the network can have a higher resolution for the area with a higher terrain change frequency, while another part can have a lower resolution for the area with a lower terrain change frequency. Similar to other univariate functions, a TIN network also lacks the ability to represent vertical object features (such as side facets of buildings).

Once a TIN network is generated, its topological relationship can be built. Topological relationships can be, for example, the relationship of a given triangle with the three vertices of the triangle and of its three neighbor triangles. Such topological relationships can be used for spatial queries, spatial analysis, etc. A disadvantage of storing topological relationships for a TIN model is on its heavy storage cost and computation load.

8.3 ERROR SOURCES OF A DEM MODEL

The imprecision of a generated DEM model can be caused by (a) the original sample point error, and (b) the interpolation methods used to generate the model from the original sample points.

Before digital photogrammetry was well developed and widely applied, elevation points (either from contours or spot heights) from maps were used as the original sample points for generating a DEM, either through interpolation toward a regular grid or by triangulation generation. Spot heights can also be captured via a field surveying method, for example, using a total station. Contours are then interpolated based on the field measured spot heights or together with other feature points. In such

DEM generation processes, error sources include field surveying, map digitization, and interpolation error.

Sampling density and sample size is one factor that directly affects DEM accuracy. In general, the model generated from more sample points is of higher quality and can represent the natural world more realistically.

Sampling schemes are other factors affecting DEM accuracy. The sampling point distribution can follow (a) a regular pattern, and (b) an irregular pattern. With a regular pattern, the sampling points are evenly distributed (such as, as squares, rectangle, regular triangles, and regular hexagons). With a regular distribution pattern, the feature points or lines may not be represented by the sample points, hence generating an error source of a DEM.

Another factor that affects the effectiveness of a regular gridded DEM is the inflexibility in resolution. In principle, the resolution of a DEM should be changeable, depending on the terrain to be described. The DEM resolution should be lower for an area with lower terrain change frequency, and higher for an area with higher terrain change frequency. As mentioned in Section 8.2.3, a TIN model can cope with this dynamic change of resolution within a DEM model, whereas, the regular grid model is not sufficiently flexible in this regard. Dynamic generation of multi-resolution TIN models (Yang et al., 2005) is an interesting issue, especially for three-dimensional digital city modeling.

A DEM model is a simplification and abstraction of a surface, such as a terrain surface in the real world. A set of limit discrete sample points is used in a DEM to represent a continuous surface that has an infinite number of points. This is a fundamental cause of uncertainty in such models. Therefore, no matter how high the sampling density chosen, this model is still a simplification of the surface in the real world, but without being as detailed.

Positional errors in the sample points also cause uncertainty in a DEM. The positional characteristics of the model are determined by the position of its sample points, thus positional error can propagate within the model. The quantitative relationship between the sample errors and the errors in the generated DEM are given in Sections 8.4 and 8.5 below.

A continuous surface is usually generated from discrete sample points by applying an interpolation method. The quality and characteristics of the interpolation method is, therefore, another factor that affects the accuracy of a DEM. Much research has been done on analyzing interpolation algorithms (Akima, 1996; Caselles et al., 1998; Almansa et al., 2002). The related research issues are: (a) local interpolation methods, (b) directional interpolation methods, and (c) assessment of the performance of the interpolation methods.

As indicated, each interpolation algorithm introduces error, generally different in type, in the generation of a DEM. However, these errors can be estimated and even reduced to a certain level, although it is not possible to totally avoid them. Research has been carried out to study interpolation algorithms in order to improve interpolation accuracy. During interpolation, sufficient consideration regarding the characteristics of raw data sets and the nature of the terrain concerned in interpolation method design can potentially improve the quality of the interpolated DEM. Based on this consideration, two interpolation methods aiming to improve the quality of the generated DEM model are proposed and introduced in Chapter 14.

8.4   ACCURACY ESTIMATION FOR A TIN MODEL

The error sources of DEM models have been introduced in the above section. A further issue is to quantitatively estimate the effect of these error sources on the generated DEM.

Of the two types of DEM models, the regular gridded DEM and the irregular TIN, less research has focused on the accuracy estimation of the latter, possibly, owing to the fact that it is more mathematically complex. A method for estimating positional accuracy of TIN (Zhu, et al., 2005) is introduced in this section and an accuracy estimation of regular grid DEM (Shi et al., 2005) is also given in Section 8.5.

For a TIN model, three given points (usually three vertices of a triangle in a three-dimensional space) are used to form a plane. Let A, B, and C indicate the three given points with coordinates denoted by $A(x_1, y_1, z_1)$, $B(x_2, y_2, z_2)$, and $C(x_3, y_3, z_3)$. The triangle surface is then expressed as

$$z = ax + by + c, \tag{8.12}$$

where coefficients a, b, and c are

$$a = \frac{\begin{vmatrix} z_1 & y_1 & 1 \\ z_2 & y_2 & 1 \\ z_3 & y_3 & 1 \end{vmatrix}}{\begin{vmatrix} x_1 & y_1 & 1 \\ x_2 & y_2 & 1 \\ x_3 & y_3 & 1 \end{vmatrix}}, \quad b = \frac{\begin{vmatrix} x_1 & z_1 & 1 \\ x_2 & z_2 & 1 \\ x_3 & z_3 & 1 \end{vmatrix}}{\begin{vmatrix} x_1 & y_1 & 1 \\ x_2 & y_2 & 1 \\ x_3 & y_3 & 1 \end{vmatrix}}, \quad c = \frac{\begin{vmatrix} x_1 & y_1 & z_1 \\ x_2 & y_2 & z_2 \\ x_3 & y_3 & z_3 \end{vmatrix}}{\begin{vmatrix} x_1 & y_1 & 1 \\ x_2 & y_2 & 1 \\ x_3 & y_3 & 1 \end{vmatrix}}. \tag{8.13}$$

A TIN model is expressed by a set of equations. The z-value of a point on the TIN surface can be obtained by an interpolation based on the set of equations and given x- and y-coordinates of the point.

Equations 8.12 and 8.13 state that the linear interpolation model of a point on a TIN surface is solely determined by the initial points A, B, and C. Errors in these initial points are propagated to the points on the TIN surface through the interpolation process, and these propagated errors exist in the surface of the DEM.

To quantify the error in the model, an error propagation formula for the TIN generated from linear interpolation based on probability theory has currently been derived and is described in the following sections.

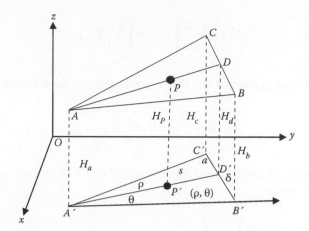

FIGURE 8.5 Linear model for a triangle. (From Zhu, C. Q. et al., 2005. *Int. J. Remote Sens*, 26(24): 5509–5523. With permission.)

8.4.1 MATHEMATICAL FORMULA OF TIN ACCURACY ESTIMATION

The mean elevation error is identified here as an error indicator for a TIN. A mathematical expression for the mean elevation error is then given and numerical experiments accomplished to validate the mathematical expression derived. In addition to the experimental validation, a rigorous theoretical proof is also provided.

Consider $\triangle ABC$ is a triangle in a TIN, and any point in this triangle is acquired by a linear interpolation. Triangle $\triangle ABC$ is then one of the faces of the DEM model (see Figure 8.5). Suppose H_a, H_b, and H_c are elevations of the three vertices of the triangle and their errors are independent and identical, which are denoted by σ^2_{node}.

Let P be any point in $\triangle ABC$. A line passing through A and P intersects another line passing through B and C at point D. On the other hand, suppose the projection of $\triangle ABC$ on the plane x and y is denoted by $\triangle A'B'C'$. Three interior angles of $\triangle A'B'C'$ are $\angle B'A'C' = \alpha$, $\angle A'B'C' = \beta$, $\angle B'C'A' = \gamma$.

The coordinate system is then converted to the polar coordinate system defined by the origin A' and the semi-infinite line leading from A' via B'. Let $B'D' = \delta$, $A'D' = s$, $A'P' = \rho$, $B'C' = a$.

From the linear interpolation algorithm, the following equation is obtained:

$$H_d = \frac{a-\delta}{a} H_b + \frac{\delta}{a} H_c. \tag{8.14}$$

This shows that errors in points B and C are propagated to point D. Since the errors in B and C are independent from the property of variance, the error in D is given by

$$\sigma_d^2 = \left(\frac{a-\delta}{a}\right)^2 \sigma_{node}^2 + \left(\frac{\delta}{a}\right)^2 \sigma_{node}^2. \tag{8.15}$$

Similarly, the errors in points A and D are propagated to point P because

$$H_p = \frac{s-\rho}{s}H_a + \frac{\rho}{s}H_d. \tag{8.16}$$

From the property of variance and Equations 8.15 and 8.16,, the elevation error in point P is expressed as

$$
\begin{aligned}
\sigma_P^2 &= \left(\frac{s-\rho}{s}\right)^2 \sigma_a^2 + \left(\frac{\rho}{s}\right)^2 \sigma_d^2 \\
&= \left(\frac{s-\rho}{s}\right)^2 \sigma_{node}^2 + \left(\frac{\rho}{s}\right)^2 \left(\left(\frac{a-\delta}{a}\right)^2 \sigma_{node}^2 + \left(\frac{\delta}{a}\right)^2 \sigma_{node}^2\right).
\end{aligned} \tag{8.17}
$$

This mathematical expression has a direct impact on the precision of the location of point P within the triangle.

The mean elevation error in $\triangle ABC$ (denoted by σ_H^2) is the integration of σ_P^2 over $\triangle A'B'C'$,

$$
\begin{aligned}
\sigma_H^2 &= \frac{1}{S_{A'B'C'}} \int_0^\alpha \int_0^s \left[\left(\frac{s-\rho}{s}\right)^2 \sigma_{node}^2 + \left(\frac{\rho}{s}\right)^2 \left(\left(\frac{a-\delta}{a}\right)^2 \sigma_{node}^2 + \left(\frac{\delta}{a}\right)^2 \sigma_{node}^2\right)\right] \rho \, d\rho \, d\theta \\
&= \frac{\sigma_{node}^2}{S_{A'B'C'}} \int_0^\alpha \left(\frac{s^2}{3} - \frac{s^2\delta}{2a} + \frac{s^2\delta^2}{2a^2}\right) d\theta = \frac{I_1 - I_2 + I_3}{S_{A'B'C'}} \sigma_{node}^2,
\end{aligned} \tag{8.18}
$$

where $S_{A'B'C'}$ is the area of $\triangle A'B'C'$.

For $\triangle A'B'D'$, we have

$$\delta = \frac{c\sin\theta}{\sin(180° - (\theta+\beta))} = \frac{c\sin\theta}{\sin(\theta+\beta)}, \tag{8.19}$$

and

$$s = \frac{c\sin\beta}{\sin(180° - (\theta+\beta))} = \frac{c\sin\beta}{\sin(\theta+\beta)}. \tag{8.20}$$

From Equations 8.18 and 8.20, we have

$$I_1 = \int_0^\alpha \frac{s^2}{3} d\theta = \frac{c^2 (\sin\beta)^2}{3} \int_0^\alpha \frac{1}{\sin^2(\theta+\beta)} d\theta \equiv \frac{c^2 (\sin\beta)^2}{3} I_{11},$$

$$I_1 = \int_0^\alpha \frac{s^2}{3} d\theta = \frac{c^2 (\sin\beta)^2}{3} \int_0^\alpha \frac{1}{\sin^2(\theta+\beta)} d\theta \equiv \frac{c^2 (\sin\beta)^2}{3} I_{11}, \qquad (8.21)$$

and

$$I_{11} = \int_0^\alpha \frac{1}{\sin^2(\theta+\beta)} d(\theta+\beta) = \frac{\sin\alpha}{\sin\gamma\sin\beta}. \qquad (8.22)$$

Putting Equation 8.22 into Equation 8.21, we gain the value of I_1. From Equations 8.18–8.20, we have

$$I_2 = \int_0^\alpha \frac{\delta s^2}{2a} d\theta = \frac{c^3 (\sin\beta)^2}{2a} \int_0^\alpha \frac{\sin\theta}{\sin^3(\theta+\beta)} d\theta \equiv \frac{c^3 (\sin\beta)^2}{2a} I_{21}. \qquad (8.23)$$

Let $\theta + \beta = t$. Then,

$$I_{21} = \int_\beta^{\beta+\alpha} \frac{\sin(t-\beta)}{\sin^3 t} dt = \cos\beta \int_\beta^{\beta+\alpha} \frac{1}{\sin^2 t} dt - \sin\beta \int_\beta^{\beta+\alpha} \frac{1}{\sin^3 t} d\sin t \qquad (8.24)$$

$$\equiv \cos\beta I_{11} - \sin\beta I_{211}$$

and

$$I_{211} = \int_\beta^{\beta+\alpha} \frac{1}{\sin^3 t} d\sin t = -\frac{1}{2\sin^2 \gamma} + \frac{1}{2\sin^2 \beta}. \qquad (8.25)$$

The value of I_{21} can be obtained by putting Equations 8.22 and 8.25 into Equation 8.24, while the value of I_2 is attained by solving Equations 8.23 and 8.24.

Also, from Equations 8.18–8.20, we have

$$I_3 = \int_0^\alpha \frac{s^2\delta^2}{2a^2} d\theta = \frac{c^4 (\sin\beta)^2}{2a^2} \int_0^\alpha \frac{\sin^2\theta}{\sin^4(\theta+\beta)} d\theta \equiv \frac{c^4 (\sin\beta)^2}{2a^2} I_{31}. \qquad (8.26)$$

Let $\theta + \beta = t$. Then,

$$I_{31} = \int_{\beta}^{\alpha+\beta} \frac{\sin^2(t-\beta)}{\sin^4 t} dt \equiv \cos(2\beta)I_{11} - \sin(2\beta)I_{211} + \sin^2\beta I_{313}. \qquad (8.27)$$

Also,

$$I_{313} = \int_{\beta}^{\alpha+\beta} \frac{1}{\sin^4 t} dt = \frac{\cos\gamma}{3\sin^3\gamma} + \frac{\cos\beta}{3\sin^3\beta} + \frac{2}{3}I_{11}. \qquad (8.28)$$

The value of I_{31} is obtained by solving Equations 8.22, 8.25, 8.27, and 8.28, and the value of I_3 is obtained by solving the Equations 8.26 and 8.27, respectively. Finally, combining Equations 8.21, 8.23, and 8.26 will gain the result of Equation 8.18. This result indicates elevation error σ_H^2 in $\triangle ABC$.

8.4.2 Experimental Validation of the Formula

To further validate the mathematical formula of the mean elevation error given in Equation 8.18, experiments applying the model to several triangles are conducted. The coordinates of these triangles are given in Table 8.1 where each vertex of the individual triangles is written as (a, b, c). This table also lists the experimental result for the coefficient

$$\frac{I_1 - I_2 + I_3}{S_{A'B'C'}}$$

in Equation 8.18. The triangles in a three-dimensional space are displayed in Figure 8.10

Table 8.1 shows that coefficient

$$\frac{I_1 - I_2 + I_3}{S_{A'B'C'}}$$

TABLE 8.1

The Experimental Result of the Mean Error in Elevation

Triangle	Vertex Coordinate of the Triangles			$\dfrac{I_1 - I_2 + I_3}{S_{A'B'C'}}$
1	(13,14,13)	(22,33,44)	(13,42,56)	0.5003
2	(1,3,4)	(12,33,51)	(23,45,43)	0.4998
3	(12,55,62)	(23,3,55)	(13,44,37)	0.4999
4	(1,3,5)	(22,55,78)	(66,33,95)	0.5001

is approaching 0.5. Therefore, potentially the value 0.5 is the true value of this coefficient. A rigorous mathematical proof for this hypothesis is given as follows:

8.4.3　THEORETICAL PROOF FOR THE MATHEMATICAL FORMULA

Assume that the true value of the coefficient in Equation 8.18 is 0.5,

$$\sigma_H^2 = \frac{I_1 - I_2 + I_3}{S_{A'B'C'}} \sigma_{node}^2 = \frac{1}{2}\sigma_{node}^2. \tag{8.29}$$

This assumption is correct only if the following equality satisfies

$$\frac{I_1 - I_2 + I_3}{S_{A'B'C'}} = \frac{1}{2}. \tag{8.30}$$

The following aims to prove this equality.

For $\Delta A'B'C'$, we have

$$S_{A'B'C'} = \frac{1}{2}ac\sin\beta. \tag{8.31}$$

From Equations 8.21 and 8.31, we have

$$\frac{I_1}{S_{A'B'C'}} = \frac{c^2(\sin\beta)^2}{3 \bullet \frac{1}{2}ac\sin\beta}I_{11} = \frac{2c\sin\beta}{3a}I_{11}. \tag{8.32}$$

From Equations 8.23, 8.24, and 8.31, the following equation is obtained:

$$\frac{I_2}{S_{A'B'C'}} = \frac{c^3(\sin\beta)^2}{2a \bullet \frac{1}{2}ac\sin\beta}I_{21} = \frac{c^2\sin\beta}{a^2}(\cos\beta I_{11} - \sin\beta I_{211}). \tag{8.33}$$

From Equations 8.26 to 8.28, and 8.31, we have

$$\frac{I_3}{S_{A'B'C'}} = \frac{c^4(\sin\beta)^2}{2a^2 \bullet \frac{1}{2}ac\sin\beta}I_{31}$$

$$= \frac{c^3\sin\beta}{a^3}\left(\cos(2\beta)I_{11} - \sin(2\beta)I_{211} + \sin^2\beta I_{313}\right) \tag{8.34}$$

$$= \frac{c^3\sin\beta}{a^3}\left[(\cos(2\beta) + \frac{2}{3}\sin^2\beta)I_{11} - \sin(2\beta)I_{211} + \frac{\sin^2\beta\cos\gamma}{3\sin^3\gamma} + \frac{\cos\beta}{3\sin\beta}\right].$$

Let a, b, and c be ΔABC's sides opposite to vertices A, B, and C, respectively, and α, β, and γ be the interior angles at these three vertices. By the sine law and the cosine law,

$$\frac{a}{\sin\alpha} = \frac{b}{\sin\beta} = \frac{c}{\sin\gamma} \qquad (8.35)$$

and

$$\cos\beta = \frac{a^2 + c^2 - b^2}{2ac} . \qquad (8.36)$$

From 8.32 to 8.36, the following equation is obtained:

$$
\begin{aligned}
\frac{I_1 - I_2 + I_3}{S_{A'B'C'}} &= \frac{2c\sin\beta}{3a} I_{11} - \frac{c^2\sin\beta}{a^2}(\cos\beta I_{11} - \sin\beta I_{211}) \\
&\quad + \frac{c^3\sin\beta}{a^3}\left[(\cos(2\beta) + \frac{2}{3}\sin^2\beta)I_{11} - \sin(2\beta)I_{211} + \sin^2\beta\left(\frac{\cos\gamma}{3\sin^3\gamma} + \frac{\cos\beta}{3\sin^3\beta}\right)\right] \\
&= \left[\frac{2c\sin\beta}{3a} - \frac{c^2\sin\beta\cos\beta}{a^2} + \frac{c^3\sin\beta}{a^3}\left(\cos(2\beta) + \frac{2}{3}\sin^2\beta\right)\right]I_{11} \qquad (8.37) \\
&\quad + \left[\frac{c^2\sin\beta}{a^2}\sin\beta - \frac{c^3\sin\beta}{a^3}\sin(2\beta)\right]I_{211} \\
&\quad + \frac{c^3\sin\beta}{a^3}\left(\frac{\sin^2\beta\cos\gamma}{3\sin^3\gamma} + \frac{\cos\beta}{3\sin\beta}\right) \equiv W_1 + W_2 + W_3 .
\end{aligned}
$$

According to 8.22 and 8.37, W_1 is then given as

$$
\begin{aligned}
W_1 &= \left[\frac{2c\sin\beta}{3a} - \frac{c^2\sin\beta\cos\beta}{a^2} + \frac{c^3\sin\beta}{a^3}\left(\cos(2\beta) + \frac{2}{3}\sin^2\beta\right)\right]\frac{\sin\alpha}{\sin\gamma\sin\beta} \quad (8.38) \\
&= \frac{2c\sin\alpha}{3a\sin\gamma} - \frac{c^2\cos\beta\sin\alpha}{a^2\sin\gamma} + \frac{c^3\sin\alpha}{a^3\sin\gamma}\left(1 - 2\sin^2\beta + \frac{2}{3}\sin^2\beta\right) \\
&= \frac{2}{3} - \frac{c}{a}\cos\beta + \frac{c^2}{a^2} - \frac{4}{3}\cdot\frac{c^2}{a^2}\sin^2\beta .
\end{aligned}
$$

From Equations 8.25 and 8.37, W_2 is given as

$$
W_2 = \left[\frac{c^2\sin\beta}{a^2}\sin\beta - \frac{c^3\sin\beta}{a^3}\sin(2\beta)\right]\left[-\frac{1}{2\sin^2\gamma} + \frac{1}{2\sin^2\beta}\right] \qquad (8.39)
$$

$$= -\frac{c^2 \sin^2 \beta}{a^2 2 \sin^2 \gamma} + \frac{c^3 \sin \beta}{a^3 2 \sin^2 \gamma} \sin(2\beta) + \frac{c^2 \sin^2 \beta}{a^2 2 \sin^2 \beta} - \frac{c^3 \sin \beta}{a^3 2 \sin^2 \beta} \sin(2\beta)$$

$$= -\frac{\sin^2 \beta}{2 \sin^2 \alpha} + \frac{c^3 \sin \beta}{a^3 \sin^2 \gamma} \sin \beta \cos \beta + \frac{c^2}{2a^2} - \frac{c^3 \sin \beta}{a^3 \sin^2 \beta} \sin \beta \cos \beta$$

$$= -\frac{b^2}{2a^2} = \frac{b^2}{a^3} \cos \beta + \frac{c^2}{2a^2} - \frac{c^3}{a^3} \cos \beta.$$

Also, from Equation 8.37, we have

$$W_3 = \frac{c^3 \sin \beta}{a^3} \left(\frac{\sin^2 \beta \cos \gamma}{3 \sin^3 \gamma} + \frac{\cos \beta}{3 \sin \beta} \right) \tag{8.40}$$

$$= \frac{b^3}{3a^3} \cos \gamma + \frac{c^3}{3a^3} \cos \beta.$$

Based on the above three equations, the sum of W_1, W_2, and W_3 is expressed as (8.41)

$$W_1 + W_2 + W_3 = \frac{2}{3} - \frac{c}{a} \cos \beta + \frac{c^2}{a^2} - \frac{4}{3} \cdot \frac{c^2}{a^2} \sin^2 \beta$$

$$- \frac{b^2}{2a^2} + \frac{cb^2}{a^3} \cos \beta + \frac{c^2}{2a^2} - \frac{c^3}{a^3} \cos \beta + \frac{b^3}{3a^3} \cos \gamma + \frac{c^3}{3a^3} \cos \beta$$

$$= \frac{2}{3} + \left(-\frac{c}{a} + \frac{cb^2}{a^3} - \frac{2c^3}{3a^3} \right) \cos \beta + \frac{c^2}{6a^2} + \frac{4}{3} \cdot \frac{c^2}{a^2} \cos^2 \beta - \frac{b^2}{2a^2} + \frac{b^3}{3a^3} \cos \gamma$$

$$= \frac{2}{3} - \frac{1}{2} - \frac{c^2}{2a^2} + \frac{b^2}{2a^2} + \frac{b^2}{2a^2} + \frac{b^2 c^2}{2a^4} - \frac{b^4}{2a^4} - \frac{c^2}{3a^2} - \frac{c^4}{3a^4} + \frac{c^2 b^2}{3a^4}$$

$$+ \frac{c^2}{6a^2} + \frac{1}{3} + \frac{c^4}{3a^4} + \frac{b^4}{3a^4} + \frac{2c^2}{3a^2} - \frac{2b^2}{3a^2} - \frac{2c^2 b^2}{3a^4}$$

$$- \frac{b^2}{2a^2} + \frac{b^2}{6a^2} + \frac{b^4}{6a^4} - \frac{b^2 c^2}{6a^4} = \frac{1}{2}.$$

Putting it into 8.37 will gain Equation 8.30. Therefore, the assumption given in Equation 8.29 is correct, that is,

$$\sigma_H^2 = \frac{1}{2} \sigma_{node}^2.$$

8.4.4   A NOTE ON THE DERIVED MATHEMATICAL FORMULA

This section gives the mathematical formula used for estimating the TIN model, based on the error of the vertices, by using the linear interpolation. Below are several notes to the formula:

- This formula, by giving the average accuracy of a triangle surface of the TIN model, enables a description of the error in the entire TIN surface, rather than the error of a particular point on that surface.
- The average accuracy of any triangle of the TIN given by the formula is identical and is independent of the location and shape of an incorporated particular triangle. Therefore, the accuracy of the overall TIN model is also estimated by the formula.
- The error at an arbitrary point inside a triangle of the TIN model can be estimated by the Equation 8.17.

8.5   ACCURACY ESTIMATION OF A REGULAR GRID DEM

In the above section, the accuracy estimation of a TIN model, generated from a linear interpolation has been introduced. This section addresses accuracy estimation for a regular grid DEM model (Shi et al., 2005), specifically, interpolated using the bicubic interpolation method.

As introduced in Section 8.2.1, a number of interpolation methods can be used to generate a regular grid DEM model where the bicubic interpolation is one of the commonly used nonlinear model. Thus, such a regular grid model, generated through the representative bicubic interpolation, is taken as an example for accuracy assessment. In such a case, the accuracy of an interpolated DEM model is measured by (a) the surface error and (b) the elevation error. As illustrated by Table 8.2, existing studies have demonstrated that a bicubic interpolation has a higher accuracy ($0.8a$) than that of a linear interpolation (a). However, the elevation error of a linear interpolation is $4/9a$. Hence, of interest is the need to estimate an elevation error of a higher-order interpolation. A method to estimate elevation error of a DEM is introduced in this section. This, together with the surface error estimation described in previous studies, can enable the realization of the estimation of overall accuracy of higher-order interpolation model.

TABLE 8.2
Surface Error and Elevation Error for the Linear
and the Higher-Ordered (Bicubic) Interpolations

Interpolation	Linear	Higher-Order Interpolation (Bicubic)
Surface error	a	$0.8a$
Elevation error	4/9	?

8.5.1 THE BICUBIC INTERPOLATION

In a two-dimensional case, the bicubic interpolation is given as

$$
\begin{aligned}
z(x,y) = a_{00} &+ a_{01}x + a_{10}y + a_{20}x^2 + a_{11}xy + a_{02}y^2 + a_{30}x^3 \\
&+ a_{21}x^2y + a_{12}xy^2 + a_{03}y^3 + a_{31}x^3y + a_{32}x^2y^2 + a_{13}xy^3 \\
&+ a_{32}x^3y^2 + a_{23}x^2y^3 + a_{33}x^3y^3,
\end{aligned}
\tag{8.42}
$$

where the coefficients a_{ij} ($i, j = 0, 1, 2, 3$) are calculated by solving the above equation with the values of the nodes on a DEM. An alternative method to obtain the bicubic interpolation is to perform the cubic interpolation along the x- and y-axes, respectively.

8.5.1.1 Two-Dimensional Cubic Interpolation

The two-dimensional cubic interpolation is similar to the bilinear interpolation. The former method, however, uses a cubic polynomial for the interpolation, whereas the bilinear interpolation is based on a linear polynomial. Let A and B be two given points with coordinates $A(x_0, h_0)$ and $B(x_1, h_1)$ and let their first derivatives be denoted by h'_0 and h'_1. The two-dimensional cubic interpolation is as follows:

$$
\begin{aligned}
H_3(x) = \left(1 - 2\frac{x - x_0}{x_0 - x_1}\right)\left(\frac{x - x_1}{x_0 - x_1}\right)^2 h_0 + \left(1 - 2\frac{x - x_1}{x_1 - x_0}\right)\left(\frac{x - x_0}{x_1 - x_0}\right)^2 h_1 \\
+ \left(x - x_0\right)\left(\frac{x - x_1}{x_0 - x_1}\right)^2 h'_0 + \left(x - x_1\right)\left(\frac{x - x_0}{x_1 - x_0}\right)^2 h'_1.
\end{aligned}
\tag{8.43}
$$

The Equation 8.43 satisfies two conditions:

$$
\begin{aligned}
H_3(x_i) &= h_i, \\
H'_3(x_i) &= h'_i, \quad i = 0, 1.
\end{aligned}
\tag{8.44}
$$

Figure 8.6 shows the cubic interpolation on the two points.

In general, the values of h'_0 and h'_1 are unknown. They have to be estimated based on known data. One common method for this estimation is given as follows

$$
h'_1 = \frac{h_1 - h_i}{x_1 - x_i}
\tag{8.45}
$$

and

$$
h'_1 = \frac{h_1 - h_i}{x_1 - x_i}.
\tag{8.46}
$$

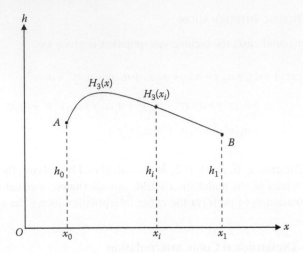

FIGURE 8.6   The bicubic interpolation on two-dimensional two points.

Usually, the derivatives are replaced by gradient functions. Substituting them into Equation 8.44 gives

$$H_3(x) = \left(1 - 2\frac{x - x_0}{x_0 - x_1}\right)\left(\frac{x - x_1}{x_0 - x_1}\right)^2 h_0 + \left(1 - 2\frac{x - x_1}{x_1 - x_0}\right)\left(\frac{x - x_0}{x_1 - x_0}\right)^2 h_1$$

$$+ (x - x_0)\left(\frac{x - x_1}{x_0 - x_1}\right)^2\left(\frac{h_1 - h_0}{x_1 - x_0}\right) + (x - x_1)\left(\frac{x - x_0}{x_1 - x_0}\right)^2\left(\frac{h_1 - h_0}{x_1 - x_0}\right)$$

$$= \left[1 - 2\left(\frac{x - x_0}{x_0 - x_1}\right)\right]\left(\frac{x - x_1}{x_0 - x_1}\right)^2 h_0 + \left[1 - 2\left(\frac{x - x_1}{x_1 - x_0}\right)\right]\left(\frac{x - x_0}{x_1 - x_0}\right)^2 h_1$$

$$+ \left[(x - x_0)\left(\frac{x - x_1}{x_0 - x_1}\right)^2 + (x - x_1)\left(\frac{x - x_0}{x_1 - x_0}\right)^2\right]\left(\frac{h_1 - h_0}{x_1 - x_0}\right).$$

(8.47)

This shows that $H_3(x)$ is defined based on the given coordinates of the two points A and B. When $x_0 = 0$ and $x_1 = d$, Equation 8.47 is simplified to be

$$H_3(x) = \left(1 + 2\frac{x}{d}\right)\left(\frac{x - d}{d}\right)^2 h_0 + \left(1 - 2\frac{x - d}{d}\right)\left(\frac{x}{d}\right)^2 h_1$$

$$+ \left[x\left(\frac{x - d}{d}\right)^2 + (x - d)\left(\frac{x}{d}\right)^2\right]\left(\frac{h_1 - h_0}{d}\right)$$

(8.48)

$$= \frac{1}{d^2}\left(\frac{2x^3}{d} - 3x^2 + d^2\right)h_0 + \frac{1}{d^2}\left(3 - \frac{2x}{d}\right)x^2 h_1$$

$$+ \left(2x^3 - 3x^2 d + xd^2\right)\left(\frac{h_1 - h_0}{d^3}\right)$$

$$= \left[\frac{1}{d^2}\left(\frac{2x^3}{d} - 3x^2 + d^2\right) - \frac{\left(2x^3 - 3x^2 d + xd^2\right)}{d^3}\right]h_0$$

$$+ \left[\frac{1}{d^2}\left(3 - \frac{2x}{d}\right)x^2 + \frac{1}{d^3}\left(2x^3 - 3x^2 d + xd^2\right)\right]h_1$$

$$= \left[S_1(x) - S_3(x)\right]h_0 + \left[S_2(x) + S_3(x)\right]h_1$$

$$= Q_1(x)h_0 + Q_2(x)h_1,$$

where

$$\begin{cases} S_1(x) = \dfrac{1}{d^2}\left(\dfrac{2x^3}{d} - 3x^2 + d^2\right) \\[2ex] S_2(x) = \dfrac{1}{d^2}\left(3 - \dfrac{2x}{d}\right)x^2 \\[2ex] S_3(x) = \dfrac{1}{d^3}\left(2x^3 - 3x^2 d + xd^2\right) \end{cases} \tag{8.49}$$

and

$$\begin{cases} Q_1(x) = S_1(x) - S_3(x) \\[1ex] Q_2(x) = S_2(x) + S_3(x) \end{cases} \tag{8.50}$$

Equation 8.48 is the mathematical expression of the elevation derived from the cubic interpolation for a two-dimensional case.

8.5.1.2 Three-Dimensional Cubic Interpolation

In a DEM surface interpolation, the bicubic interpolation is three-dimensional. The three-dimensional bicubic interpolation is extended from the two-dimensional method mentioned previously. That is, two-dimensional interpolations along the x- and y-axes are realized.

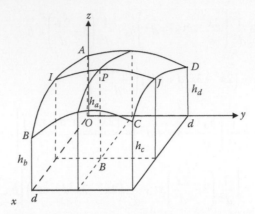

FIGURE 8.7   A bilinear interpolation on four nodes (From Shi, W. Z. et al., 2005. *Int. J. Remote Sens.* 26(4): 3069–3084. With permission.)

Suppose the polygon *ABCD* consists of four nodes *A*, *B*, *C*, and *D*, and this is one of the polygons from which a surface on a DEM has to be constructed and let its projection on the plane *xOy* be a square. Let $A = (0, 0, h_a)$, $B = (d, 0, h_b)$, $C = (d, d, h_c)$, and $D = (0, d, h_d)$ as shown in Figure 8.7.

A mathematical expression of the surface *ABCD* is denoted by $h(x, y)$. That is

$$\begin{cases} h(0,0) = h_a \\ h(d,0) = h_b \\ h(d,d) = h_c \\ h(0,d) = h_d \end{cases} \qquad (8.51)$$

A bicubic interpolation is to perform the cubic interpolation along the *x*- and *y*-axes, respectively. Function $h(x, y)$ can then, be gained from the product of the two cubic interpolation functions.

From Equation 8.48, the cubic interpolation along the x-axis is given as

$$H_x(x,y) = Q_1(x) h(0,y) + Q_2(x) h(d,y), \qquad (8.52)$$

where $Q_1(x)$ and $Q_2(x)$ are computed from Equation 8.50. Furthermore, performing the cubic interpolation on $H_x(x, y)$ along the y-axis gives $h(x, y)$,

$$\begin{aligned} h(x,y) \equiv Q_1(x) \Big[Q_1(y) h(0,0) + Q_2(y) h(0,d) \Big] \\ + Q_2(x) \Big[Q_1(y) h(d,0) + Q_2(y) h(d,d) \Big] \end{aligned} \qquad (8.53)$$

$$= Q_1(x)Q_1(y)h_a + Q_1(x)Q_2(y)h_d + Q_2(x)Q_1(y)h_b + Q_2(x)Q_2(y)h_c$$

$$\equiv T_1 h_a + T_2 h_d + T_3 h_b + T_4 h_c ,$$

where $Q_1(y)$ and $Q_2(y)$ can be calculated from Equation 8.50 by replacing x with y. For any point on the DEM surface, its elevation can then be computed from Equation 8.53.

8.5.2 ACCURACY ESTIMATION FOR THE DEM GENERATED FROM THE BICUBIC INTERPOLATION

An analysis of the elevation error propagation from the source data through the bicubic interpolation is now presented. Firstly, a mathematical expression of the elevation error in a regular grid DEM, generated by the bicubic interpolation is derived. This expression is simplified from a mathematical theory. It is then applied to measure surface error and elevation on the DEM surface. As the regular grid is composed of squares, the accuracy estimation of the whole DEM surface is equivalent to an accuracy estimation of a single square on the surface. Therefore, the following computation and proofs are performed on a single square, as illustrated by Figure 8.7.

8.5.2.1 The Average Elevation Error

The regular grid shown in Figure 8.7 consists of four nodes A, B, C, and D. The surface of these four nodes is a part of the whole DEM surface. Suppose the elevations of the nodes are h_a, h_b, h_c, and h_d, and their elevation errors are independent and identical, and written by σ_{node}^2.

For any point P on the surface $ABCD$, the elevation of this point is expressed as

$$h_p = T_1 h_a + T_2 h_d + T_3 h_b + T_4 h_c , \tag{8.54}$$

where T_1, T_2, T_3, and T_4 are derived from Equation 8.53. This represents the value of the point P generated from the bicubic interpolation, based on the four nodes A, B, C, and D.

When the node errors are independent, the variance of the elevation of point P is given by

$$\sigma_p^2 \equiv T_1^2 \sigma_{node}^2 + T_2^2 \sigma_{node}^2 + T_3^2 \sigma_{node}^2 + T_4^2 \sigma_{node}^2 . \tag{8.55}$$

This equation can also be applied to estimate the average elevation error in the surface $ABCD$. The average elevation error is computed by integrating Equation 8.55 over the projection of $ABCD$ on the plane xOy. That is,

$$\sigma_H^2 = \frac{1}{d^2} \int_0^d \int_0^d \left(T_1^2 \sigma_{node}^2 + T_2^2 \sigma_{node}^2 + T_3^2 \sigma_{node}^2 + T_4^2 \sigma_{node}^2 \right) dxdy$$

$$(8.56)$$

$$= \frac{\sigma_{node}^2}{d^2} \int_0^d \int_0^d \left(T_1^2 + T_2^2 + T_3^2 + T_4^2 \right) dxdy .$$

From Equations 8.48, 8.50, 8.53, and 8.54, we have

$$\sigma_H^2 = \frac{\sigma_{node}^2}{d^2} \int_0^d \int_0^d \left[\left(S_1(x) - S_3(x) \right) \right]^2 \left[S_1(y) - S_3(y) \right]^2$$

$$+ \left[S_1(x) - S_3(x) \right]^2 \left[S_2(y) + S_3(y) \right]^2$$

$$(8.57)$$

$$+ \left[S_2(x) + S_3(x) \right]^2 \left[S_1(y) - S_3(y) \right]^2$$

$$+ \left[S_2(x) + S_3(x) \right]^2 \left[S_2(y) + S_3(y) \right]^2 dxdy$$

$$= \frac{\sigma_{node}^2}{d^2} \left[\int_0^d \left[S_1(x) - S_3(x) \right]^2 dx \int_0^d \left[S_1(y) - S_3(y) \right]^2 dy \right.$$

$$+ \int_0^d \left[S_1(x) - S_3(x) \right]^2 dx \int_0^d \left[S_2(y) - S_3(y) \right]^2 dy$$

$$+ \int_0^d \left[S_2(x) + S_3(x) \right]^2 dx \int_0^d \left[S_1(y) - S_3(y) \right]^2 dy$$

$$+ \left. \int_0^d \left[S_2(x) + S_3(x) \right]^2 dx \int_0^d \left[S_2(y) - S_3(y) \right]^2 dy \right]$$

$$\equiv \frac{\sigma_{node}^2}{d^2} \left[M_1 N_1 + M_1 N_2 + M_2 N_1 + M_2 N_2 \right]$$

$$= \frac{\sigma_{node}^2}{d^2} \left(M_1 + M_2 \right) \left(N_1 + N_2 \right) .$$

From the integration M_1, N_1, M_2, and N_2, it is found that

$$\begin{cases} M_1 = N_1 \\ M_2 = N_2 \end{cases}.$$

(8.58)

Therefore, the average elevation error is expressed as

$$\sigma_H^2 = \frac{\sigma_{node}^2}{d^2}\left(M_1 + M_2\right)^2.$$

(8.59)

8.5.2.2 A Formula for the Average Elevation Error

Equation 8.59 is then simplified as follows. Equation 8.57 gives

$$\begin{aligned}
M_1 &= \int_0^d \left[S_1(x) - S_3(x)\right]^2 dx \\
&= \int_0^d \left[S_1^2(x) - 2S_1(x)S_3(x) + S_3^2(x)\right] dx \\
&= \int_0^d S_1^2(x)dx - 2\int_0^d S_1(x)S_3(x)dx + \int_0^d S_3^2(x)dx \\
&\equiv K_1 - 2K_2 + K_3.
\end{aligned}$$

(8.60)

Based on Equation 8.49, K_1, K_2, and K_3 in Equation 8.60 are expressed as

$$K_1 = \int_0^d S_1^2(x)dx = \int_0^d \frac{1}{d^4}\left(\frac{2x^3}{d} - 3x^2 + d^2\right)^2 dx = d\left(\frac{4}{7} + \frac{9}{5} - 2\right),$$

(8.61)

$$K_2 = \frac{1}{d^5}\int_0^d \left(\frac{2x^3}{d} - 3x^2 + d^2\right)\left(2x^3 - 3x^2d + xd^2\right)dx = d\left(\frac{4}{7} + \frac{11}{5} + \frac{1}{4} - 3\right)$$

(8.62)

and

$$K_3 = \frac{1}{d^6}\int_0^d \left(2x^3 - 3x^2d + xd^2\right)^2 dx = d\left(\frac{4}{7} + \frac{13}{5} + \frac{1}{3} - 2 - \frac{3}{2}\right).$$

(8.63)

Substituting Equations 8.61–8.63 into Equation 8.60 gives

$$M_1 = d\left(\frac{4}{7}+\frac{9}{5}-2\right) - 2d\left(\frac{4}{7}+\frac{11}{5}+\frac{1}{4}-3\right) + d\left(\frac{4}{7}+\frac{13}{5}+\frac{1}{3}-2-\frac{3}{2}\right) = \frac{1}{3}d . \quad (8.64)$$

From Equation 8.63,

$$
\begin{aligned}
M_2 &= \int_0^d \left[S_2(x) + S_3(x) \right]^2 dx \\[6pt]
&= \int_0^d \left[S_2^2(x) + 2S_2(x)\,S_3(x) + S_3^2(x) \right] dx \\[6pt]
&= \int_0^d S_2^2(x)\,dx + 2\int_0^d S_2(x)\,S_3(x)\,dx + \int_0^d S_3^2(x)\,dx \\[6pt]
&\equiv K_4 + 2K_5 + K_3 .
\end{aligned}
\qquad (8.65)
$$

Putting Equation 8.51 into the above equation gives

$$K_4 = \int_0^d \frac{1}{d^4}\left[\left(3-\frac{2x}{d}\right)\cdot x^2\right]^2 dx = d\left(\frac{9}{5}-2+\frac{4}{7}\right) \qquad (8.66)$$

and

$$K_5 = \frac{1}{d^5}\int_0^d \left(3-\frac{2}{d}x\right)x^2\left(2x^3 - 3x^2 d + xd^2\right)dx = d\left(2-\frac{11}{5}+\frac{3}{4}-\frac{4}{7}\right). \qquad (8.67)$$

Based on Equations 8.63, 8.66, and 8.67, Equation 8.65 then becomes

$$M_2 = d\left(\frac{9}{5}-2+\frac{4}{7}\right) + 2d\left(2-\frac{11}{5}+\frac{3}{4}-\frac{4}{7}\right) + d\left(\frac{4}{7}+\frac{13}{5}+\frac{1}{3}-2-\frac{3}{2}\right) = \frac{1}{3}d . \qquad (8.68)$$

From Equation 8.59, 8.64, and 8.68, the following formula is achieved:

$$\sigma_H^2 = \frac{\sigma_{node}^2}{d^2}\left(N_1 + N_2\right)^2 = \frac{\sigma_{node}^2}{d^2}\left(\frac{1}{3}d + \frac{1}{3}d\right)^2 = \frac{4}{9}\sigma_{node}^2 . \qquad (8.69)$$

This is the formula for estimating the average elevation error in a regular grid DEM generated from the bicubic interpolation.

TABLE 8.2a
Surface Error and Elevation Error for the Linear
and the Higher-Ordered (Bicubic) Interpolations

Interpolation	Linear	Higher-Order Interpolation (Bicubic)
Surface error	a	$0.8a$
Elevation error	4/9	4/9

8.5.2.3 Error Measure for the Bicubic Interpolation

In early studies, it was proved that the average error in the linear interpolation was

$$\sigma_H^2 = \frac{4}{9}\sigma_{node}^2. \tag{8.70}$$

This equation is identical to Equation 8.69. Therefore, Table 8.2 can be updated as Table 8.2a, as follows,

Combining (a) the average elevation error (4/9a) result in a regular grid DEM, generated from the bicubic interpolation presented in this subsection, and (b) the surface error (0.8a) revealed in early studies, it can be concluded that the bicubic interpolation not only has a higher surface accuracy but also has a higher overall DEM accuracy. This includes both surface accuracy and elevation accuracy. In general, a DEM model generated from the bicubic interpolation is more accurate than that from the bilinear interpolation. This was also proved by Kidner (2003), who found that the former was 20% more accurate than the latter.

8.5.2.4 A Note

The average elevation accuracy can be estimated by Formula 8.69, for a regular grid DEM generated from the bicubic interpolation.

- Since the average elevation accuracy in each regular grid is identical, the average elevation accuracy calculated by Equation 8.69 is in fact equal to the elevation accuracy of the entire DEM model, rather than for a single point.
- The average elevation accuracy is invariant with locations of the nodes in the DEM.
- The accuracy of a single interpolated point can be estimated by the Equation 8.55, depending on the location of the point.

8.6 SUMMARY

Uncertainty in a DEM model has been addressed in this chapter. A brief introduction of error sources, the existing methods of generating a DEM model, plus accuracy assessment have been given. Accuracy assessment methods have been further developed for both the TIN and regular grid DEM models. An average accuracy estimation formula for the TIN model was established and mathematically proved. Accuracy estimation for regular grid DEM, generated by a bicubic model has been given, and a comparison with a bilinear model made. The conclusion is that, considering both the elevation accuracy and the surface accuracy, the bicubic interpolation model has a higher overall interpolation accuracy than the bilinear interpolation model made.

REFERENCES

Akima, H., 1996. Algorithm 760: rectangular grid data surface fitting that has the accuracy of a bi-cubic polynomial. *ACM Trans. Math. Softw.* 22: 357–361.

Almansa, A., Cao, F., Gousseau, Y., and Rougé, B., 2002. Interpolation of digital elevation models using AMLE and related methods. *IEEE Trans. Geosci. Rem. Sens.* 40: 314–325.

Caselles, V., J. M. Morel, and C. Sbert, 1998. An axiomatic approach to image interpolation. *IEEE Trans. Image Process.* 7: 376–386.

Kidner, D. B., 2003. Higher-order interpolation of regular grid digital elevation models. *Int. J. Rem. Sens.* 24: 2981–2987.

Shi, W. Z., Li, Q. Q., and Zhu, C. Q., 2005, Estimating the propagation error of DEM from higher-order interpolation algorithms, *Int. J. Rem. Sens.* 26(4): 3069–3084.

Yang, B. S., Shi, W. Z., and Li, Q. Q., 2005. A Dynamic Method for Generating Multi-Resolution TIN Models. *Photogram. Eng. Rem. Sens.* 71(8): 917–926.

Zhu, C. Q., Shi, W. Z., Li, Q. Q., Wang, G. X., Cheung, C. K., Dai, E. F., and Shea, G. Y. K., 2005. Estimation of average DEM accuracy under linear interpolation considering random error at the nodes of TIN model. *Int. J. Rem. Sens*, 26(24): 5509–5523.

Section IV

Modeling Uncertainties in Spatial Analyses

9 Modeling Positional Uncertainties in Overlay Analysis

Uncertainty modeling for static spatial data and models has been introduced in Section II and Section III, respectively. We now address methods for modeling uncertainties in dynamic spatial analyses in Section IV, from Chapter 9 to Chapter 11. These methods cover uncertainty modeling for buffer, overlay, and line simplification spatial analyses. This particular chapter, Chapter 9, focuses on positional uncertainty modeling for overlay analysis. Two approaches are presented: the analytical method and the simulation method.

9.1 INTRODUCTION

A spatial analysis in GISci provides an automatic or semiautomatic process for solving a spatial problem by executing a spatial data computational process. Spatial data, however, are always subject to different uncertainties, and the uncertainties are likely to propagate through the spatial analysis. A consequence of this situation is that spatial analysis results contain inaccuracies, and may even be incorrect in terms of the users' expectations. This being the case, the user's objective must be first clearly identified in terms of the degree of concern.

The focus of this chapter is on modeling positional uncertainty in a vector overlay analysis. Vector overlay analysis is a fundamental spatial analysis in GISci. It combines two or more spatial data collections, for example, a simultaneous observation of spatial data for analyzing the corresponding geographical phenomenon on the Earth's surface. The vector overlay analysis is classified, based on the categories of the spatial feature concerned, point overlay, line overlay, or polygon overlay. Uncertainty modeling for polygon overlay is described in this chapter.

A point overlay refers to a combination of two or more map layers of point features. The result is a point layer that contains the attributes and full extent of both source map layers. The computational processes for the point overlay include (a) determination of whether a point on one map layer is identical to a point in another map layer; (b) merging duplicate points on the source map layers and adding the merged point to the resultant map layer; and (c) adding unique points from the source map layers to the resultant map layer.

A line overlay is a combination of two or more line layers to obtain the resultant line layer. Its computational processes are similar to those of the point overlay. Two additional processes include (a) determination of whether a line in one map layer intersects with a line in another, and (b) adding the intersectional line to the resultant line layer if this line does exist.

A combination of two or more polygon layers to form a new polygon layer is called a *polygon overlay*. This operation involves the vital parts of both the computation processes for the point overlay and the line overlay; in addition, it is necessary to (a) decide whether two polygons in different map layers intersect, and (b) determine any intersectional or uncrossed parts of the polygons.

Positional uncertainty in a vector overlay analysis addresses the following: whether the accuracy of new spatial features generated in the analysis is correct, and how, in newly generated features, to quantify errors found in an overlay analysis.

Overlaying polygon layers may generate new polygons. These generated polygons are the intersection of two polygons or the difference between two polygons, each from different layers. For a quantitative description, existing uncertainty models for a polygon's feature (such as area and perimeter) are derived under different assumptions. Uncertainty indicators are defined for the perimeter and area of the generated polygon, and for intersection points of the source polygons. The uncertainty indicators are then estimated based on the error propagation law and simulation technique given in Section 9.4.1, and Section 9.4.2, respectively.

9.2   REVIEW OF THE EXISTING MODELS

Positional uncertainty of a generated polygon may be studied based upon its characteristics, including major measurements such as perimeter and area. The uncertainty of these measurements is quantified based on assumptions regarding source polygon vertices uncertainty.

Prisley et al. (1989) made three assumptions to derive the mean and variance of a generated polygon area: (a) uncertainty in the vertices of the source polygons has equal variance in both the x- and y-directions, (b) uncertainty in the x- and y-directions at a given vertex of each source polygon is independent, and (c) uncertainty in the x-direction (or y-direction) at two neighbor vertices are interrelated.

Caspary and Scheduring (1992) proposed different uncertainty measures for the generated polygon, based on the assumption that the vertices uncertainty in each source polygon is random and independent with a known variance. Kraus and Kager (1994) developed uncertainty models to describe the area accuracy of a generated polygon by considering that the vertices of each source polygon in the x- and y-directions are correlated, but that the uncertainty in the x-direction (or the y-direction) at two neighbor vertices is uncorrelated.

The above existing uncertainty models for polygon measurements are derived under different assumptions. Uncertainty indicators for the perimeter and area of a generated polygon are therefore also derived for the intersection point of source polygons. Below in a later section of this chapter, the uncertainty modeling is based upon a more relaxed assumption that both correlation of the uncertainty in the x- and y-direction at each vertex of the source polygons and correlation between the uncertainty in the x-direction (or y-direction) at all of the vertices of the source polygons might exist.

9.3 UNCERTAINTY MODELING FOR THE GENERATED POLYGON

In this section, positional uncertainty in a generated polygon from an overlay spatial analysis is described by a set of uncertainty indicators. These indicators can be classified into three categories: (a) the covariance matrix of any intersection point of source polygons; (b) uncertainty of the measurements of the generated polygon, such as the area and perimeter; and (c) a radial interval for describing the uncertainty in all vertices of the generated polygon.

Two approaches are introduced below, one, in Section 9.3.1, is based on the error propagation law and, the other, in Section 9.3.2, is based on simulation (Shi et al., 2004).

9.3.1 ANALYTICAL APPROACH

Uncertainty indicators for positional uncertainty in the generated polygon described below are derived by using the error propagation law. These uncertainty indicators are expressed as a function of vertices of the source polygons.

9.3.1.1 Covariance Matrix of an Intersection Point

To derive the intersection point of two line segments from two polygons, the following are considered: a line segment Q_iQ_j is of endpoints $Q_i = [x_i, y_i]^T$ and $Q_j = [x_j, y_j]^T$, and a line segment Q_sQ_t is of endpoints $Q_s = [x_s, y_s]^T$ and $Q_t = [x_t, y_t]^T$. Equations of these two line segments are given by

$$\begin{cases} Q_iQ_j : y - y_i = \dfrac{y_j - y_i}{x_j - x_i}\left(x - x_i\right) \\[4mm] Q_sQ_t : y - y_s = \dfrac{y_t - y_s}{x_t - x_s}\left(x - x_s\right) \end{cases}. \tag{9.1}$$

Suppose these two line segments intersect at point P, denoted by $[x_p, y_p]^T$ (see Figure 9.1). Since the intersection point must be a point passing through Q_iQ_j and Q_sQ_t, its coordinate is determined by solving Equation 9.1,

$$\begin{cases} x_p = x_i + \dfrac{\Delta x_{ij}\Delta y_{st}\Delta x_{si}}{\Delta x_{st}\Delta y_{ij} - \Delta y_{st}\Delta x_{ij}} - \dfrac{\Delta x_{ij}\Delta x_{st}\Delta y_{si}}{\Delta x_{st}\Delta y_{ij} - \Delta y_{st}\Delta x_{ij}} \\[4mm] y_p = y_i + \dfrac{\Delta y_{ij}\Delta y_{st}\Delta x_{si}}{\Delta x_{st}\Delta y_{ij} - \Delta y_{st}\Delta x_{ij}} - \dfrac{\Delta y_{ij}\Delta x_{st}\Delta y_{si}}{\Delta x_{st}\Delta y_{ij} - \Delta y_{st}\Delta x_{ij}} \end{cases} \tag{9.2}$$

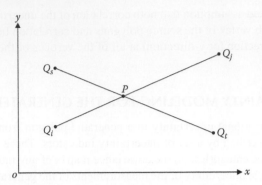

FIGURE 9.1   Intersection of two line segments: Q_iQ_j and Q_sQ_t at point $P(x_p, y_p)$.

satisfying the following two conditions:

$$\begin{cases} 0 \le r_1 \le 1 \\ 0 \le r_2 \le 1 \end{cases}, \tag{9.3}$$

where $\Delta x_{pq} = x_q - x_p$ and $\Delta y_{pq} = y_q - y_p$ ($p, q \in \{i, j, s, t\}$),

$$\begin{cases} r_1 = \dfrac{\sqrt{\left(x_p - x_i\right)^2 + \left(y_p - y_i\right)^2}}{\sqrt{\left(\Delta x_{ij}\right)^2 + \left(\Delta y_{ij}\right)^2}} \\[2em] r_2 = \dfrac{\sqrt{\left(x_p - x_s\right)^2 + \left(y_p - y_s\right)^2}}{\sqrt{\left(\Delta x_{st}\right)^2 + \left(\Delta y_{st}\right)^2}} \end{cases}. \tag{9.4}$$

Equation 9.2 shows the mathematical relationship between the intersection point and the two line segments. Positional uncertainty for this intersection point is subject to positional uncertainty of the line segments. Based on the error propagation law, this mathematical relationship is differentiated and expresses the uncertainty of the intersection point, analytically, in the form of a covariance matrix. Let $\Sigma_{Q(i)Q(j)}$ and $\Sigma_{Q(s)Q(t)}$ denote the covariance matrix for the random endpoints of line segment Q_iQ_j, and that for the random endpoints of line segment Q_sQ_t

$$\Sigma_{Q_iQ_j} = \begin{bmatrix} \sigma_{x_i,x_i} & \sigma_{x_i,y_i} & \sigma_{x_i,x_j} & \sigma_{x_i,y_j} \\ \sigma_{y_i,x_i} & \sigma_{y_i,y_i} & \sigma_{y_i,x_j} & \sigma_{y_i,y_j} \\ \sigma_{x_j,x_i} & \sigma_{x_j,y_i} & \sigma_{x_j,x_j} & \sigma_{x_j,y_j} \\ \sigma_{y_j,x_i} & \sigma_{y_j,y_i} & \sigma_{y_j,x_j} & \sigma_{y_j,y_j} \end{bmatrix} \tag{9.5}$$

and

$$\Sigma_{Q_sQ_t} = \begin{bmatrix} \sigma_{x_s,x_s} & \sigma_{x_s,y_s} & \sigma_{x_s,x_t} & \sigma_{x_s,y_t} \\ \sigma_{y_s,x_s} & \sigma_{y_s,y_s} & \sigma_{y_s,x_t} & \sigma_{y_s,y_t} \\ \sigma_{x_t,x_s} & \sigma_{x_t,y_s} & \sigma_{x_t,x_t} & \sigma_{x_t,y_t} \\ \sigma_{y_t,x_s} & \sigma_{y_t,y_s} & \sigma_{y_t,x_t} & \sigma_{y_t,y_t} \end{bmatrix}. \tag{9.6}$$

And let $\Sigma_{Q_iQ_j,Q_sQ_t}$ denote the covariance matrix of elements from a set of the endpoints of Q_iQ_j and a set of the endpoints of Q_sQ_t, which gives all the covariance coefficients between an endpoint of Q_iQ_j and an endpoint of Q_sQ_t,

$$\Sigma_{Q_iQ_j,Q_sQ_t} = \begin{bmatrix} \sigma_{x_i,x_s} & \sigma_{x_i,y_s} & \sigma_{x_i,x_t} & \sigma_{x_i,y_t} \\ \sigma_{y_i,x_s} & \sigma_{y_i,y_s} & \sigma_{y_i,x_t} & \sigma_{y_i,y_t} \\ \sigma_{x_j,x_s} & \sigma_{x_j,y_s} & \sigma_{x_j,x_t} & \sigma_{x_j,y_t} \\ \sigma_{y_j,x} & \sigma_{y_j,y_s} & \sigma_{y_j,x_t} & \sigma_{y_j,y_t} \end{bmatrix}, \tag{9.7}$$

where $\sigma_{p,q}$ is the covariance coefficient between variables p and q.

Matrix $\Sigma_{Q_iQ_j,Q_sQ_t}$ is not necessarily symmetric. The covariance matrix for all random endpoints of Q_iQ_j and Q_sQ_t is then expressed as

$$\Sigma_{Q_iQ_jQ_sQ_t} = \begin{bmatrix} \Sigma_{Q_iQ_j} & \Sigma_{Q_iQ_j,Q_sQ_t} \\ \Sigma_{Q_iQ_j,Q_sQ_t} & \Sigma_{Q_sQ_t} \end{bmatrix}. \tag{9.8}$$

Let $\zeta_P = [x_p, y_p]^T$. Differentiating Equation 9.2 gives

$$d\zeta_P = \begin{bmatrix} dx_p \\ dy_p \end{bmatrix} = \frac{\partial \zeta_P}{\partial \zeta_{Q_iQ_j,Q_sQ_t}} d\zeta_{Q_iQ_j,Q_sQ_t} = P_\zeta \cdot d\zeta_{Q_iQ_j,Q_sQ_t}, \tag{9.9}$$

where

$$\frac{\partial \zeta_P}{\partial \zeta_{Q_iQ_j,Q_sQ_t}} = \begin{bmatrix} \dfrac{\partial x_p}{\partial x_i} & \dfrac{\partial x_p}{\partial y_i} & \dfrac{\partial x_p}{\partial x_j} & \dfrac{\partial x_p}{\partial y_j} & \dfrac{\partial x_p}{\partial x_s} & \dfrac{\partial x_p}{\partial y_s} & \dfrac{\partial x_p}{\partial x_t} & \dfrac{\partial x_p}{\partial y_t} \\ \dfrac{\partial y_p}{\partial x_i} & \dfrac{\partial y_p}{\partial y_i} & \dfrac{\partial y_p}{\partial x_j} & \dfrac{\partial y_p}{\partial y_j} & \dfrac{\partial y_p}{\partial x_s} & \dfrac{\partial y_p}{\partial y_s} & \dfrac{\partial y_p}{\partial x_t} & \dfrac{\partial y_p}{\partial y_t} \end{bmatrix} \tag{9.10}$$

and

$$\partial \zeta_{Q_iQ_j,Q_sQ_t} = \begin{bmatrix} dx_i, dy_i, dx_j, dy_j, dx_s, dy_s, dx_t, dy_t \end{bmatrix}^T. \tag{9.11}$$

According to the error propagation law, the covariance matrix of the intersection point P is

$$\Sigma_P = \begin{bmatrix} \sigma_{x_p,x_p} & \sigma_{x_p,y_q} \\ \sigma_{x_p,y_q} & \sigma_{y_p,y_p} \end{bmatrix} = P_\zeta \Sigma_{Q_i Q_j Q_s Q_t} P_\zeta^T. \tag{9.12}$$

9.3.1.2 Covariance Matrix of Vertices of the Generated Polygon

The covariance matrix for the random vertices of the generated polygon is defined as a square matrix. This gives all the covariance between two vertices of the generated polygon. Suppose polygon A has Na vertices, $[Ax_1, Ay_1]^T$, $[Ax_2, Ay_2]^T$, ..., $[Ax_{Na}, Ay_{Na}]^T$, and polygon B has N_b vertices, $[Bx_1, By_1]^T$, $[Bx_2, By_2]^T$, ..., $[Bx_{Nb}, By_{Nb}]^T$. A set of the vertices of polygon A and that of the vertices of polygon B are then denoted by ζ_A and ζ_B,

$$\begin{cases} \zeta_A = \begin{bmatrix} Ax_1, Ay_1, Ax_2, Ay_2, ..., Ax_{Na}, Ay_{Na} \end{bmatrix}^T \\ \zeta_B = \begin{bmatrix} Bx_1, By_1, Bx_2, By_2, ..., Bx_{Nb}, By_{Nb} \end{bmatrix}^T \end{cases}. \tag{9.13}$$

The covariance matrix Σ_{AB} for all elements of ζ_A and ζ_B is

$$\Sigma_{AB} = \begin{bmatrix} \Sigma_{A,A} & \Sigma_{A,B} \\ \Sigma_{A,B} & \Sigma_{B,B} \end{bmatrix}, \tag{9.14}$$

where matrix $\Sigma_{A,A}$ is the covariance matrix of ζ_A, matrix $\Sigma_{B,B}$ is the covariance matrix of ζ_B, and matrix $\Sigma_{A,B}$ is the covariance matrix of components from ζ_A and ζ_B.

Suppose polygon C of N_c vertices is a generated polygon after polygons A and B overlay. A set of its vertices is expressed as

$$\zeta_C = \begin{bmatrix} Cx_1, Cy_1, Cx_2, Cy_2, ..., Cx_{Nc}, Cy_{Nc} \end{bmatrix}^T, \tag{9.15}$$

where for $k = 1, 2, ..., N_c$, point $[Cx_k, Cy_k]^T$ is a vertex of polygon C. All components of ζ_C are functions of $[Ax_i, Ay_i]^T$ and $[Bx_j, By_j]^T$ (for $i = 1, 2, ..., N_a$ and $j = 1, 2, ..., N_b$). We, therefore, differentiate it with

$$\zeta_{AB} = [Ax_1, Ay_1, Ax_2, Ay_2, ..., Ax_{Na}, Ay_{Na}, Bx_1, By_1, ..., Bx_{Nb}, By_{Nb}]^T, \tag{9.16}$$

that is, a collection of vertices of polygons A and B,

$$d\zeta_C = \frac{\partial \zeta_C}{\partial \zeta_{AB}} d\zeta_{AB} = C_\zeta d\zeta_{AB}, \tag{9.17}$$

where $d\zeta_C = \begin{bmatrix} dCx_1, dCy_1, dCx_2, dCy_2, ..., dCx_{Nc}, dCy_{Nc} \end{bmatrix}^T$,

$$C_\zeta = \frac{\partial \zeta_C}{\partial \zeta_{AB}}$$

$$
= \begin{bmatrix}
\frac{\partial Cx_1}{\partial Ax_1} & \frac{\partial Cx_1}{\partial Ay_1} & \cdots & \frac{\partial Cx_1}{\partial Ay_{Na}} & \frac{\partial Cx_1}{\partial Bx_1} & \cdots & \frac{\partial Cx_1}{\partial Bx_{Nb}} & \frac{\partial Cx_1}{\partial By_{Nb}} \\[2mm]
\frac{\partial Cy_1}{\partial Ax_1} & \frac{\partial Cy_1}{\partial Ay_1} & \cdots & \frac{\partial Cy_1}{\partial Ay_{Na}} & \frac{\partial Cy_1}{\partial Bx_1} & \cdots & \frac{\partial Cy_1}{\partial Bx_{Nb}} & \frac{\partial Cy_1}{\partial By_{Nb}} \\[2mm]
& & & & & & & \\[2mm]
\frac{\partial Cx_{Nc}}{\partial Ax_1} & \frac{\partial Cx_{Nc}}{\partial Ay_1} & & \frac{\partial Cx_{Nc}}{\partial Ay_{Na}} & \frac{\partial Cx_{Nc}}{\partial Bx_1} & & \frac{\partial Cx_{Nc}}{\partial Bx_{Nb}} & \frac{\partial Cy_{Nc}}{\partial By_{Nb}} \\[2mm]
\frac{\partial Cy_{Nc}}{\partial Ax_1} & \frac{\partial Cy_{Nc}}{\partial Ay_1} & & \frac{\partial Cy_{Nc}}{\partial Ay_{Na}} & \frac{\partial Cy_{Nc}}{\partial Bx_1} & & \frac{\partial Cy_{Nc}}{\partial Bx_{Nb}} & \frac{\partial Cy_{Nc}}{\partial By_{Nb}}
\end{bmatrix},
$$

$$d\zeta_{AB} = \left[dAx_1, dy_1, \ldots, dAx_{Na}, dAy_{Na}, dBx_1, dBy_1, \ldots, dBx_{Nb}, dBy_{Nb} \right]^T.$$

Based on the error propagation law, the covariance matrix for polygon C is then represented as

$$\Sigma_{C,C} = C_\zeta \Sigma_{AB} C_\zeta^T. \tag{9.18}$$

9.3.1.3 Variances of Measurements of the Generated Polygon

A generated polygon can be characterized by its parameters, such as its perimeter and its area. Uncertainties of these two parameters are described as follows.

It is considered that a generated polygon C, consisting of N_c vertices: $[Cx_1, Cy_1]^T$, $[Cx_2, Cy_2]^T, \ldots, [Cx_{Nc}, Cy_{Nc}]^T$, where $[Cx_0, Cy_0]^T = [Cx_{Nc}, Cy_{Nc}]^T$ and $[Cx_{Nc+1}, Cy_{Nc+1}]^T = [Cx_1, Cy_1]^T$. Its perimeter L and area R are expressed as

$$L = \sum_{k=1}^{Nc} \sqrt{\left(Cx_k - Cx_{k+1}\right)^2 + \left(Cy_k - Cy_{k+1}\right)^2}, \tag{9.19}$$

$$R = \frac{1}{2} \sum_{k=1}^{Nc} Cx_k \left(Cy_{k+1} - Cy_{k-1}\right). \tag{9.20}$$

Differentiating these two equations yields:

$$dL = \left[\frac{\partial L}{\partial Cx_1}, \frac{\partial L}{\partial Cy_1}, \cdots, \frac{\partial L}{\partial Cx_{Nc}}, \frac{\partial L}{\partial Cx_{Nc}} \right] \left[dCx_1, dCy_1, \cdots, dCx_{Nc}, dCy_{Nc} \right]^T$$

$$= \frac{\partial L}{\partial \zeta_C} d\zeta_C, \tag{9.21}$$

$$dR = \left[\frac{\partial R}{\partial Cx_1}, \frac{\partial R}{\partial Cy_1}, \cdots, \frac{\partial R}{\partial Cx_{Nc}}, \frac{\partial R}{\partial Cy_{Nc}} \right] \left[dCx_1, dCy_1, \cdots, dCx_{Nc}, dCy_{Nc} \right]^T$$

(9.22)

$$= \frac{\partial R}{\partial \zeta_C} d\zeta_C .$$

Based on the error propagation law, the variance of the perimeter and area are given as

$$\sigma_L^2 = \frac{\partial L}{\partial \zeta_C} \Sigma_{C,C} \left(\frac{\partial L}{\partial \zeta_C} \right)^T ,$$

(9.23)

$$\sigma_R^2 = \frac{\partial R}{\partial \zeta_C} \Sigma_{C,C} \left(\frac{\partial R}{\partial \zeta_C} \right)^T ,$$

(9.24)

where $\Sigma_{C,C}$ is the covariance matrix of the vertices of the generated polygon C.

9.3.1.4 Uncertainty Interval for the Vertices of the Generated Polygon

Positional uncertainties for individual vertices of a generated polygon might be different. Therefore, (a) uncertainty intervals in the x- and y-directions of the shape point and (b) radial uncertainty interval for the generated polygon are introduced below. The interval indicator describes uncertainty by giving a possible range of the uncertainty for the vertices in the generated polygon.

For a generated polygon C of N_c vertices, variances of each vertex in the x- and y-directions (i.e., σ_{Cx_k,Cx_k} and σ_{Cy_k,Cy_k} for $k = 1, 2, ..., N_c$) are derived from Equation 9.18.

The maximum and the minimum standard deviations for the vertices of generated polygon C are then defined as

$$\begin{cases} \min \sigma_x = \min \left\{ \sigma_{Cx_1,Cx_1}, \sigma_{Cx_2,Cx_2}, \cdots, \sigma_{Cx_{N_c},Cx_{N_c}} \right\} \\ \max \sigma_x = \max \left\{ \sigma_{Cx_1,Cx_1}, \sigma_{Cx_2,Cx_2}, \cdots, \sigma_{Cx_{N_c},Cx_{N_c}} \right\} \\ \min \sigma_y = \min \left\{ \sigma_{Cy_1,Cy_1}, \sigma_{Cy_2,Cy_2}, \cdots, \sigma_{Cy_{N_c},Cy_{N_c}} \right\} \\ \max \sigma_y = \max \left\{ \sigma_{Cy_1,Cy_1}, \sigma_{Cy_2,Cy_2}, \cdots, \sigma_{Cy_{N_c},Cy_{N_c}} \right\} \end{cases} .$$

(9.25)

In addition, the x- and y-directional uncertainty intervals for C are

$$I_x = [\min \sigma_x, \max \sigma_x]$$

(9.26)

and

$$I_y = [\min \sigma_y, \max \sigma_y],$$

(9.27)

respectively.

An interval of the radial positional uncertainty for the vertices of the generated polygon is represented by

$$I_p = [\min \sigma_p, \max \sigma_p], \tag{9.28}$$

where for $k = 1, 2, ..., N_c$,

$$\sigma_{p_k} = \sqrt{\sigma_{Cx_k,Cx_k} + \sigma_{Cy_k,Cy_k}}, \tag{9.29}$$

$$\min \sigma_p = \min\left\{\sigma_{p_1}, \sigma_{p_2}, \cdots, \sigma_{p_{N_c}}\right\}, \tag{9.30}$$

$$\max \sigma_p = \max\left\{\sigma_{p_1}, \sigma_{p_2}, \cdots, \sigma_{p_{N_c}}\right\}. \tag{9.31}$$

The x- and y-directional uncertainty intervals for a generated polygon are adopted for the overall x- and y-directional uncertainties, respectively. However, the x- and y-directional uncertainties might not be proportional. A generated polygon, for example, may have a small x-directional uncertainty but a large y-directional uncertainty; another generated polygon has a large x-directional uncertainty but a small y-directional uncertainty. In this case, it is difficult to conclude which generated polygon is more accurate as the x- and y-directional uncertainties are considered separately. Therefore, in many cases, the radial positional uncertainty is more appropriate for describing the uncertainty in the vertices of the generated polygon. Both the x- and y-directional uncertainties are considered, simultaneously, for individual vertices of the generated polygon.

9.3.1.5 A Note on the Analytical Model

In this section, an analytical solution for modeling the uncertainty propagation in a vector overlay analysis, based on the error propagation law, has been introduced. Involved are several uncertainty indicators for a polygon. They include the covariance matrix of the vertices of the polygon, the variances of the perimeter and area of the polygon, and the x- and y-directional, as well as the radial positional error intervals. These uncertainty indicators not only measure positional uncertainty in a generated polygon, but also measure positional uncertainty in the source polygons.

Uncertainty indicators for the generated polygon derived from the analytical approach are computed under several assumptions: (a) that the given location of two edges of source polygons intersect if and only if their true locations intersect; and (b) that the given location of two edges of the source polygons are disjointed if and only if their true locations are also disjointed. Such assumptions may not always be valid, especially when intersection points of the source polygons are close to any source polygons vertices or when the boundaries of the source polygons are close. This is illustrated in Figure 9.2. In this figure, A_i and $\bar{A}_i$ are a measured location and the true location of a vertex of source polygon A (for $i = 1, 2,$ or 3), B_j and $\bar{B}_j$ are a measured location and the true location of a vertex of source polygon B (for $j = 1, 2, 3,$ or 4), point $P_{i,j,s,t}$ is the intersection point of edges A_iA_j and B_sB_t (for $i, j = 1, 2,$ or

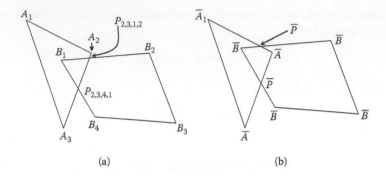

FIGURE 9.2 Intersection of the measured locations of two source polygons in (a) and that of the corresponding expected locations of the source polygons in (b). (From Shi, W. Z., Cheung, C. K., and Tong, X. H., 2004. *ISPRS J. Photogram. Rem. Sens.* 59: 47–59. With permission.)

3 and $s, t = 1, 2, 3,$ or 4), and point $\bar{P}_{i,j,s,t}$ is the intersection point of edges $\bar{A}_i\bar{A}_j$ and $\bar{B}_s\bar{B}_t$ (for $i, j = 1, 2,$ or 3 and $s, t = 1, 2, 3,$ or 4).

The measured location of the overlaid region of polygons A and B (which is the measured location of a generated polygon) is composed of three vertices B_1, $P_{2,3,1,2}$, and $P_{2,3,4,1}$, whereas the true location of the generated polygon is composed of four vertices $\bar{B}_1$, $\bar{P}_{1,2,1,2}$, $\bar{A}_2$, and $\bar{P}_{2,3,4,1}$. Both $P_{2,3,4,1}$ and $\bar{P}_{2,3,4,1}$ are derived from the same edges of the source polygons and the covariance matrix of this intersection point is able to be computed. However, $P_{2,3,1,2}$ and $\bar{P}_{1,2,1,2}$ are points derived from different edges of the source polygons so that the covariance matrix is unknown The generated polygon in Figure 9.2(a) and that in Figure 9.2(b) are also different in shape. This implies that the covariance matrix of the vertices of this generated polygon cannot be computed, based on the error propagation law. The analytical solution, therefore, is restricted by the assumptions.

9.3.2 SIMULATION APPROACH

To avoid the limitation of the analytical model, the uncertainty indicators are now estimated with a simulation technique. In the simulation approach, positional uncertainty of the source polygons is simulated, based on the assumption of uncertainty distributions of the vertices of the source polygons. After the simulation is repeated for n times, the simulated results are applied to compute a sample covariance matrix of the intersection point of the source polygons, sample variances of the perimeter and area of a generated (or source) polygon, and a sample radial positional uncertainty interval for the generated (or source) polygon.

Suppose that two source polygons A and B of N_a and N_b vertices, respectively, have undergone simulation NP times. This yields NP random locations of the source polygons, identified as a realization of the true locations of the source polygons. Figure 9.3 gives the logic flow of the simulation method.

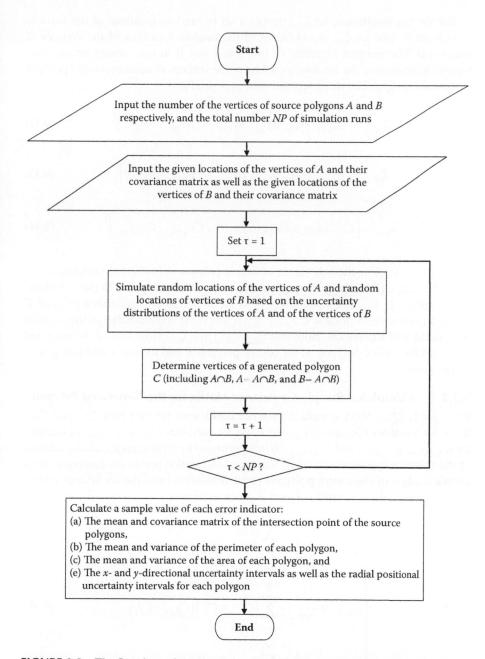

FIGURE 9.3 The flowchart of the simulation method for modeling uncertainty in a vector overlay analysis. (From Shi, W. Z., Cheung, C. K., and Tong, X. H., 2004. *ISPRS J. Photogram. Rem. Sens.* 59: 47–59. With permission.)

For the τth simulation, let $\zeta_{A(\tau)}$ denote a set of random locations of the vertices of polygon A, and let $\zeta_{B(\tau)}$ denote a set of the random locations of the vertices of polygon B. The random locations of polygon A and B in this simulation are then applied to determine the random locations of the vertices of each generated polygon C, denoted by $\zeta_{C(\tau)}$. The three sets are then expressed as

$$\zeta_{A(\tau)} = \left[Ax_{1,\tau},\, Ay_{1,\tau},\, Ax_{2,\tau},\, Ay_{2,\tau},\, \cdots,\, Ax_{Na,\tau},\, Ay_{Na,\tau} \right]^T, \qquad (9.32)$$

$$\zeta_{B(\tau)} = \left[Bx_{1,\tau},\, By_{1,\tau},\, Bx_{2,\tau},\, By_{2,\tau},\, \cdots,\, Bx_{Nb,\tau},\, By_{Nb,\tau} \right]^T, \qquad (9.33)$$

$$\zeta_{C(\tau)} = \left[Cx_{1,\tau},\, Cy_{1,\tau},\, Cx_{2,\tau},\, Cy_{2,\tau},\, \cdots,\, Cx_{Nc(\tau),\tau},\, Cy_{Nc(\tau),\tau} \right]^T, \qquad (9.34)$$

where $[Ax_{i,\tau}, Ay_{i,\tau}]^T$ is the i-th vertex of source polygon A in the τ-th simulation (for $i = 1, 2, \ldots, N_a$), $[Bx_{j,\tau}, By_{j,\tau}]^T$ is the j-th vertex of source polygon B in the τ-th simulation (for $j = 1, 2, \ldots, N_b$), $[Cx_{k,\tau}, Cy_{k,\tau}]^T$ is the k-th vertex of generated polygon C in the τ-th simulation (for $k = 1, 2, \ldots, N_c(\tau)$), and $N_c(\tau)$ is the number of the vertices of C in the τ-th simulation. Note that $\zeta_{A(0)}$, $\zeta_{B(0)}$, and $\zeta_{C(0)}$ store sets of the expected locations for source polygon A, for source polygon B and for generated polygon C, respectively.

9.3.2.1   A Simulated Sample Covariance Matrix for the Generated Polygon

For $i, j = 1, 2, \ldots, N_c(\tau)$, sample covariance coefficients for variables $Cx_{i,\tau}$ and $Cx_{j,\tau}$, those for variables $Cx_{i,\tau}$ and $Cy_{j,\tau}$, and those for variables $Cy_{i,\tau}$ and $Cy_{j,\tau}$ are denoted by $s_{Cx_{i,\tau},Cx_{i,\tau}}$, $s_{Cx_{i,\tau},Cy_{j,\tau}}$, and $s_{Cy_{i,\tau},Cy_{i,\tau}}$. When positional uncertainties for all the vertices of the source polygons are so small that their intersection points are computed from identical edges of the source polygons in all simulation runs, the covariance matrix of the vertices of a generated polygon is given as follows:

$$
\left\{
\begin{aligned}
s_{Cx_i,Cx_j} &= \frac{1}{NP-1} \sum_{\tau=1}^{NP} \left(Cx_{i,\tau} - \overline{Cx}_i \right)\left(Cx_{j,\tau} - \overline{Cx}_j \right) \\
s_{Cx_i,Cy_j} &= \frac{1}{NP-1} \sum_{\tau=1}^{NP} \left(Cx_{i,\tau} - \overline{Cx}_i \right)\left(Cy_{j,\tau} - \overline{Cy}_j \right), \\
s_{Cy_i,Cy_j} &= \frac{1}{NP-1} \sum_{\tau=1}^{NP} \left(Cy_{i,\tau} - \overline{Cy}_i \right)\left(Cy_{j,\tau} - \overline{Cy}_j \right)
\end{aligned}
\right. \qquad (9.35)
$$

where for $k = i$ or j,

$$\begin{cases} C\overline{x}_k = \dfrac{1}{NP} \sum_{\tau=1}^{NP} Cx_{k,\tau} \\[2em] C\overline{y}_k = \dfrac{1}{NP} \sum_{\tau=1}^{NP} Cy_{k,\tau} \end{cases}. \qquad (9.36)$$

A sample covariance matrix of the vertices of the generated polygon C is then,

$$S_{C,C} = \begin{bmatrix} S_{Cx_1,Cx_1} & S_{Cx_1,Cy_1} & \cdots & S_{Cx_1,Cx_{N_c}} & S_{Cx_1,Cy_{N_c}} \\ S_{Cx_1,Cy_1} & S_{Cy_1,Cy_1} & \cdots & S_{Cy_1,Cx_{N_c}} & S_{Cy_1,Cy_{N_c}} \\ & & \ddots & & \\ S_{Cx_1,Cx_{N_c}} & S_{Cy_1,Cx_{N_c}} & \cdots & S_{Cx_{N_c},Cx_{N_c}} & S_{Cx_{N_c},Cy_{N_c}} \\ S_{Cx_1,Cy_{N_c}} & S_{Cy_1,Cy_{N_c}} & \cdots & S_{Cx_{N_c},Cy_{N_c}} & S_{Cy_{N_c},Cy_{N_c}} \end{bmatrix}, \qquad (9.37)$$

where $N_c(1) = N_c(2) = \ldots = N_c(NP) = N_c$ is the number of the vertices of the generated polygon C.

9.3.2.2 Simulated Sample Variances of Measurements of the Generated Polygon

Given the vertices of the source polygons in the τ-th simulation, vertices of all corresponding generated polygons are computed. Suppose a generated polygon C in this simulation is of $N_c(\tau)$ vertices $[Cx_{1,\tau}, Cy_{1,\tau}]^T$, $[Cx_{2,\tau}, Cy_{2,\tau}]^T$, ..., $[Cx_{Nc(\tau),\tau}, Cy_{Nc(\tau),\tau}]^T$. The mathematical formula for the perimeter and area are already shown in Equations 9.19 and 9.20.

Let $L(\tau)$ and $R(\tau)$ denote the perimeter and area of the generated polygon in the τ-th simulation. Their variances are derived as follows:

$$s_{L,L} = \frac{1}{NP-1} \sum_{\tau=1}^{NP} \left(L(\tau) - \overline{L}\right)^2, \qquad (9.38)$$

$$s_{R,R} = \frac{1}{NP-1} \sum_{\tau=1}^{NP} \left(R(\tau) - \overline{R}\right)^2, \qquad (9.39)$$

where $\overline{L}$ and $\overline{R}$ are mean values of the perimeter and area of the generated polygon, respectively. That is,

$$\bar{L} = \frac{1}{NP} \sum_{\tau=1}^{NP} L(\tau), \tag{9.40}$$

$$\bar{R} = \frac{1}{NP} \sum_{\tau=1}^{NP} R(\tau). \tag{9.41}$$

9.3.2.3 Simulated Uncertainty Intervals of the Generated Polygon

Similar to the analytical model, a simulation-based uncertainty interval for the vertices of a generated polygon is defined as follows. Let $N_c(\tau)$ denote the number of the vertices of generated polygon C in the τ-th simulation where the vertices are $[Cx_{1,\tau}, Cy_{1,\tau}]^T$, $[Cx_{2,\tau}, Cy_{2,\tau}]^T$, ..., $[Cx_{Nc(\tau),\tau}, Cy_{Nc(\tau),\tau}]^T$.

The maximum and minimum standard deviations for C are given as

$$\begin{cases} \min \sigma_x = \min_{\tau=1,2,\ldots,NP} \left(\min\left\{ \sigma_{Cx_{1,\tau}}, \sigma_{Cx_{2,\tau}}, \cdots, \sigma_{Cx_{N_c,\tau}} \right\} \right) \\[2ex] \max \sigma_x = \max_{\tau=1,2,\ldots,NP} \left(\max\left\{ \sigma_{Cx_{1,\tau}}, \sigma_{Cx_{2,\tau}}, \cdots, \sigma_{Cx_{N_c,\tau}} \right\} \right) \\[2ex] \min \sigma_y = \min_{\tau=1,2,\ldots,NP} \left(\min\left\{ \sigma_{Cy_{1,\tau}}, \sigma_{Cy_{2,\tau}}, \cdots, \sigma_{Cy_{N_c,\tau}} \right\} \right) \\[2ex] \max \sigma_y = \max_{\tau=1,2,\ldots,NP} \left(\max\left\{ \sigma_{Cy_{1,\tau}}, \sigma_{Cy_{2,\tau}}, \cdots, \sigma_{Cy_{N_c,\tau}} \right\} \right) \end{cases}, \tag{9.42}$$

where $\delta_{Cx_{i,\tau}}$ and $\delta_{Cy_{i,\tau}}$ are standard deviations of $Cx_{i,\tau}$ and $Cy_{i,\tau}$ for $i = 1, 2, ..., Nc(\tau)$.

The x- and y-directional uncertainty intervals are

$$I_x = [\min \sigma_x, \max \sigma_x] \tag{9.43}$$

and

$$I_y = [\min \sigma_y, \max \sigma_y]. \tag{9.44}$$

Furthermore, the radial positional uncertainty interval for C is

$$I_p = [\min \sigma_p, \max \sigma_p], \tag{9.45}$$

where $k = 1, 2, ..., N_c(\tau)$,

$$\sigma_{p_{k,\tau}} = \sqrt{\sigma_{Cx_{k,\tau},Cx_{k,\tau}} + \sigma_{Cy_{k,\tau},Cy_{k,\tau}}}, \tag{9.46}$$

$$\min \sigma_p = \min_{\tau=1,2,\ldots,NP} \left(\min\left\{ \sigma_{p_{1,\tau}}, \sigma_{p_{2,\tau}}, \cdots, \sigma_{p_{N_c,\tau}} \right\} \right), \tag{9.47}$$

$$\max \sigma_p = \max_{\tau=1,2,\ldots,NP} \left(\max\left\{ \sigma_{p_{1,\tau}}, \sigma_{p_{2,\tau}}, \cdots, \sigma_{p_{N_c,\tau}} \right\} \right). \tag{9.48}$$

9.4 SUMMARY AND COMMENT

In this chapter, analytical and simulation methods for modeling positional uncertainty in a vector overlay analysis have been introduced and compared. The analytical method is able to assess the uncertainty in an overlay analysis under the following two assumptions: (a) The intersection point with uncertainty is further away from the vertices of the source polygons with the uncertainty; (b) the true location of the source polygons intersects when the boundaries of the two given source polygons are disjointed. Therefore, if the boundaries of the given source polygons are disjointed and not close, the true locations of the boundaries will be disjointed. Such assumptions restrict the applicability of the analytical model in the uncertainty assessment for some overlay analyses.

However, the simulation method can model the uncertainty in the overlay analysis for more generic cases, when the uncertainty distributions of the vertices of the source polygons are given. However, the computational load in the simulation method is higher than that in the analytical method. In the simulation model, generated polygons are formed for each simulation and the vertices of the generated polygons are computed. Alternatively, the analytical method initially computes the vertices of the generated polygons. Therefore, the simulation method is more time-consuming.

REFERENCES

Caspary, W. and Scheuring, R., 1992. Error-bands as measures of geometrical accuracy. In *Proceedings of the Third European Conference on GIS (EGIS'92)*, Munich, Germany, pp. 227–233.

Kraus, K. and Kager, H., 1994. Accuracy of derived data in a geographic information system. *Comput. Environ. Urban Syst.* 18: 87–94.

Prisley, S., Gregoire, T., and Smith, J., 1989. The mean and variance of area estimates computed in an arc-node geographic information system. *Photogram. Eng. Rem. Sens.* 55:1601–1612.

Shi, W. Z., Cheung, C. K., and Tong, X. H., 2004. Modeling error propagation in GIS overlay spatial analysis. *ISPRS J. Photogram. Rem. Sens.* 59: 47–59.

10 Modeling Positional Uncertainty in Buffer Analysis

An uncertainty model for buffer spatial analysis is presented and discussed in this chapter. Three uncertainty indices for describing positional uncertainty—an error of commission, an error of omission, and a normalized discrepant area—are presented. The first two indices are applicable if the "true" and measured locations of a buffer overlap; the latter is not subject to this restriction. A rigorous mathematical expression for each of the uncertainty indices is derived, based upon multivariate integration. In contrast to existing research on uncertainty modeling for buffer spatial analysis, the proposed model relaxes the assumption of positional uncertainty distribution for a spatial feature surrounded by its buffer. Hence, this model has the added advantage of being suitable for dealing with more generic application situations.

10.1 INTRODUCTION

The discussion on uncertainty propagation problems in spatial analyses, conducted in previous chapters continues in this chapter, but with the focus more specifically on uncertainty modeling in relation to buffer analysis, and in particular, vector buffer analysis.

Buffer analysis is a form of spatial analysis in GISci, in which a buffer of a predefined width is delineated around a specific spatial feature (such as a point, a line, or a polygon). Any spatial features located inside this buffer are identified as being, in terms of specific degrees, close to that particular spatial feature. The extent of closeness is quantified using a predefined width. For example, if a forest fire breaks out, regions probably devastated by the fire can be predicted in a buffer analysis. Firemen can take action to stop the fire spreading, and neighboring residents can be evacuated at an early stage. A similar buffer analysis can be implemented in areas such as environment protection, urban planning, and resource allocation.

A buffer analysis is conducted on either object-based or field-based spatial data. In field-based spatial data, a buffer zone comprises a set of pixels within a specified distance threshold (the buffer width) from a set of pixels that represents a source spatial feature. In object-based spatial data, a buffer around a spatial feature is a geometric object with a boundary, offset around the spatial feature by the buffer width. It is a circle for a point and a corridor for a line. Given a location of a spatial feature and a buffer width value, the corresponding buffer can be drawn to identify or select nearby spatial features.

A buffer around a source spatial feature can be either static or dynamic, depending on a characteristic of the source spatial feature, relative to its neighbor. If the buffer characteristics remain constant, that buffer is static. However, if the characteristics vary with distance from the source spatial feature, the buffer is dynamic. The boundary of a dynamic buffer is a line that connects all points of the characteristics of equal value. An example is provided by an environment protector wanting to detect air pollution caused by a factory located among nearby buildings. In such a case, as the air pollutant concentration in a nearby building is higher than in a more remote one, this characteristic is inversely proportional to distance. A buffer around a pollutant source, determined from the concentration of the air pollutant, is therefore dynamic.

In fact, a buffer analysis contains positional uncertainty. Positional uncertainty for a vector buffer is caused by (a) positional uncertainty of the source spatial feature and (b) buffer width uncertainty.

Positional uncertainty of a spatial feature might arise from the moment of the first data capture step to the later spatial feature decision-making operations in GISci. Data can be captured through, for example, ground surveys, remote-sensing data collection, and map digitization, each of which may introduce positional uncertainty in spatial features. (Details concerning data capture errors are given in Section 2.4.3.) Uncertainty for the source spatial feature therefore propagates toward that of the buffer. In addition, buffer uncertainty might arise from inputting the buffer width. However, uncertainty does not exist (given that there is no typing error) when the buffer width is a keyboard input, but uncertainty might exist if the buffer width is input by other methods, such as digitization.

Various studies have been carried out on modeling uncertainty propagation in a raster buffer analysis in GISci (Veregin, 1994, 1996; de Genst and Canters, 2000). Less attention has been paid to uncertainty propagation in a vector buffer analysis, probably because few standard positional uncertainty models for vector spatial features exist in this area. Zhang et al. (1998) derived an uncertainty propagation model for a vector buffer based on the epsilon band and E-band models, two error band models for assessing positional uncertainty in spatial features. This model is restricted to the assumption of error band models, in which positional uncertainty in all spatial features vertices, is identically and independently distributed. Uncertainty in the buffer width is ignored. Details of this model are reviewed below in Section 10.2.

A geometrical uncertainty indicator to assess positional uncertainty in a vector buffer in a more generic case is presented in Section 10.3. This uncertainty model considers the two uncertainty sources of the buffer analysis and relaxes the assumption of the uncertainty in the source spatial feature.

10.2 EXISTING BUFFER ANALYSIS ERROR MODEL

Zhang et al. (1998) derived an uncertainty propagation buffer model in a vector environment, based on the epsilon band (ε-band) and the E-band models. For each error band, an absolute uncertainty and a relative uncertainty for the buffer around a spatial feature, based on positional uncertainty at all the spatial feature vertices, were derived. The absolute uncertainty is defined as the buffer area around an error band of a spatial feature minus the buffer area around that feature. The relative uncertainty in the buffer is defined as the absolute uncertainty value divided by the buffer area around the spatial feature.

This buffer spatial analysis model provides a solution to assess corresponding positional uncertainty. In both the ε-band and E-band models, however, the positional uncertainty for all source spatial feature vertices is assumed to be equal, without autocorrelation and cross-correlation. It is possible to further develop the existing error band models in order to derive a more generic model for buffer analysis uncertainty propagation by considering the correlation between the spatial feature vertices.

The confidence region model for a line (see Section 4.5) has been proposed to describe the line's positional uncertainty under the assumption of independent positional uncertainty for all its vertices. This confidence region around the given line location is a region, inside which the true location of the line feature is most probably located. A further developed stochastic error band model (see Section 4.7) by considering the case in which the positional uncertainty for the vertices is correlated.

The confidence region model and stochastic error band model realize the assumption made in the ε-band model; namely, positional uncertainty for each line vertex might be of different value.

The error band approach for vector buffer analysis only deals with buffer uncertainty if it is the result of the source spatial feature propagated uncertainty. This is a special case, because buffer width uncertainty might exist and also affect the buffer accuracy. An uncertainty model for buffer analysis is introduced in Section 10.3. In this model, the source spatial feature uncertainty and the buffer width uncertainty are taken into account (Shi et al., 2003).

10.3 A PROBABILITY-BASED UNCERTAINTY MODEL FOR BUFFER ANALYSIS

Positional uncertainty in a buffer is measured using three uncertainty measures: an error of commission, an error of omission, and a normalized discrepant area between the "true" and measured locations of the buffer. The measured location of a spatial feature might be far from its true location. This leads to a deviation between the measured and true locations. In addition, buffer width uncertainty might arise in the buffer analysis. Based on the assumptions of an uncertainty source spatial feature distribution and that for the buffer width, the mean values of the error of commission, the error of omission, and the normalized discrepant area are expressed as multiple integrals. These mathematical expressions can be solved by using a numerical integration technique or a simulation technique.

10.3.1 ERROR OF COMMISSION AND ERROR OF OMISSION

A measured location of a buffer around a spatial feature may be far from the true location of the buffer. For the sake of convenience, the measured location and the

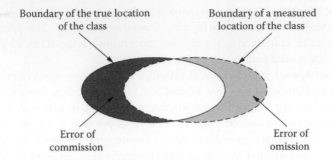

Boundary of the true location of the class

Boundary of a measured location of the class

Error of commission

Error of omission

FIGURE 10.1 The error of commission and the error of omission in remote sensing classification.

true location of the buffer are called *measured buffer* and *true buffer*, respectively. The measured buffer might partially or entirely overlap with the true buffer. Two uncertainty indicators for buffer analysis—error of commission and error of omission—to assess positional uncertainty of the buffer are now defined. These definitions have been adopted, initially, from remote sensing classification accuracy assessments.

In remote sensing, the error of commission is uncertainty due to assigning an area of one class to another class; the error of omission is the omission of an area of one class from the data set. Figure 10.1 shows the error of commission, shaded in dark gray, and the error of omission in light gray. The true location of a class is bound by a solid line and a measured location of the class by a line of dashes.

In a buffer analysis, the error of commission is defined as uncertainty consisting of the misidentified part of the measured buffer as a part of the true buffer. The error of omission is a missing part of the true buffer in the measured buffer. From these definitions, it follows that

$$\begin{cases} \text{Area of the error of commission} \leq \text{Area of the measured buffer} \\ \text{Area of the error of omission} \leq \text{Area of the true buffer} \end{cases} . \quad (10.1)$$

The question that is now raised is, How can the error of commission and the error of omission be calculated when the true buffer is unknown? In an experimental study, "reference data" or "more accurate data" are commonly used as the "true" value for investigation. In most real situations, such reference data or more accurate data are not available, owing to the expense of collection. Given an observation L and the corresponding true value $\tilde{L}$, the numerical difference is described by

$$\Delta = \tilde{L} - L . \quad (10.2)$$

From a statistical point of view, when the observation has a random error, its mean value is considered as the "true" value. This implies that $E(L) = \tilde{L}$ and thus that

$$E(L) = \tilde{L}. \qquad (10.3)$$

The mean location of a spatial feature is explored in the sections below.

10.3.1.1　Buffer Around a Point

The error of commission in a buffer around a point is calculated by subtracting (a) the overlaid part of the "true" and measured buffers from (b) the measured buffer. The error of omission is the result of the subtraction of (a) the "true" buffer and (b) the overlaid part of the "true" and measured buffers. Figure 10.2 shows the "true" and measured locations of a point denoted by $\tilde{Q}_1 = [\mu_{x_1}, \mu_{y_1}]^T$ and $Q_1 = [x_1, y_1]^T$, respectively. The circles centered around $\tilde{Q}_1$ and Q_1 are the "true" buffer of the buffer width μ_w, and the measured buffer of the buffer width w.

Let $A_{\tilde{Q}_1}$ denote a collection of locations in the buffer around the "true" point and A_{Q_1} denote a collection of locations in the buffer around the measured point. That is,

$$\begin{cases} A_{\tilde{Q}_1} = \left\{ [x,y]^T : (x - \mu_{x_1})^2 + (y - \mu_{y_1})^2 \leq \mu_w^2 \right\} \\ A_{Q_1} = \left\{ [x,y]^T : (x - x_1)^2 + (y - y_1)^2 \leq w^2 \right\} \end{cases}. \qquad (10.4)$$

Let $A_{C\text{-}Error}$ denote a collection of locations in which the error of commission occurs, and let $A_{O\text{-}Error}$ denote a collection of locations in which the error of omission occurs. These two collections are expressed as

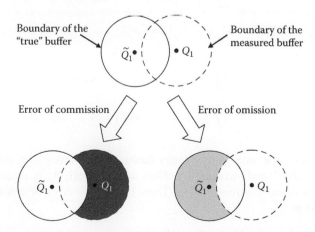

FIGURE 10.2　The error of commission and the error of omission in a buffer around a point, represented in dark gray and light gray, respectively. (From Shi, W. Z. et al., 2003. *Int. J. Geog. Inform. Sci.* 17: 251–271. With permission.)

$$\begin{cases} A_{C\text{-}Error} = A_{Q_1} - \left(A_{Q_1} \cap A_{\tilde{Q}_1}\right) \\ A_{O\text{-}Error} = A_{\tilde{Q}_1} - \left(A_{Q_1} \cap A_{\tilde{Q}_1}\right) \end{cases}. \tag{10.5}$$

The error of commission and the error of omission are then measured by the area of $A_{C\text{-}Error}$ and by the area of $A_{O\text{-}Error}$. Both are functions of the "true" and measured points as well as the "true" and measured values of the buffer width,

$$\begin{cases} Area\left(A_{C\text{-}Error}\right) = g\left(x_1, y_1, w, \mu_{x_1}, \mu_{y_1}, \mu_w\right) \\ Area\left(A_{O\text{-}Error}\right) = h\left(x_1, y_1, w, \mu_{x_1}, \mu_{y_1}, \mu_w\right) \end{cases}, \tag{10.6}$$

where $Area(X)$ represents the area of the region that contains all elements in set X. If the "true" point and the "true" value of the buffer width are given, Equation (10.6) becomes

$$\begin{cases} Area\left(A_{C\text{-}Error}\right) = \hat{g}\left(x_1, y_1, w\right) \\ Area\left(A_{O\text{-}Error}\right) = \hat{h}\left(x_1, y_1, w\right) \end{cases}. \tag{10.7}$$

The mean values of the error of commission and the error of omission are then

$$\begin{aligned} E\left(A_{C\text{-}Error}\right) &= \iint f\left(x_1, y_1, w\right) \hat{g}\left(x_1, y_1, w\right) dx_1 dy_1 dw \\ &= \iint f\left(x_1, y_1, w\right) Area\left(A_{C\text{-}Error}\right) dx_1 dy_1 dw \end{aligned} \tag{10.8}$$

and

$$\begin{aligned} E\left(A_{O\text{-}Error}\right) &= \iint f\left(x_1, y_1, w\right) \hat{h}\left(x_1, y_1, w\right) dx_1 dy_1 dw \\ &= \iint f\left(x_1, y_1, w\right) Area\left(A_{O\text{-}Error}\right) dx_1 dy_1 dw, \end{aligned} \tag{10.9}$$

where $f(x_1, y_1, w)$ is the probability density function of variables x_1, y_1 and w.

When the "true" and measured buffers are identical, $A_{\tilde{Q}}$ is equal to A_Q and $Area(A_{C\text{-}Error})$ is equal to $Area(A_{O\text{-}Error})$. Then, there is no error of commission and error of omission.

10.3.1.2 Buffer Around a Straight-Line Segment

A straight-line segment is composed of two endpoints, and a buffer around this segment is created by rolling a circle along that segment. The center of each circle is a point on the segment. The buffer errors of commission and omission are

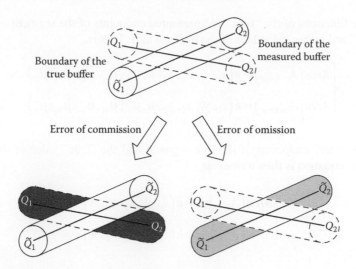

FIGURE 10.3 The errors of commission and omission in a buffer around a straight-line segment is represented in dark gray and light gray, respectively.

defined and shown in Figure 10.3. The "true" location $\tilde{Q}_1\tilde{Q}_2$ and the measured location Q_1Q_2 of a straight-line segment are in solid thick lines, whereas their buffers are bound by a solid line and a dashed line, respectively. Here, $\tilde{Q}_1 = [\mu_{x_1}, \mu_{y_1}]^T$, $\tilde{Q}_2 = [\mu_{x_2}, \mu_{y_2}]^T$, $Q_1 = [x_1, y_1]^T$, and $Q_2 = [x_2, y_2]^T$. The shaded region in dark gray represents the error of commission and that in light gray represents the error of omission.

Let $A_{\tilde{Q}_1\tilde{Q}_2}$ denote a collection of locations in the buffer around the "true" line $\tilde{Q}_1\tilde{Q}_2$ and $A_{Q_1Q_2}$ denote a collection of locations in the buffer around the measured line Q_1Q_2.

$$\begin{cases} A_{\tilde{Q}_1\tilde{Q}_2} = \left\{ [x,y]^T : \left(x - \mu_{xr} \right)^2 + \left(y - \mu_{yr} \right)^2 \leq \mu_w^2, \right. \\ \left. \mu_{xr} = r\mu_{x(1)} + \left(1 - r \right)\mu_{x(2)}, \mu_{yr} = r\mu_{y(1)} + \left(1 - r \right)\mu_{y(2)}, 0 \leq r \leq 1 \right\} \\ A_{Q_1Q_2} = \left\{ [x,y]^T : \left(x - x_r \right)^2 + \left(y - y_r \right)^2 \leq w^2, \right. \\ \left. x_r = rx_1 + \left(1 - r \right)x_2, y_r = ry_1 + \left(1 - r \right)y_2, 0 \leq r \leq 1 \right\} \end{cases} \qquad (10.10)$$

The error of commission and the error of omission are then described by

$$\begin{cases} Area\left(A_{C-Error} \right) = Area\left(A_{Q_1Q_2} - \left(A_{Q_1Q_2} \cap A_{\tilde{Q}_1\tilde{Q}_2} \right) \right) \\ Area\left(A_{O-Error} \right) = Area\left(A_{\tilde{Q}_1\tilde{Q}_2} - \left(A_{Q_1Q_2} \cap A_{\tilde{Q}_1\tilde{Q}_2} \right) \right) \end{cases}, \qquad (10.11)$$

which are functions of the "true" and measured endpoints of the segment as well as the "true" and measured values of the buffer width. That is,

$$
\begin{cases}
Area\left(A_{C-Error}\right) = g\left(x_1, y_1, x_2, y_2, w, \mu_{x_1}, \mu_{y_1}, \mu_{x_2}, \mu_{y_2}, \mu_w\right) \\
Area\left(A_{O-Error}\right) = h\left(x_1, y_1, x_2, y_2, w, \mu_{x_1}, \mu_{y_1}, \mu_{x_2}, \mu_{y_2}, \mu_w\right)
\end{cases}.
\tag{10.12}
$$

Given the "true" endpoints of the line segment and the "true" value of the buffer width, this equation is then written as

$$
\begin{cases}
Area\left(A_{C-Error}\right) = \hat{g}\left(x_1, y_1, x_2, y_2, w\right) \\
Area\left(A_{O-Error}\right) = \hat{h}\left(x_1, y_1, x_2, y_2, w\right)
\end{cases}.
\tag{10.13}
$$

Therefore, the mean values of the error of commission and the error of omission are

$$
\begin{aligned}
E\left(A_{C-Error}\right) &= \iint f\left(x_1, y_1, x_2, y_2, w\right)\hat{g}\left(x_1, y_1, x_2, y_2, w\right)dx_1 dy_1 dx_2 dy_2 dw \\
&= \iint f\left(x_1, y_1, x_2, y_2, w\right) Area\left(A_{C-Error}\right)dx_1 dy_1 dx_2 dy_2 dw
\end{aligned}
\tag{10.14}
$$

and

$$
\begin{aligned}
E\left(A_{O-Error}\right) &= \iint f\left(x_1, y_1, x_2, y_2, w\right)\hat{h}\left(x_1, y_1, x_2, y_2, w\right)dx_1 dy_1 dx_2 dy_2 dw \\
&= \iint f\left(x_1, y_1, x_2, y_2, w\right) Area\left(A_{O-Error}\right)dx_1 dy_1 dx_2 dy_2 dw
\end{aligned},
\tag{10.15}
$$

where $f(x_1, y_1, x_2, y_2, w)$ is the probability density function of variables x_1, y_1, x_2, y_2 and w.

10.3.1.3   Buffer Around a Polyline

Joining more than one straight-line segment forms a polyline. A buffer of this polyline line is the union of the buffers around the composed straight-line segments. Figure 10.4 gives an example of a polyline: the "true" location $\tilde{Q}_{1..3}$ of this polyline is composed of vertices $\tilde{Q}_1 = [\mu_{x_1}, \mu_{y_1}]^T$, $\tilde{Q}_2 = [\mu_{x_2}, \mu_{y_2}]^T$, and $\tilde{Q}_3 = [\mu_{x_3}, \mu_{y_3}]^T$; the measured polyline $Q_{1..3}$ is composed of $Q_1 = [x_1, y_1]^T$, $Q_2 = [x_2, y_2]^T$ and $Q_3 = [x_3, y_3]^T$. The dark gray and the light gray regions in this figure represent the errors of commission and omission, respectively.

Suppose a polyline is composed of m vertices where the "true" locations of the vertices are $\tilde{Q}_1 = [\mu_{x_1}, \mu_{y_1}]^T$, $\tilde{Q}_2 = [\mu_{x_2}, \mu_{y_2}]^T$, ..., $\tilde{Q}_m = [\mu_{x_m}, \mu_{y_m}]^T$, and the measured locations of the vertices are $Q_1 = [x_1, y_1]^T$, $Q_2 = [x_2, y_2]^T$, ..., $Q_m = [x_m, y_m]^T$.

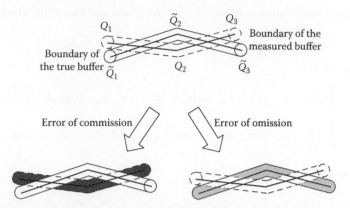

FIGURE 10.4 The errors of commission and omission in a buffer around a polyline segment are represented in dark gray and light gray, respectively.

Let $A_{\tilde{Q}_{1\ldots m}}$ denote a collection of locations in the buffer around the "true" line $\tilde{Q}_{1\ldots m}$, and let $A_{Q_{1\ldots m}}$ denote a collection of locations in the buffer around the measured line $Q_{1\ldots m}$. These two sets are

$$\begin{cases} A_{\tilde{Q}_{1\ldots m}} = A_{\tilde{Q}_1\tilde{Q}_2} \cup A_{\tilde{Q}_2\tilde{Q}_3} \cup \cdots \cup A_{\tilde{Q}_{m-1}\tilde{Q}_m} \\ A_{Q_{1\ldots m}} = A_{Q_1Q_2} \cup A_{Q_2Q_3} \cup \cdots \cup A_{Q_{m-1}Q_m} \end{cases}. \tag{10.16}$$

The error of commission and the error of omission are, then, defined by

$$\begin{cases} Area\left(A_{C-Error}\right) = Area\left(A_{Q_{1\ldots m}} - \left(A_{Q_{1\ldots m}} \cap A_{\tilde{Q}_{1\ldots m}}\right)\right) \\ Area\left(A_{O-Error}\right) = Area\left(A_{\tilde{Q}_{1\ldots m}} - \left(A_{Q_{1\ldots m}} \cap A_{\tilde{Q}_{1\ldots m}}\right)\right) \end{cases}, \tag{10.17}$$

which are functions of the "true" and measured vertices of the line as well as the "true" and measured values of the buffer width. That is,

$$\begin{cases} Area\left(A_{C-Error}\right) = g\left(x_1, y_1, \cdots, x_m, y_m, w, \mu_{x_1}, \mu_{y_1}, \cdots, \mu_{x_m}, \mu_{y_m}, \mu_w\right) \\ Area\left(A_{O-Error}\right) = h\left(x_1, y_1, \cdots, x_m, y_m, w, \mu_{x_1}, \mu_{y_1}, \cdots, \mu_{x_m}, \mu_{y_m}, \mu_w\right) \end{cases}. \tag{10.18}$$

When the "true" vertices of the line and the "true" value of the buffer width are given, Equation 10.18 becomes

$$\begin{cases} Area\left(A_{C-Error}\right) = \hat{g}\left(x_1, y_1, \cdots, x_m, y_m, w\right) \\ Area\left(A_{O-Error}\right) = \hat{h}\left(x_1, y_1, \cdots, x_m, y_m, w\right) \end{cases}. \tag{10.19}$$

Therefore, the mean value of the error of commission and that of the error of omission are

$$
\begin{cases}
E\left(A_{C-Error}\right) \\
\quad = \int \cdots \int f\left(x_1, y_1, \cdots, x_m, y_m, w\right)\hat{g}\left(x_1, y_1, \cdots, x_m, y_m, w\right)dx_1 dy_1 \cdots dx_m dy_m dw \\
\quad = \int \cdots \int f\left(x_1, y_1, \cdots, x_m, y_m, w\right) Area\left(A_{C-Error}\right)dx_1 dy_1 \cdots dx_m dy_m dw \\
E\left(A_{O-Error}\right) \\
\quad = \int \cdots \int f\left(x_1, y_1, \cdots, x_m, y_m, w\right)\hat{h}\left(x_1, y_1, \cdots, x_m, y_m, w\right)dx_1 dy_1 \cdots dx_m dy_m dw \\
\quad = \int \cdots \int f\left(x_1, y_1, \cdots, x_m, y_m, w\right) Area\left(A_{O-Error}\right)dx_1 dy_1 \cdots dx_m dy_m dw
\end{cases}
\tag{10.20}
$$

where $f(x_1, y_1, \ldots, x_m, y_m, w)$ is the probability density function of variables $x_1, y_1, \ldots, x_m, y_m$ and w.

10.3.1.4 Buffer Around a Polygon

The error of commission and the error of omission for a buffer around a polygon are shown in Figure 10.5. The polygon in this example consists of four vertices: the "true" locations of the vertices are $\tilde{Q}_1 = [\mu_{x_1}, \mu_{y_1}]^T$, $\tilde{Q}_2 = [\mu_{x_2}, \mu_{y_2}]^T$, $\tilde{Q}_3 = [\mu_{x_3}, \mu_{y_3}]^T$

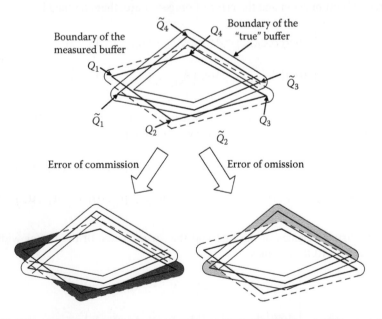

FIGURE 10.5 The error of commission and the error of omission in a buffer around a line, are represented in dark gray and light gray, respectively.

and $\tilde{Q}_4 = [\mu_{x_4}, \mu_{y_4}]^T$, and the measured locations are $Q_1 = [x_1, y_1]^T$, $Q_2 = [x_2, y_2]^T$, $Q_3 = [x_3, y_3]^T$ and $Q_4 = [x_4, y_4]^T$. Similar to the errors of commission and omission for the buffer around a point or a polyline, let $A_{\tilde{Q}_{1<->m}}$ denote a collection of locations in the buffer around the "true" boundary $\tilde{Q}_{1<->m}$ of the polygon, and let $A_{Q_{1<->m}}$ denote a collection of locations in the buffer around the measured boundary $Q_{1<->m}$ of the polygon. Then,

$$
\begin{cases}
Area\left(A_{C-Error}\right) = Area\left(A_{Q_{1<->m}} - \left(A_{\tilde{Q}_{1<->m}} \cap A_{Q_{1<->m}}\right)\right) \\
Area\left(A_{O-Error}\right) = Area\left(A_{\tilde{Q}_{1<->m}} - \left(A_{\tilde{Q}_{1<->m}} \cap A_{Q_{1<->m}}\right)\right)
\end{cases}, \qquad (10.21)
$$

where $Area(A_{C-Error})$ is the area of the region in which the error of commission occurs (the dark gray region in Figure 10.5), and $Area(A_{O-Error})$ is the area of the region in which the error of omission occurs (the light gray region in Figure 10.5).

Mean values of the error of commission and the error of omission in a buffer around a polygon of m vertices are

$$
\begin{cases}
E\left(A_{C-Error}\right) \\
= \int \cdots \int f\left(x_1, y_1, \cdots, x_m, y_m, w\right) Area\left(A_{C-Error}\right) dx_1 dy_1 \cdots dx_m dy_m dw \\
E\left(A_{O-Error}\right) \\
= \int \cdots \int f\left(x_1, y_1, \cdots, x_m, y_m, w\right) Area\left(A_{O-Error}\right) dx_1 dy_1 \cdots dx_m dy_m dw
\end{cases}. \qquad (10.22)
$$

10.3.1.5 Discussion of the Error of Commission and the Error of Omission

The errors of commission and omission are uncertainty indicators for positional uncertainty in a buffer analysis, especially in the case where the "true" and measured locations of the buffer overlap. However, if two measured locations of the buffer do not intersect with the "true" buffer, it is difficult to determine, from the errors of commission and omission, which measured buffer has the higher accuracy. Figure 10.6 gives an example of two measured locations of a buffer, one of which is further from the "true" buffer. Theoretically, positional uncertainty in (a) is greater than that in (b). The error of commission and the error of omission for both measured buffers, however, are equal. No differences exist in both cases. This means that measurement of the uncertainty in the buffer cannot be enabled by these two uncertainty indicators, when the "true" and the measured buffers are mutually exclusive.

It can be argued that a measured location of a spatial feature is most likely to be close to its "true" location, that the corresponding measured buffer always intersects with the "true" buffer, and that the closeness of the "true" buffer and the measured buffer are subject to positional uncertainty of the source spatial feature and the buffer width. However, if the uncertainty in the spatial feature is larger than the sum of the measured and "true" values of the buffer width, the "true" and the measured

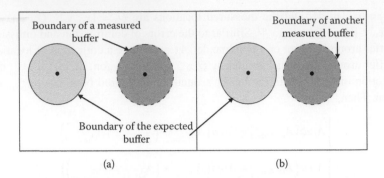

Boundary of a measured buffer

Boundary of another measured buffer

Boundary of the expected buffer

(a)   (b)

FIGURE 10.6 Limitations of the error of commission and the error of omission in buffer analysis.

buffers will most likely be disjointed. If positional uncertainty of the source spatial feature is large, another uncertainty indicator—normalized discrepant area—is proposed to assess the uncertainty in the buffer spatial analysis.

10.3.2 NORMALIZED DISCREPANT AREA

A normalized discrepant area, to measure buffer positional uncertainty in a generic case, where the "true" and the measured buffers either overlap or do not overlap, presents another uncertainty indicator. This uncertainty indicator is derived based on the concept of a discrepant area, originally used to describe discrepancies between spatial features.

A simple method for estimating positional uncertainty in a buffer around a point is based on minimum and maximum distances between the "true" buffer and the measured buffer. The minimum and maximum distances are

$$\begin{cases} Min\_d = \min\limits_{p \in A_{\tilde{Q}_1}, q \in A_{Q_1}} d(p,q) \\ Max\_d = \max\limits_{p \in A_{\tilde{Q}_1}, q \in A_{Q_1}} d(p,q) \end{cases}, \qquad (10.23)$$

where $A_{\tilde{Q}_1}$ is a collection of locations in the buffer around the "true" point $\tilde{Q}_1$, A_{Q_1} is a collection of locations in the buffer around the measured point Q_1, and $d(p, q)$ is the Euclidean distance between points p and q. The value of Min_d is the shortest distance between the "true" and the measured buffers, and the value of Max_d is the farthest distance between the buffers (see Figure 10.7).

This simple approach does not consider positional uncertainty in all points inside the buffer. Mathematically, a line and a region contain an infinite number of points. Likewise, both the boundary and the interior part of a buffer also contain an infinite number of points. The minimum and the maximum distances between the points are relatively simple measures of the uncertainty in the buffer.

To provide a full description of buffer uncertainty, the distance discrepancy is proposed, as the area difference between the "true" buffer and the measured buffer.

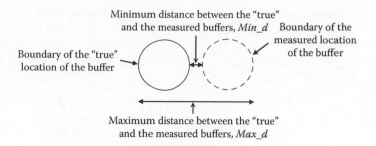

FIGURE 10.7   Uncertainty indicators for a point buffer—minimum and maximum distance between the "true" and the measured locations of the buffer. (From Shi, W. Z. et al., 2003. *Int. J. Geog. Inform. Sci.*, 17: 251–271. With permission.)

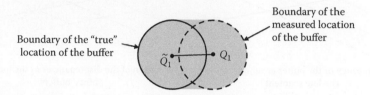

FIGURE 10.8   Discrepancy of the "true" and the measured locations of a buffer around a point, which is shaded.

Positional uncertainty for each point within the buffer is measured as the length of a line segment joining the "true" and the measured locations of the point. Therefore, the "sum" of the uncertainty for all the points within the buffer defines buffer positional uncertainty as a whole (including the boundary and the interior part of the buffer). The buffer discrepancy is shaded in Figure 10.8, where the solid line segment joining $\tilde{Q}_1 = [\mu_{x_1}, \mu_{y_1}]^T$, and $Q_1 = [x_1, y_1]^T$ is a measure of the discrepancy for Q_1.

Let $D_{Q_1 \tilde{Q}_1}$ denote a set of locations inside the discrepancy of the buffer around a point,

$$D_{Q_1 \tilde{Q}_1} = \left\{ [x, y]^T : (x - x_r)^2 + (y - y_r)^2 \le w_r^2, x_r = r\mu_{x_1} + (1 - r)\mu_{x_2}, \right.$$

$$y_r = r\mu_{y_1} + (1 - r)\mu_{y_2}, w_r = r\mu_w + (1 - r)w, 0 \le r \le 1 \right\}. \tag{10.24}$$

Buffer discrepancies around all points on a straight-line segment, are used to determine the buffer discrepancy around the segment. A straight-line segment, as in Figure 10.9, contains an infinite number of points, given by

$$(x, y) = r(\mu_{x_1}, \mu_{y_1}) + (1 - r)(\mu_{x_2}, \mu_{y_2}), \tag{10.25}$$

where $0 \le r \le 1$.

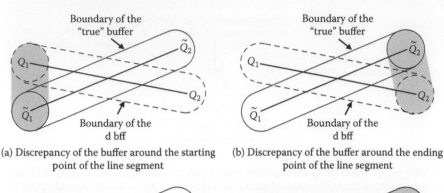

(a) Discrepancy of the buffer around the starting point of the line segment

(b) Discrepancy of the buffer around the ending point of the line segment

(c) Discrepancy of the buffer around a point of the line segment

(d) Union of the discrepancies of the individual points' buffers

FIGURE 10.9   The discrepancy of the buffer around a straight-line segment.

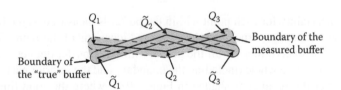

FIGURE 10.10   The discrepancy of the buffer around a straight line.

The buffer around this segment is obtained by rolling a buffer along the segment, uniting all buffers and, hence, enabling the creation of a buffer around an arbitrary segment point, as shown in Figure 10.9(d).

In the same manner, the buffer discrepancy around a polyline is defined as a discrepancy union, composed of straight-line segments of that straight line (Figure 10.10).

Let $D_{Q_{i...i+1}\tilde{Q}_{i...i+1}}$ denote a set of locations within the buffer discrepancy around a straight-line segment Q_iQ_{i+1}, and let $D_{Q_{1...m}\tilde{Q}_{1...m}}$ denote a set of locations within the buffer discrepancy around the straight line composed of m vertices. The relation between these two sets is expressed as

$$D_{Q_{1...m}\tilde{Q}_{1...m}} = \bigcup_{i=1}^{m-1} D_{Q_{i...i+1}\tilde{Q}_{i...i+1}}. \tag{10.26}$$

The buffer discrepancy around a polygon is defined similarly. Let $D_{Q_{1<->m}\tilde{Q}_{1<->m}}$ denote a set of locations within the buffer discrepancy, giving:

$$D_{Q_{1<->m}\tilde{Q}_{1<->m}} = \left(\bigcup_{i=1}^{m} D_{Q_{i<->i+1}\tilde{Q}_{i<->i+1}} \right) \bigcup A_{Q_{1<->m}} \qquad (10.27)$$

where $D_{Q_{m<->m+1}\tilde{Q}_{m<->m+1}} = D_{Q_{m<->1}\tilde{Q}_{m<->1}}$ as $Q_{m+1} = Q_1$, and $A_{Q_{1<->m}}$ is the set of locations in the buffer around the measured boundary $Q_{1<->m}$ of the polygon.

The buffer area discrepancy can be interpreted as a buffer uncertainty indicator, because the discrepancy is the difference between the "true" buffer and its measured locations. It is noted that when the "true" and the measured buffer locations are identical, the area of the discrepancy is equal to either the area of the "true" buffer or that of the measured buffer. This contradicts the fact that an uncertainty indicator should be zero when there is no uncertainty in the data. The distance function discrepant area—the *normalized discrepant area*—is therefore introduced so that the distance function will be zero when there is no positional difference between the "true" buffer and the measured buffer.

From the definition of distance, a function $d: X \times X \to \Re_+$ for all i, j in a set X is called a distance defined in X, if it satisfies the following axioms:

$$\begin{cases} \text{For all } i, j \in X, & D(i,j) = D(j,i) \\ \text{For all } i \in X, & D(i,i) = 0 \\ \text{For all } i, j, k \in X, & D(i,k) \le D(i,j) + D(i,k) \end{cases} \qquad (10.28)$$

The normalized discrepant area A_{D_Error} is then defined by

$$A_{D\_Error} = Area(D) - \min\left\{ Area(X), Area(\tilde{X}) \right\}, \qquad (10.29)$$

where $Area(D)$ is the area of the buffer discrepancy, $Area(\tilde{X})$ is the area of the "true" buffer, and $Area(X)$ is the measured buffer area; it is a function of the "true" and the measured locations of the source spatial feature, as well as the "true" and the measured values of the buffer width,

$$A_{D\_Error} = l\left(x_1, y_1, \cdots, x_m, y_m, w, \mu_{x_1}, \mu_{y_1}, \cdots, \mu_{x_m}, \mu_{y_m}, \mu_w \right). \qquad (10.30)$$

If the "true" locations of the source spatial feature and the "true buffer width" value are given, the normalized discrepant area becomes

$$A_{D\_Error} = \tilde{l}\left(x_1, y_1, \cdots, x_m, y_m, w \right). \qquad (10.31)$$

Its mean value is given by

$$E\left(A_{D\_Error}\right) \tag{10.32}$$

$$= \int \cdots \int f\left(x_1, y_1, \cdots, x_m, y_m, w\right) \tilde{l}\left(x_1, y_1, \cdots, x_m, y_m, w\right) dx_1 dy_1 \cdots dx_m dy_m dw$$

$$= \int \cdots \int f\left(x_1, y_1, \cdots, x_m, y_m, w\right) Area\left(A_{D\_Error}\right) dx_1 dy_1 \cdots dx_m dy_m dw,$$

where $f(x_1, y_1, \ldots, x_m, y_m, w)$ is the probability density function of variables $x_1, y_1, \ldots, x_m, y_m,$ and W.

The analytical expression of the mean value of each uncertainty indicator (including the error of commission, the error of omission, and the normalized discrepant area) involves the measured vertices probability density function of the source spatial feature and the measured buffer width value. If the uncertainty in the vertices of the spatial feature and that in the buffer width are independent, the probability density function becomes

$$f\left(x_1, y_1, \cdots, x_m, y_m, w\right) = f_1\left(x_1, y_1, \cdots, x_m, y_m\right) f_2\left(w\right), \tag{10.33}$$

where $f_1(x_1, y_1, \cdots, x_m, y_m)$ is the probability density function of variables $x_1, y_1, \ldots, x_m$ and y_m, and $f_2(w)$ is the probability density function of variable w. The mean values of the error of commission, the error of omission, and the normalized discrepant area are

$$E\left(A_{C\_Error}\right) \tag{10.34}$$

$$= \int \cdots \int f_1\left(x_1, y_1, \cdots, x_m, y_m\right) f_2\left(w\right) Area\left(A_{C\_Error}\right) dx_1 dy_1 \cdots dx_m dy_m dw,$$

$$E\left(A_{O\_Error}\right)$$

$$= \int \cdots \int f_1\left(x_1, y_1, \cdots, x_m, y_m\right) f_2\left(w\right) Area\left(A_{O\_Error}\right) dx_1 dy_1 \cdots dx_m dy_m dw,$$

$$E\left(A_{D\_Error}\right)$$

$$= \int \cdots \int f_1\left(x_1, y_1, \cdots, x_m, y_m\right) f_2\left(w\right) Area\left(A_{D\_Error}\right) dx_1 dy_1 \cdots dx_m dy_m dw.$$

In the case where the buffer width is free from any uncertainties, Equation 10.33 is simplified to

$$f\left(x_1, y_1, \cdots, x_m, y_m, w\right) = f_1\left(x_1, y_1, \cdots, x_m, y_m\right). \tag{10.35}$$

Therefore, the mean values of the uncertainty indicators are

$$E\left(A_{C\_Error}\right) \tag{10.36}$$
$$= \int \cdots \int f_1\left(x_1, y_1, \cdots, x_m, y_m\right) Area\left(A_{C\_Error}\right) dx_1 dy_1 \cdots dx_m dy_m,$$

$$E\left(A_{O\_Error}\right)$$
$$= \int \cdots \int f_1\left(x_1, y_1, \cdots, x_m, y_m\right) Area\left(A_{O\_Error}\right) dx_1 dy_1 \cdots dx_m dy_m,$$

$$E\left(A_{D\_Error}\right)$$
$$= \int \cdots \int f_1\left(x_1, y_1, \cdots, x_m, y_m\right) Area\left(A_{D\_Error}\right) dx_1 dy_1 \cdots dx_m dy_m.$$

10.4 SUMMARY AND COMMENT

Within the framework of spatial analysis uncertainty modeling, a method for modeling positional uncertainty propagation in a vector buffer analysis has been presented in this chapter. Buffer analysis positional uncertainty is defined as the difference between the "true" buffer location and the measured buffer location. The buffer can be for spatial features of points, line segments, polylines, and polygons. The buffer analysis uncertainty sources include positional uncertainty in vertices of the source spatial features and buffer width uncertainty.

Three uncertainty indicators and corresponding mathematical models, expressed as multiple integrals, have been derived for describing buffer uncertainty. These include the error of commission, the error of omission, and the normalized discrepant area. These indicators are used to describe various buffer analysis uncertainty situations. Both the error of commission and the error of omission are suitable for describing a situation where the "true" location and the measured location of the buffer overlap. The definition of these two uncertainty indicators is consistent with the uncertainty measures used for assessing classified images in remote sensing. The normalized discrepant area is used to consider the case that the "true" buffer location and the measured buffer location are mutually exclusive. The normalized discrepant area satisfies the conditions of algebraic distance definition. Both conceptual definitions for the three uncertainty indicators and the corresponding rigorous mathematical models for those indicators have been given.

REFERENCES

de Genst, W. and Canters, F., 2000. Handling uncertainty propagation through the buffer operation in a raster environment. In *Proceedings of Accuracy 2000*. Delft University Press, Delft, Holland: 145–152.

Shi, W. Z., Cheung, C. K., and Zhu, C. Q., 2003. Modeling error propagation of buffer spatial analysis in vector-based GIS. *Int. J. Geog. Inform. Sci.* 17: 251–271.

Veregin, H., 1994. Integration of simulation modeling and error propagation for the buffer operation in GIS. *Photogram. Eng. Rem. Sens.* 60: 427–435.

Veregin, H., 1996. Error propagation through the buffer operation for probability surface. *Photogram. Eng. Rem. Sens.* 62: 419–428.

Zhang, B., Zhu, L., and Zhu, G., 1998. The uncertainty propagation model of vector data on buffer operation in GIS. *ACTA Geodaetica et Cartographic Sinica* 27: 259–266 (in Chinese).

11 Modeling Positional Uncertainty in Line Simplification

An analytical method for quantifying positional uncertainties in line simplification spatial analysis is presented in this chapter. The total error in a line simplification is quantified as the sum of the measurement error in the initial line and the line simplification model error.

11.1 INTRODUCTION

Line simplification is a process of eliminating unnecessary or less important shape points from an initial line. The corresponding simplified line should be as similar as possible to the original line, in terms of its geometric shape.

Line simplification is widely used in cartographic generalization. It is achieved by means of the following four steps: (a) the selection of geographic features and their attributes to be represented on a map, (b) the simplification of the retained objects, (c) the classification and grouping of similar retained objects into a limited number of classes, and (d) the representation of the retained objects by symbolization. In map generalization, line simplification is also used for vector data compression. A simplified version of the original line is representative of that line. The simplified line is composed of a set of selected points from the point set of the initial line. Hence, the data volume of the simplified line is smaller than that of the initial line. Normally, line generalization is required when a larger scale map is generalized into a smaller scale map. For example, a line generalization process can generate a coarse coastline from the original detailed coastline.

Basic geometric elements in GIS include points, lines, polygon, areas, and volumetric objects. The focus in this chapter is line simplification error modeling. This restriction arises from the fact that most geographic objects in GIS and in particular medium-scale topographic maps (Müller, 1991) are represented as lines. Simplification of other complex objects (such as polygons) is based on simplification of composition boundaries/lines. Object simplification, including points, lines, polygons, areas, and volumetric object simplifications—in fact, line simplification—is a fundamental issue for overall object simplification.

The quality of line simplification is affected by two major factors: (a) error in the original line and (b) uncertainty generated from the line simplification operation. Existing error modeling methods mainly concentrate on errors from line simplification, without considering positional error in the original lines.

Two approaches can be used for modeling uncertainties in the line simplification: (a) linear attribute measurements and (b) displacement measurement. Several studies are reported in the literature on line simplification error modeling. These studies are based on the linear attribute measurement approach (Buttenfield, 1985, 1991; McMaster, 1986; Jasinski, 1990; Veregin, 2000). Additional, research has been conducted on error object simplification modeling, based on displacement measurement. These studies include, for example, methods based on polygon distortion (White, 1985; McMaster, 1987; Buttenfield, 1991), methods based on uniform distance distortion (Veregin, 2000; Cheung and Shi, 2004), methods based on displacement vector (McMaster, 1986), and critical distance methods (Little, 1989; Veregin, 2000).

This chapter offers comprehensive uncertainty modeling in line simplification. Covered are both types of error: error in the initial line and error generated through a line simplification process.

11.2 UNCERTAINTIES IN LINE SIMPLIFICATION

The uncertainty of a line simplification method can be quantified by measuring the difference between the initial line and the simplified version of the line. The uncertainty in a simplified line is measured by considering two error sources: (a) uncertainty of an original line and (b) uncertainty introduced when generating the simplified line from the original line by applying a line simplification method.

Uncertainties in line simplification are defined as three categories of error as follows: (a) the measurement error in the original line, (b) the model error generated by a line simplification algorithm, and (c) the total error, namely, the sum of (a) and (b).

Measurement error is defined as the deviation between the "true" and the measured line. This error indicator can be used to measure the difference between the two versions of the line, yielding the uncertainty of the error in the original line.

As described in Section 2.4, a line in GIS can be captured by either map digitization from a paper map, field survey by GPS, or total station, photogrammetric, or other survey methods. In such line capture processes, line measurement error is unavoidable. The nature of these measurement errors can be random, systematic, or gross. In the real case, the error is largely random. Furthermore, these line measurement errors will propagate to a simplified line by means of a line simplification process.

Model error is defined as the deviation of a simplified line from the original line, due to applying a line simplification operation on the original line. The line simplification error can be measured by comparing the difference between the original line and the simplified line after a line simplification operation is applied.

The nature of the line simplification model error is complex and depends on the characteristics of the specific line simplification method, applied. Model error varies according to different simplification algorithms: some may have relatively smaller model error, while others may be significant.

Total error is defined as the overall error, which covers both measurement error and line simplification model error. The total error can be measured by comparing the "true" original line with the measured simplified line.

11.3   LINEAR ATTRIBUTE MEASUREMENT-BASED ERROR MODEL

Having defined the three types of errors in line simplification (measurement error, model error, and total error), the models to quantify these errors are now introduced.

Uncertainties in line simplification spatial analysis can be quantified by either (a) an analytical approach or (b) a simulation approach. The focus of this chapter is on the analytical approach. The simulation approach has been introduced in Cheung and Shi (2002). Within the framework of analytical modeling of line simplification uncertainty, simplification errors can be modeled either by (1) linear attribute measurements or (2) distortion measurements. A distortion measurement, measuring the shape similarity before and after applying a line simplification, is given by Cheung and Shi (2006). Presented in this chapter is a linear attribute measurement-based analytical method for modeling errors in line simplification (Cheung and Shi, 2004; Shi and Cheung, 2006).

11.3.1   Modeling Measurement Error of the Initial Line

The measurement error of the initial line is defined as the difference between a line in GIS at its truth in the real world. The factors causing this error may be due to the technological limitations of the data capture methods, such as GPS or total station, map digitization, or laser scanning technologies, as discussed in Section 2.4.3.

Mathematically, an infinite number of points constitute a line and their positional error can be correlated. In a GIS, the initial line is represented by a set of that line's shape points. Usually an assumption of point random error normality (especially those lines from map digitization) is made (Dutton, 1992).

A line error can be described by two methods: (a) treating the overall line as a whole and (b) treating the line in terms of its component points. The second approach is used below.

The probability density functions of the individual shape points are, in fact, different from the probability density function of the initial line. They are used to form the multivariate probability density function of all the shape points of the initial line. This multivariate probability density function is then used to measure error in sample points of that initial line.

An error ellipse for each shape point of the initial line describes the error in the shape point. The error ellipse is one measure of the error region of each shape point that follows from a normal distribution. The error ellipse of a shape point (x, y) satisfies

$$\left(\left(x,y\right)-\left(\mu_x,\mu_y\right)\right)^T \Sigma^{-1}\left(\left(x,y\right)-\left(\mu_x,\mu_y\right)\right) \leq \chi_p^2\left(\alpha\right), \tag{11.1}$$

where (μ_x, μ_y) is the mean location of the shape point, Σ is the variance-covariance matrix of the shape point, and $\chi_p^2(\alpha)$ is the α percentile of the chi-square distribution with p degrees of freedom. Figure 11.1 shows the error ellipses of the shape points of lines.

As described in Section 4.7, the G-band model can describe positional uncertainty in a line using error ellipses for the shape points and for all implicit points on

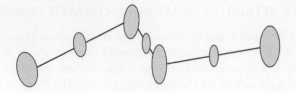

FIGURE 11.1   Error ellipses of the shape points of lines.

the line. The model covers the cases of two shape points, the independent and the dependent. The following discussion relating to initial line error modeling in line simplification is limited to the assumption that the adjacent shape points are independent in the G-band model.

Some points on the initial line are sampled randomly; the sample points and the shape points of the initial line form a point set—the *symbolic points*. An error ellipse of a symbolic point on the initial line is determined by the variance–covariance of the points. In line data capture, the shape points are selected differently from one time measurement to another, except for the starting and ending shape points.

Let Σ denote the variance-covariance matrix of a point on the line,

$$\Sigma = \begin{pmatrix} \sigma_x^2 & \sigma_{xy} \\ \sigma_{xy} & \sigma_y^2 \end{pmatrix}. \tag{11.2}$$

The semi-major and semi-minor axes of the error ellipse of the corresponding point are square roots of two solutions to the following equation

$$\lambda^2 - \left(\sigma_x^2 + \sigma_y^2\right)\lambda + \left(\sigma_x^2\sigma_y^2 - \sigma_{xy}^2\right) = 0. \tag{11.3}$$

The orientation of this standard error ellipse is given by the angle θ between the x-axis and the semi-major axis

$$\theta = \frac{1}{2}\alpha\tan\left[2\sigma_{xy}\Big/\left(\sigma_x^2 - \sigma_y^2\right)\right]. \tag{11.4}$$

Let there be n symbolic points on the line. When n tends to infinity, the integration of the error ellipses of the symbolic points forms the G-bands—as illustrated by the dashed lines in Figure 11.2.

11.3.2   MODELING LINE SIMPLIFICATION ERROR

Three indices are used to measure the error in a line simplification process: the normalized difference, the median difference, and the maximum difference between an initial line and its simplified version.

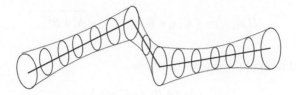

FIGURE 11.2 Modeling the measurement error of the original line by the error ellipses of symbolic the points of the line.

The *normalized (or mean) difference* measures the general uncertainty of the simplified line. It is a method used to locate a central distribution point in the measurement error for the central tendency. The normalized value can anticipate additional statistical analysis, such as variance of the measurements. The advantage of this index is that all available information is used. When a distribution has a few extreme cases, however, the mean difference may become misleading.

The *median* difference measures average error, even for a highly skewed distribution. The advantage of this index is that it is based on the middle value for points arranged in ascending order and is not affected by extreme cases in measurements.

In fact, both the normalized and the median uncertainty indices measure the central tendency of the discrepancy. However, their values may be different when the distribution of the discrepancy is skewed. For example, if the general value is required, the mean is used; if minimization of the average value of a positively skewed distribution of the discrepancy is required, the median is used.

The *maximum* difference measures the maximum error of all measurements. It is a measure of the extreme error measurement case and does not affect other differences in the error distribution.

A line simplification model error is measured by the maximum distortion of a simplified line from its initial line. A distortion is defined as the shortest distances of the individual shape points of the initial line to its simplified line. Figure 11.3 illustrates the concept of the maximum distortion error. Mathematically, the distortion error is given by

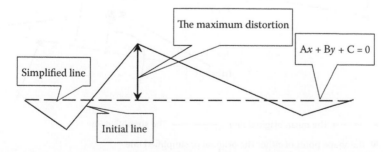

FIGURE 11.3 The line simplification model error measured by the maximum distortion between the initial line and its simplified version.

$$d(x_1, y_1) = \left| (A \cdot x_1 + B \cdot y_1 + C) \big/ \sqrt{A^2 + B^2} \right|,$$ (11.5)

where a general form of the simplified line is

$$A \cdot x + B \cdot y + C = 0$$ (11.6)

and (x_1, y_1) is a shape point on the initial line. The maximum of the distortions (the shortest distances) is defined as the maximum distortion.

11.3.3 MODELING THE TOTAL ERROR

The *total error* includes both the measurement error and line simplification model error. The total error can be measured by comparing the "true" original line with the measured simplified line.

Figure 11.4 illustrates the above concept by representing the mean locations of the initial line, the simplified lines, and the integrated band of the initial line from error ellipses.

Mathematically, the total error of the line simplification is defined as follows. Let n denote the total number of the symbolic points of an initial line and (x_i, y_i) a variable symbolic point of the initial line with mean $\mu_M = (\mu_{x_i}, \mu_{y_i})$ and variance matrix

$$\Sigma = \begin{pmatrix} \sigma_{x_i x_i} & \sigma_{x_i y_i} \\ \sigma_{y_i x_i} & \sigma_{y_i y_i} \end{pmatrix}, i = 1, 2, \ldots, n.$$

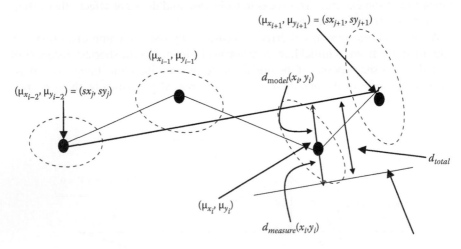

FIGURE 11.4   The concept of the total error model.

Let $n_s(w)$ denote the total number of the shape points of the simplified line of a tolerance value w for line simplification, and $(sx_{w,j}, sy_{w,j})$ a shape point of the simplified line, $j = 1, 2, ..., n_s(w)$. The tolerance is a threshold value in line simplification, where error, caused by the simplified line, can be ignored if it is smaller than the tolerance. It is assumed that a symbolic point (x_i, y_i) reaches a straight line $SL_{j,j+1}$ that links two adjacent shape points $(sx_{w,j}, sy_{w,j})$ and $(sx_{w,j+1}, sy_{w,j+1})$ of the simplified line.

The impact of the uncertainty in the initial line on the simplification process is estimated by calculating the deviation of the measured shape point on the original line $P_i(x_i, y_i)$ to its true value (μ_{x_i}, μ_{y_i}). The ellipse around (μ_{x_i}, μ_{y_i}) in Figure 11.4 is the region containing (x_i, y_i) with 95% certainty. The deviation is then defined as the largest perpendicular distance of (x_i, y_i) to (μ_{x_i}, μ_{y_i}), being a distance between (μ_{x_i}, μ_{y_i}) and the ellipse tangent of (x_i, y_i) parallel to $SL_{j,j+1}$. If the symbolic point of the initial line (x_i, y_i) is error free, $d_{Measure}(x_i, y_i)$ is equal to zero.

The deviation between the mean locations of the initial and simplified lines is derived above (Section 11.3.2). Let $d_{Model}(x_i, y_i|w)$ denote the shortest distance of a symbolic point (μ_{x_i}, μ_{y_i}) of the initial line to a simplified line $SL_{j,j+1}$. This describes the uncertainty in those points that are projections of the symbolic points on the simplified line when the initial line is error free. This deviation is then maximized with the uncertainty measures for the uncertainty in the initial line in order to assess the uncertainty in the simplified line.

The *total error* of the line simplification is the sum of $d_{Measure}(x_i, y_i)$ and $d_{Model}(x_i, y_i|w)$. It measures the deviation of a symbolic point (x_i, y_i) of the initial line from simplified line $SL_{j,j+1}$ under uncertainty. Maximizing the sum, with respect to the symbolic point results in an uncertainty indicator of the simplified line, namely, the uncertainty indicator of the simplified line d_{Total} is

$$d_{Total} = \underset{i}{Max}\left(d_{Measure}(x_i, y_i) + d_{Model}(x_i, y_i|w)\right). \tag{11.7}$$

The maximum deviation of the simplified line from the initial line is defined as the maximization of the uncertainties of all the points on the simplified line, in comparison with those on the initial line. This uncertainty indicator measures the maximum uncertainty in the simplified line.

11.3.4 AN EXAMPLE

To illustrate analyzing uncertainty in line simplification, an example of a portion of a 1:1000 base map was taken to demonstrate the application of the proposed error modeling method. In this case, road centerlines were taken for a line simplification process.

By repeatedly digitizing the lines on the map, the hypothesis that the error of the symbolic points comes from a bivariate normal distribution was tested, using a normal probability plot and the Kolmogorov–Smirnov statistic test. For example, the symbolic points of Line 1 on the ground (with unit in meter) in Figure 11.5 are given in Table 11.1.

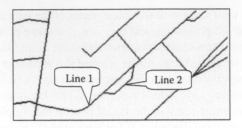

FIGURE 11.5   The road centerline to be simplified in GIS.

TABLE 11.1
Coordinates of the Symbolic
Points of Line 1 (Unit: m)

Point No. (i)	(μ_{x_i}, μ_{y_i})
1	(836261.058, 817614.464)
2	(836229.864, 817588.316)
3	(836220.873, 817583.298)
4	(836211.464, 817579.953)
5	(836204.146, 817578.907)
6	(836193.482, 817579.534)
7	(836142.256, 817589.152)
8	(836128.669, 817593.937)

TABLE 11.2
Error of the Symbolic Points of Line 1 (Unit: m)

Point No. (i)	$d_{Measure}$ (m)	d_{Model} (m)	$d_{Measure} + d_{Model}$ (m)
1	1.057	0.000	1.057
2	9.726	4.036	13.762
3	4.339	4.008	8.348
4	4.679	2.276	6.955
5	5.024	0.000	5.024
6	1.640	2.064	3.704
7	1.688	2.377	4.065
8	1.319	0.000	1.319

Based on the models described in the previous section, the measurement error, line simplification error, and total error are computed and listed in the Table 11.2. Here, the Douglas–Peucker line simplification method (Douglas and Peucker, 1973) was applied on Line 1 with the weed threshold at 5 m.

The measurement error ranges from 1.057 m to 9.726 m on the ground, while the line simplification model ranges from 0.000 m to 4.036 m on the ground. From this, it can be seen that the measurement error makes a larger contribution to the total error than the model error. If the measurement error was neglected, the total error would be 4.04 m, rather than 13.762 m, and would thus underestimate the measurement error being considered. Therefore, it is essential to take the measurement error of the original line into consideration when measuring the overall error of a line simplification process.

11.4 SUMMARY

This chapter has addressed line simplification error modeling, within the framework of modeling uncertainties in spatial object simplification. Uncertainty modeling for object simplification can be studied through two approaches: (a) analytical solutions and (b) simulation solutions. This chapter has presented an analytical solution. However, it should be realized that the simulation solution is, in fact, another effective approach to solve the error modeling in line simplifications, especially for complex geometric objects.

REFERENCES

Buttenfield, B. P., 1985. Treatment of the cartographic line. *Cartographica* 22(2): 1–26.
Buttenfield, B. P., 1991. A rule for describing line feature geometry. In *Map Generalization: Making Rules for Knowledge Representation*, ed. McMaster, R. B. and Buttenfield, B. P. Longman, Harlow. pp. 150–171.
Cheung, C. K. and Shi, W. Z., 2002. A simulation based approach to measure uncertainty of line simplification in GIS. *Fifth International Symposium on Spatial Accuracy Assessment in Nature Resources and Environmental Science*. Australia, July 10–12: 389–396.
Cheung, C. K. and Shi, W. Z., 2004. Estimation of the positional uncertainty in line simplification in GIS. *The Cartography J.*, 41(1): 37–45.
Cheung C. K. and Shi W. Z., 2006. Positional error modeling for line simplification based on automatic shape similarity analysis in GIS. *Comput. Geosci.* 32(4): 462–475 May.
Douglas, D. and Peucker, T., 1973. Algorithms for the reduction of the number of points required to represent a digitized line or its caricature. *Can. Cartographer* 10: 112–122.
Dutton, G., 1992. Handling positional uncertainty in spatial databases. In *Proceedings of Fifth International Symposium on Spatial Data Handling*, Charleston, South Carolina, pp. 460–469.
Jasinski, M. J., 1990. Comparison of complexity measures for cartographic lines. *NCGIA Technical Report* 90-1.
Little, A. R. 1989. An evaluation of selected computer-assisted line simplification algorithms in the context of map accuracy standards. Technical papers: 1989 ASPRS/ACSM Annual Conventions, *American Society for Photogrammetry and Remote Sensing and American Congress on Surveying and Mapping* 5: 122–132.
McMaster, R. B., 1986. A statistical analysis of mathematical measures for linear simplification. *The Amer. Cartographer* 13(2):103–117.
McMaster, R. B., 1987. The geometric properties of numerical generalization. *Geog. Anal.* 19: 330–346.
Müller, J. C., 1991. Generalization of spatial databases. In *Geographical Information Systems: Principles and Applications* 1 ed. Maguire, D., Goodchild, M., and Rhind, D. Longman, Harlow. pp. 457–475.

Shi, W. Z. and Cheung, C. K., 2006. Performance evaluation of line simplification algorithms for vector-mode generalization. *The Cartographic J.*, 43(1): 27–44.

Veregin, H., 2000. Quantifying positional error induced by line simplification. *Int. J. Geog. Inform. Sci.* 14: 113–130.

White, E. R., 1985. Assessment of line-generalization algorithms using characteristic points. *The Amer. Cartographer* 12: 17–27.

Section V

Quality Control of Spatial Data

12 Quality Control for Object-Based Spatial Data

The aim of spatial data quality control is to limit or even reduce overall spatial data uncertainty. Methods for describing uncertainties in spatial data have been given in the previous chapters. Examples provided include the methods for modeling positional uncertainty (Chapter 4) and the methods for modeling attribute uncertainty (Chapter 5). Both these chapters are introduced in Section II of this book. A further theoretical development beyond the uncertainty description is the creation of methods for controlling the quality of spatial data. These methods are introduced in Section V, and are found in Chapters 12 to 14.

A new systematic strategy for controlling the present quality of object-based spatial data is proposed and presented in this chapter. Cadastral data, which requires high-quality standards, are highly representative object-based data in GIS and hence are appropriate for use as an example when exploring and developing methods for spatial data quality control for object-based spatial data. A land parcel is used as a base upon which data inconsistencies are solved, using least square adjustments.

The strategy involves three steps. Firstly: to constitute the adjustment conditional equations, the registered parcel area is taken as the true value and the digitized vertex coordinates are taken as the observations with random error. Secondly, both the registered area and the digitized vertex coordinates held are taken as observations with random error. The weights of these two observation types are estimated based on the variance components method. Thirdly, the scale factor is added to the adjustment model for handling systematic error due to scale error in the land parcel area.

12.1 INTRODUCTION

In cadastral information systems, land parcels are represented by polygons. The values of parcel areas are usually taken from the following two different sources: (a) the registered area value of a land parcel measured on the ground or from descriptive documents—the registered land parcel area that has legal authority—and (b) the calculated area value from the digitized vertex's coordinates of the land parcel boundary. Due to errors in the latter, the calculated parcel area in (b) normally has a lower accuracy, and is not a match with the value of the registered areas of the same land parcel. Hence, such land parcel areas have inconsistencies that need to be solved.

Chrisman and Yandell (1988, 1989) derived a statistical model on area variances based on the error in the component nodes of the polygon and their correlations. Najeh and Burkhard (1995) applied the adjustment method for processing digitized cadastral data, with an emphasis on determining the weight of the observations. Merrit and Masters (1999) explored the determination of point coordinates from observed direction, distance, and radius based on parametric adjustments in digital cadastral databases.

Liu et al. (1999) proposed the use of the least square adjustment method for processing parcel areas, considering registered parcel areas as true values. Shi and Tong (2005) discussed the case that the registered parcel possesses error. Tong et al. (2005) further studied the scale effect on land parcels, as a systematic error.

12.2   ADJUSTMENT METHODS FOR AREA OBJECT ERROR PROCESSING

12.2.1   Conditional Equations of the Parcel Area

During cadastral map digitization, a land parcel is captured by means of digitizing its component boundary vertices. Due to the data capture method nature, errors exist in the vertex's coordinates, and this error will propagate, for example, to the land parcel area. The nature of the error can be, for example, systematic error, random error, or even gross error. To process land parcel area calculation errors, based on the vertex coordinates, the least square adjustment data processing method is applied. In such cases, the coordinates of the vertices are regarded as observations with random or systematic error. A conditional adjustment method with the conditional equations accommodate the following constraints: the true value of a land parcel area, right angles, or arcs of land parcels in relation to the component vertex coordinates, is proposed.

Let (x_i, y_i) $(i = 1, 2, …, n)$ denote the digitized coordinates of the vertices of a land parcel boundary, and the corresponding adjustment value and corrective values are, $(\hat{x}_i, \hat{y}_i)$ and (v_{x_i}, y_{x_i}), giving,

$$\hat{x}_i = x_i + v_{x_i}, \tag{12.1}$$

$$\hat{y}_i = y_i + v_{y_i} \qquad (i = 1, 2, \cdots, n).$$

Figure 12.1 shows a land parcel polygon with vertices denoted by P_i $(i = 1, 2, …, n)$. Assuming that the registered area of the parcel is known as S_0, the following relationship between the registered area and the digitized coordinates of the vertices is found:

$$\hat{S} = \frac{1}{2} \sum_{i=1}^{n} \hat{x}_i (\hat{y}_{i+1} - \hat{y}_{i-1}) = S_0, \tag{12.2}$$

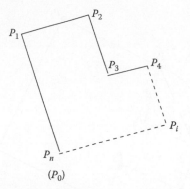

FIGURE 12.1 A polygon of land parcel.

where $P_0 = P_n$; $P_{n+1} = P_1$. Thus, the conditional equations for the area are,

$$\frac{1}{2}\sum_{i=1}^{n}\hat{x}_i(\hat{y}_{i+1}-\hat{y}_{i-1})-S_0=0.\tag{12.3}$$

The linearized form of the above conditional equation is

$$\sum_{i=1}^{n}a_iv_{x_i}+\sum_{i=1}^{n}b_iv_{y_i}+\omega=0,\tag{12.4}$$

where

$$a_i=\frac{1}{2}(y_{i+1}-y_{i-1}),$$

$$b_i=\frac{1}{2}(x_{i+1}-x_{i-1}),$$

$$\omega=S-S_0,$$

$$S=\frac{1}{2}\sum_{i=1}^{n}x_i(y_{i+1}-y_{i-1}).$$

12.2.1.1 The Rectangular Conditional Equation

Right angles in the parcel polygons are noticed in Figures 12.1 and 12.2. For example, in Figure 12.2, the three right angles are: $\angle P_1P_2P_3$, $\angle P_2P_3P_4$, and $\angle P_3P_4P_5$. Hence, at the vertices P_2, P_3, and P_4, rectangular constraints apply. Assume that an angle β is composed of three vertices, denoted as i, j, and k. A rectangular conditional equation between these three vertices is

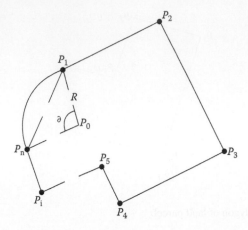

FIGURE 12.2 A parcel polygon with an arc. (From Tong, X. H. et al., 2005. *Photogram. Eng. Remote Sens.* 71(2): 189–195. With permission.)

$$\hat{\alpha}_{ik} - \hat{\alpha}_{ij} = \beta, \tag{12.5}$$

where $\hat{\partial}_{ik}$, $\hat{\partial}_{ij}$ are the adjustment values of the azimuths in directions *ik* and *ij*.
The conditional equation then equals:

$$a_{ik} v_{x_k} + b_{ik} v_{y_k} - (a_{ik} - a_{ij}) v_{x_i} - (b_{ik} - b_{ij}) v_{y_i} - a_{ij} v_{x_j} - b_{ij} v_{y_j} + \omega_i = 0, \tag{12.6}$$

where the calculations of a_{ik}, b_{ik}, a_{ij}, b_{ij}, a_{ik}, b_{ik}, ω_i are referred to Liu et al. (1999). It is easy to see that if $\beta = \pi/2$ (or $3\pi/2$), Formula 12.5 is a rectangular conditional equation.

12.2.1.2 Conditional Equation for Arcs

If a parcel polygon has several arcs as shown in Figure 12.2, the area of the parcel is the sum of the area of the polygon with vertices $P_0, P_1, ..., P_n$ and of the sector with vertices P_n, P_0, and P_1. Assuming that the radius of the sector equals R, and taking the center point $P_0(x_0, y_0)$ of the arc as the unknown parameters, the following area conditional equation is given:

$$\frac{1}{2} \sum_{i=0}^{n} \hat{x}_i (\hat{y}_{i+1} - \hat{y}_{i-1}) + \frac{1}{2} R^2 \hat{\alpha} - S_0 = 0, \tag{12.7}$$

where $P_{-1} = P_n$; $P_{n+1} = P_0$, i.e., $x_{-1} = x_n$, $y_{-1} = y_n$; $x_{n+1} = x_0$, $y_{n+1} = y_0$; where α is the central angle of the sector, calculated by

$$\sin \frac{\hat{\alpha}}{2} = \frac{1}{2R} \sqrt{(\hat{x}_2 - \hat{x}_1)^2 + (\hat{y}_2 - \hat{y}_1)^2}. \tag{12.8}$$

Linearization of the Equation 12.8 and substituting it into Equation 12.7, yields

$$\sum_{\substack{i=0 \\ i\neq1,2}}^{n} a_i v_{x_i} + \sum_{\substack{i=0 \\ i\neq1,2}}^{n} b_i v_{y_i} + (a_1 - c_1)v_{x_1} + (b_1 + d_1)v_{y_1} + (a_2 + c_1)v_{x_2} + (b_1 - d_1)v_{y_2} + w = 0, \quad (12.9)$$

where

$$c_1 = \frac{R}{2\cos\dfrac{\alpha^0}{2}}\cos\alpha_{12},$$

$$d_1 = -\frac{R}{2\cos\dfrac{\alpha^0}{2}}\sin\alpha_{12},$$

$$w = \frac{1}{2}\sum_{i=0}^{n} x_i(y_{i+1} - y_{i-1}) + \frac{1}{2}R^2\alpha^0 - S_0, \text{ and}$$

α^0 being an approximate value of α.

At least three observed points including the two end points, should be digitized to determine the arc, giving the following further conditional equation:

$$(\hat{x}_{ri} - \hat{x}_0)^2 + (\hat{y}_{ri} - \hat{y}_0)^2 - R^2 = 0. \quad (12.10)$$

Linearizing Equation 12.10 results in the following conditional equation with unknown parameters:

$$\Delta x_{0i}v_{x_{ri}} + \Delta y_{0i}v_{y_{ri}} - \Delta x_{0i}\delta\hat{x}_0 - \Delta y_{0i}\delta\hat{y}_0 + w = 0, \quad (12.11)$$

where

$$w = \frac{1}{2}[(x_{ri} - x_0^0)^2 + (y_{ri} - y_0^0)^2 - R^2].$$

Supposing that there are m arcs in a parcel polygon and that the radius of each arc is R_i ($i = 1, 2, \ldots, m$), giving the following conditional equation:

$$\sum_{i=0}^{n} a_i v_{x_i} + \sum_{i=0}^{n} b_i v_{y_i} + \sum_{i=1}^{m} \pm\frac{1}{2}R_i^2\alpha_i^0\delta\hat{\alpha}_i + \omega = 0, \quad (12.12)$$

where the positive or negative sign ($\pm$) in the above equation is determined by the shape of the arcs in the polygon. If the arc is convex the sign is positive; if it is concave the sign is negative. a_i is calculated by the endpoints and radius of the arc according to Equation 12.8. Furthermore,

$$\omega = \frac{1}{2}\sum_{i=0}^{n} x_i(y_{i+1}-y_{i-1})+\sum_{i=1}^{m}\pm\frac{R_i^2}{2}\alpha_i-S_0.$$

Usually, the parcel areas are adjusted within a region that is a block or a composition of several blocks. Assuming that the digitized coordinate observation is denoted as a vector L, and that a corresponding variance–covariance matrix is denoted as a matrix Q, the corrective value of the observation is denoted as a vector V. Therefore, the conditional equation can be represented by following generalized form:

$$AV + A_x\delta\hat{X} + w = 0, \tag{12.13}$$

where coefficients in A, A_x, and w are calculated based on Equations 12.4, 12.6, 12.9, 12.11, and 12.12, and $\delta\hat{X}$ is the unknown parameter matrix.

Based on the least squares theory, the corresponding normal equation equals:

$$\begin{bmatrix} AQA^T & A_x \\ A_x^T & 0 \end{bmatrix}\begin{bmatrix} K \\ \delta\hat{X} \end{bmatrix}+\begin{bmatrix} w \\ 0 \end{bmatrix}=0, \tag{12.14}$$

where $\delta\hat{X}$ is an unknown parameter matrix and K is a Lagrange multiplier. Thus, $\delta\hat{X}$ and K are calculated by

$$\begin{bmatrix} K \\ \delta\hat{X} \end{bmatrix}=-\begin{bmatrix} AQA^T & A_x \\ A_x^T & 0 \end{bmatrix}^{-1}\begin{bmatrix} w \\ 0 \end{bmatrix}. \tag{12.15}$$

The adjustment values of the digitized coordinates of the parcels are obtained by

$$\hat{L}=L+V=L-QA^TK. \tag{12.16}$$

The variance–covariance matrixes of the adjusted coordinates equal

$$Q_{\hat{L}} = Q-QA^T(AQA^T)^{-1}(I-A_x(A_x^T(AQA^T)^{-1}A_x)^{-1}A_x^T(AQA^T)^{-1})AQ,$$
$$D_{\hat{L}}=\hat{\sigma}_0^2 Q_{\hat{L}}, \tag{12.17}$$

and the estimator of the unit root mean square error is obtained by

$$\hat{\sigma}_0^2 = \frac{V^T P V}{r},$$ (12.18)

where r is the number of the redundant measurements.

12.2.2 ESTIMATION OF VARIANCE COMPONENTS

As stated in Section 12.0 above, the land parcel area values are from two different sources: (a) the registered parcel area value, which may be from a ground survey, and (b) the computed area value based on the digitized coordinates of the component vertices of the land parcel. The area value from source (b) contains errors due to map digitization measurement errors. The area value from source (a) may also possess errors, since it is also measured by ground survey or other methods; however, this error may be smaller than the error from source (b). If parcel areas from both sources (a) and (b) possess errors, they are treated as two types of observations with different error levels in the least square adjustment, rather than as constants, without error and as an observation with error, as was the case given in Section 12.2.1. One critical issue here is the estimation of the weight of the two types of observations. The application of the variance components method (or the Helmert method) is proposed for this purpose.

It is assumed that the area conditional equations contain two kinds of independent observations: L_1 (coordinates that are used to compute parcel area from source [b]) and L_2 (the registered area from source [a]) with weight matrices P_1 and P_2, respectively. The conditional equations are then:

$$A_1 V_1 + A_2 V_2 + w = 0,$$ (12.19)

where

$$L = \begin{pmatrix} L_1 \\ L_2 \end{pmatrix}, \quad V = \begin{pmatrix} V_1 \\ V_2 \end{pmatrix}, \quad A = \begin{pmatrix} A_1 & A_2 \end{pmatrix}, \quad P = \begin{pmatrix} P_1 & 0 \\ 0 & P_2 \end{pmatrix}.$$

Let the initial weights of two types observations be P_1 and P_2. $V_1^T P_1 V_1$ and $V_2^T P_2 V_2$ are then computed after each computation, and they are used to reestimate their unit weight variance factors $\hat{\sigma}_{01}^2$ and $\hat{\sigma}_{02}^2$ as (Cui et al., 1992):

$$\begin{cases} E(V_1^T P_1 V_1) = tr(N^{-1} N_1 N^{-1} N_1)\hat{\sigma}_{01}^2 + tr(N^{-1} N_1 N^{-1} N_2)\hat{\sigma}_{02}^2 \\ E(V_2^T P_2 V_2) = tr(N^{-1} N_1 N^{-1} N_2)\hat{\sigma}_{01}^2 + tr(N^{-1} N_2 N^{-1} N_2)\hat{\sigma}_{02}^2 \end{cases},$$ (12.20)

where $N = N_1 + N_2$, $N_1 = A_1 P_1^{-1} A_1^T$, $N_2 = A_2 P_2^{-1} A_2^T$.

The factors of the unit weight variance for the two types of observation are estimated as:

$$S_A \hat{\theta} = W_\theta , \tag{12.21}$$

where

$$S_A = \begin{pmatrix} tr(N^{-1}N_1N^{-1}N_1) & tr(N^{-1}N_1N^{-1}N_2) \\ tr(N^{-1}N_1N^{-1}N_2) & tr(N^{-1}N_2N^{-1}N_2) \end{pmatrix}, \hat{\theta} = \begin{pmatrix} \hat{\sigma}_{01}^2 \\ \hat{\sigma}_{02}^2 \end{pmatrix}, W_\theta = \begin{pmatrix} V_1^T P_1 V_1 \\ V_2^T P_2 V_2 \end{pmatrix}.$$

The obtained $\hat{\sigma}_{01}^2$ and $\hat{\sigma}_{02}^2$ are used to reestimate the weights of the two types of observations, and to carry out the adjustment calculation once more. This iterative process continues until $\hat{\sigma}_{01}^2 = \hat{\sigma}_{02}^2$ or the ratio of the unit weight variances equals 1, based on other statistic test.

Let the observations of digitized coordinates of parcel vertices be denoted as (x_i, y_i) $(i = 1, 2, ..., n)$, and the observations of registered parcel areas as $S_j (j = 1, 2, ..., m)$. It is assumed that these observations are independent and the corresponding coordinate adjustment values and corrective values are denoted as $(\hat{x}_i, \hat{y}_i)$ and (v_{x_i}, v_{y_i}); the adjustment value and corrective values of the registered area are $\hat{S}_j$ and v_{s_j}. We have that

$$\begin{aligned} \hat{x}_i &= x_i + v_{x_i} \\ \hat{y}_i &= y_i + v_{y_i} \end{aligned} \quad (i = 1, 2, ..., n), \tag{12.22}$$

and

$$\hat{S}_j = \hat{S}_j + v_{s_j} \quad (j = 1, 2, ..., m), \tag{12.23}$$

Assuming that the parcel area S_j is generated from the digitized vertices (x_i, y_i) $(i = 1, 2, t)$, conditional equation for S_j can then be constituted as

$$\frac{1}{2}\sum_{i=1}^{t} \hat{x}_i(\hat{y}_{i+1} - \hat{y}_{i-1}) - \hat{S}_j = 0 , \tag{12.24}$$

with the linearized form of this conditional equation equal to:

$$\sum_{i=0}^{t} a_i v_{x_i} + \sum_{i=1}^{t} b_i v_{y_i} - v_{s_j} + w_j = 0 , \tag{12.25}$$

where

$$a_i = \frac{1}{2}(y_{i+1} - y_{i-1}),$$

$$b_i = \frac{1}{2}(x_{i-1} - x_{i+1}),$$

$$w_j = S_0 - S_j,$$

$$S_0 = \frac{1}{2}\sum_{i=1}^{t} x_i(y_{i+1} - y_{i-1}).$$

Let the coordinate observation of a digitized vertex of a parcel be denoted as L_1, the corrective value as V; the observation of the registered area as L_2, the corrective value as V_s; and the corresponding weight matrices as P_1 and P_2. The conditional equations of the adjustment can then be expressed as:

$$AV - V_s + W = 0 \qquad (12.26)$$

leading to corresponding normal equations:

$$NK + W = 0, \qquad (12.27)$$

where K is a Lagrange multiplier, and the meaning of A and N are the same as in Formula 12.19 and Formula 12.20.

The variance components, V and V_s, are estimated and computed by using the above conditional adjustments. The weights of these two types of observations can be obtained by means of iterative computations. The final area adjustment of the parcels can be processed accordingly.

12.2.3   AREA ADJUSTMENT WITH SCALE PARAMETER

12.2.3.1   Area Conditional Equation with Scale Parameter

The observations of the digitized coordinate of a parcel vertex are denoted as (x_i, y_i) $(i = 1, 2, ..., n)$; the observations of the registered area are denoted as $S_j (j = 1, 2, ..., m)$. It is assumed that these two types of observations are independent. The corresponding coordinate vertex adjustment and corrective values are $(\hat{x}_i, \hat{y}_i)$ and (v_{x_i}, v_{y_i}), respectively, and the adjustment and corrective values of the registered area are S_j and v_{s_j}, respectively. The conditions between digitized vertex coordinates and the land parcel/ registered area gives the following conditional equation with scale parameter:

$$\frac{1}{2}\sum_{i=1}^{n} \hat{x}_i(\hat{y}_{i+1} - \hat{y}_{i-1}) - \hat{\mu} \cdot \hat{S}_j = 0 \qquad (12.28)$$

and the following conditional equation after linearization:

$$\sum_{i=0}^{n} a_i v_{xi} + \sum_{i=1}^{n} b_i v_{yi} - \mu_0 v_{s_j} - S_0 v_\mu + w = 0. \tag{12.29}$$

Here, μ is the scale parameter for the parcel area, with a user-defined approximate initial value μ_0 normally set to be 1 and,

$$a_i = \frac{1}{2}(y_{i+1} - y_{i-1}),$$

$$b_i = \frac{1}{2}(x_{i-1} - x_{i+1}),$$

$$w_j = S^0 - u_0 S_j,$$

$$S^0 = \frac{1}{2}\sum_{i=1}^{n} x_i(y_{i+1} - y_{i-1}).$$

In a region area adjustment, in which that area may include one or more street blocks, the observations of the digitized coordinate are denoted as the vector L, and that of a corresponding covariance matrix as the matrix Q; the corrective value of the observation is denoted as the vector V, the observed area values as L_S, the corresponding corrective value as V_S, and the covariance matrix as Q_S. Therefore, the conditional equations with the scale parameter are

$$AV + A_s V_s + A_\lambda \nabla\lambda + w = 0, \tag{12.30}$$

where $A_s^T = [0 \ -I]$, and the coefficients in A, A_λ, w can be obtained through relevant computations.

In practical applications, one or more scale parameters can be set according to the complexity level of the parcel area data to be processed.

12.2.3.2 The Effect of Scale Error and Its Significance Test

Adopting the conditional adjustment with parameter, the corresponding normal equation for the conditional Equation 12.30 is

$$\begin{bmatrix} AQA^T + Q_S & A_\lambda \\ A_\lambda^T & 0 \end{bmatrix} \begin{bmatrix} K \\ \nabla\lambda \end{bmatrix} + \begin{bmatrix} w \\ 0 \end{bmatrix} = 0, \tag{12.31}$$

where K is a Lagrange multiplier, which can be calculated by

$$K = -(AQA^T + Q_S)^{-1}(A_\lambda \nabla \lambda + w). \tag{12.32}$$

Therefore, the corrective value of the coordinate observations equals

$$V = QA^T K = -QA^T(AQA^T + Q_S)^{-1}(A_\lambda \nabla \lambda + w). \tag{12.33}$$

If the area scale parameter is not considered, the corrective value of the coordinate observations equals:

$$V = -Q(AQA^T + Q_S)^{-1}w. \tag{12.34}$$

The effect of the area scale to the coordinate corrective value can be estimated by

$$\nabla V = Q(AQA^T + Q_S)^{-1}(S_0 \nabla \lambda). \tag{12.35}$$

To test the significance of area scale, a hypothesis test can be applied. The null hypothesis H_0 and the alternative hypothesis H_1 is

$$\begin{aligned} H_0 &: E(\nabla \lambda) = \nabla \lambda = 0, \\ H_1 &: E(\nabla \lambda) = \nabla \lambda \neq 0. \end{aligned} \tag{12.36}$$

Using the t test, the following statistic can be constituted:

$$t = \frac{\nabla \lambda}{\hat{\sigma}_0 \sqrt{Q_\lambda}} \sim t(r), \tag{12.37}$$

where r is the number of the redundancies, Q_λ is the covariance of $\nabla \hat{\lambda}$, and

$$\hat{\sigma}_0^2 = \frac{1}{r}\Omega, \quad \text{with} \quad \Omega = V^T Q^{-1} V + V_s^T Q_s^{-1} V_s.$$

According to a given confidence level α, $t_{\alpha/2}$ is then obtained. If $|t| > t_{\alpha/2}$, H_1 is then accepted and H_0 is rejected. That is to say, $\nabla \lambda$ is significant. When r is large, F test can be applied to replace t test.

The linear hypothesis test is used to constitute the statistic

$$F = \frac{R/c}{\Omega/r} \sim F(c,r), \tag{12.38}$$

where c is the number of the scale parameter and $R = \nabla\hat{\lambda}^T Q_\lambda \nabla\hat{\lambda}$. According to a given confidence level α, $F_\alpha(c, r)$ can be calculated. If $|F| > F_\alpha(c, r)$, then H_1 is accepted and H_0 is rejected, and the scale parameter $\nabla\lambda$ is considered to be significant.

12.2.3.3   Variance Components Estimation for the Conditional Adjustment with Scale Parameter

Since the two types of observed values (registered areas and digitized vertex coordinates) are involved, and some scale parameters are unknown, the variance components estimation method for conditional adjustment with parameter is adopted to estimate the weights of these two types of observations.

If the covariance of the two given types of observations, Q and Q_s, are not given properly during the first adjustment process, this may be denoted as

$$\begin{cases} D(L) = \sigma_{01}^2 Q \\ D(L_s) = \sigma_{02}^2 Q_s \end{cases}. \tag{12.39}$$

By using $V^T Q^{-1} V$ and $V_s^T Q_s^{-1} V_s$ resulting from the first adjustment posterior and successive adjustments thereafter, the unit weight variance factors, $\hat{\sigma}_{01}^2$ and $\hat{\sigma}_{02}^2$, of the two types of observations, according to the following formula, can be reestimated:

$$\begin{pmatrix} tr(Q_{rr}N_1Q_{rr}N_1) & tr(Q_{rr}N_1Q_{rr}N_2) \\ tr(Q_{rr}N_1Q_{rr}N_2) & tr(Q_{rr}N_2Q_{rr}N_2) \end{pmatrix} \begin{bmatrix} \hat{\sigma}_{01}^2 \\ \hat{\sigma}_{02}^2 \end{bmatrix} = \begin{bmatrix} V^T Q^{-1} V \\ V_s^T Q_s^{-1} V_s \end{bmatrix}, \tag{12.40}$$

where $N = N_1 + N_2$, $N_1 = A\,Q^{-1}A^T$, $N_2 = A_\lambda Q_\lambda^{-1} A_\lambda^T$.

$$\begin{pmatrix} Q_{rr} & Q_{rt} \\ Q_{tr} & Qtt \end{pmatrix} = \begin{pmatrix} N & A_\lambda \\ A_\lambda^T & 0 \end{pmatrix}^{-1}.$$

Weights are reestimated by using the obtained more accurate $\hat{\sigma}_{01}^2$ and $\hat{\sigma}_{02}^2$. Adjustments are recalculated using the newly obtained weights. This process repeats iteratively until $\hat{\sigma}_{01}^2 = \hat{\sigma}_{02}^2$ or the ratio of the unit weight variance equals 1 through tests such as the F test.

12.3   QUALITY CONTROL METHODS FOR CADASTRAL DATA

In a cadastral map, a parcel belongs to a particular street block. A street block is usually composed of many parcels without gaps between adjacent parcels. A block is a relatively independent unit of a region. In principle, the block area is equal to the sum of the areas of its composing parcels. Blocks are usually separated by streets or other geographic features. Thus, it is pertinent to regard a single block as a basic land parcel adjustment process unit.

The parcel errors correlate within the same block. The correlations between parcels can be interpreted as constraints, which can be used as the conditions in the parcel area adjustment processing. If these correlations are not considered, the boundaries of different parcels may either overlap or lead to gaps between the parcels. In such a case, the area adjustment target cannot be achieved. Owing to the destruction of the topological relations between land parcels, it is impossible for the parcel data to meet the quality requirements.

The corresponding data quality control methods for cadastral data, with regard to various possible parcel and block relationships, are as follows:

(1) *Area adjustment for a single parcel*

This adjustment solution is suitable for the situations of independent parcels or if there is only one single parcel within a block, such as the examples shown in Figures 12.1 and 12.2. The conditional equations can be built based on Equations 12.6, 12.11, and 12.12. The area adjustment process for a land parcel can thus be carried out accordingly.

(2) *Area adjustment for parcels with "holes" or an "island"*

An "island" refers to the situation whereby a parcel is located inside another parcel. Two area conditions should be considered: one for the area of the "island" parcel and one for the parcel area containing the "island." The conditions for each of the parcels should be used for forming conditional equations. The two parcels should be adjusted simultaneously. A "hole" refers to the situation whereby part of a parcel belongs to another parcel. Similar adjustments to those made to the "island" should be made to the parcel with a "hole."

(3) *Area adjustment for multiple parcels*

In common situations, many parcels occur within one block, as shown in Figure 12.3. If the data volume of a block is within certain limits, multiple parcels within the block should be integrally and simultaneously adjusted, for example, when the total number of parcels is less than 50, and the total number of digitized vertices within a block is less than 1000. The key issue

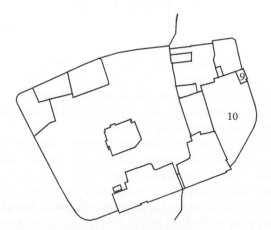

FIGURE 12.3   Several parcels are within one block.

for an integrated adjustment is to ensure that the common vertices and boundaries of correlated parcels are moved simultaneously. Therefore, the topologies between parcel polygons remain unchanged.

In the integrated adjustment algorithm for parcels, two object types are defined: *PolygonStruct* and *CombinedPolygonStruct*. The *PolygonStruct* type includes parcel number, registered parcel area, digitized parcel area, area error limit of the area adjustment, "parcels with holes" ("yes" or "no"), the total number of points contained in a parcel, and the coordinate observations of each vertex. The *CombinedPolygonStruct* typed contains all the vertices that constitute the adjustment.

According to the adjustment-enrolled points and constraints, conditional equations are constituted. Coefficients, a_i, b_i, a_{ij}, b_{ij}, a_{ik}, b_{ik}, s_{ij}, and s_{ik}, are computed according to the observations of the vertex coordinates in the parcels to which they belong. They are computed via the iterative computing process of the least square conditional adjustment. The corrective value of the digitized coordinate observations of parcel vertices and corresponding adjustment values are then obtained. The consistency between digitized parcel areas and registered areas is then warranted.

(4) *Area adjustment of multiple parcels with fixed points and parcels*

It may happen that several parcels or parcel vertices have a higher positional accuracy than the rest of the parcels or vertices. For instance, the parcels with number 9 and 10 in Figure 12.3, and all the vertices along the block boundary of the parcels have a higher positional accuracy than those of the other parcels and vertices. These fixed points and parcels should not be adjusted or moved during the area adjustment for the overall parcels in the region. The area adjustment method for the cases with fixed points or parcels should be carried out, based on the above method (3), with the following additional conditional equations:

$$\left. \begin{array}{l} v_{x_i} = 0 \\ \\ v_{y_i} = 0 \end{array} \right\} \quad (i = 1, 2, \ldots, n_0), \quad (12.41)$$

where n_0 is the total number of the fixed parcels and fixed points.

(5) *Sequential area adjustment of multiple parcels with a large data volume*

When the data volume of the parcels and blocks is large, a sequential adjustment method for multiple parcels is proposed. This method is a further development based on the solutions (3) and (4).

In a sequential adjustment of multiple parcels, the area is first divided into a series of adjustment regions (for example, blocks), and each region includes many parcels. The sequential adjustment of multiple parcels is initially applied to the region level—to adjust the vertices of the regions based on method (3). In addition, the second-level adjustment is applied to a single region by applying method (4), where the adjusted coordinates of the vertices are regarded as fixed points.

(6) *The error limit for area parcel adjustment*

The difference between the registered and digitized parcel areas is sometimes large. This may be because the digitization error is itself too large or because the value of the registered area is in doubt. For this reason, an error limit is set up for parcel area adjustments to ensure that the adjusted parcel coordinates are within an acceptable level. If the difference is below the error limit, adjustments to the parcel area are made. If not, further effort should be made to check the reason for the difference. In the following, the error limit is derived, based on the assumption that the root mean square error of the digitized coordinate is 7cm. (This example is based on previous case studies.) According to Equation 12.2, the variance of area is

$$\sigma_S^2 = AD_\zeta A^T, \tag{12.42}$$

where $A = [a_1 \ b_1 \ a_2 \ b_2 \ \dots \ a_n \ b_n]$; $\zeta = [x_1 \ y_1 \ x_2 \ y_2 \ \dots \ x_n \ y_n]^T$. σ_s is the variance in area, D_ζ is a diagonal matrix representing the variances of the digitized coordinates, and $D_\zeta = diag\{0.49, 0.49, \dots, 0.49\}_{2n \times 2n}$.

From the above equation, it can be seen that the area variance is related to the number of vertices and sides of the parcels. Therefore, $2\sigma_s$ to $3\sigma_s$ is usually chosen as the area error limit.

12.4 A CASE STUDY

The quality control methods for the cadastral data introduced below are implemented in a GIS software package with additional modules written in Visual C++ for land parcel area adjustment.

Different error modeling and quality control methods were applied to various example data sets. The operational flow of the adjustment program is illustrated in Figure 12.4. The results from the case study were then compared and analyzed. The parcel data of two typical blocks were selected as examples, as illustrated in Figure 12.5.

Table 12.1 shows the portion of results on the registered parcel area, the area from digitized vertices, and the difference between them. Relative parcel errors are also provided. It is noticed that the relative error for some parcels is rather large, such as that for parcels Number 7 and 8 of block Number 1. If they are larger than the error limit, the parcels are further checked, rather than simply included (without check) in the overall parcel adjustment process.

12.4.1 AREA ADJUSTMENT NEGLECTING ERROR OF REGISTERED PARCEL AREA

Table 12.2 shows the statistical results of the corrective values of the digitized coordinates in the x and y directions for both examples. From Table 12.2, it can be seen that the number of the vertex coordinate corrective values that have fallen into each region is different. A common point can be identified, which is that corrective values

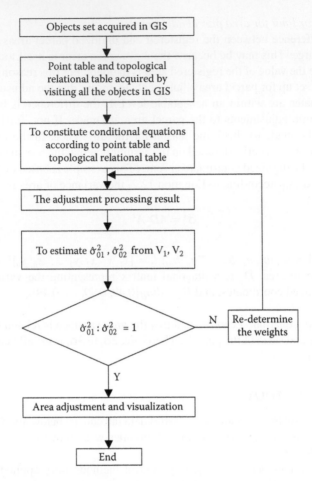

FIGURE 12.4 The operational flow of the adjustment program. (a). Example 1: parcels in block No. 1 (b). Example 2: parcels in block No. 2.

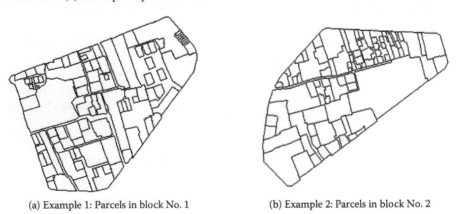

(a) Example 1: Parcels in block No. 1   (b) Example 2: Parcels in block No. 2

FIGURE 12.5 Examples of blocks with parcel to be adjusted.

TABLE 12.1

The Result of Parcel Area

No. of Samples	Data for Lock 1 (Figure 12.5a)			Data for Block 2 (Figure 12.5b)		
	Registered Area	Digitized Area	Relative Error (‰)	Registered Area	Digitized Area	Relative Error (‰)
1	1531	1522.91	5	753	750.832	3
2	265	268.373	12	482	480.409	3
3	226	224.868	5	2550	2530.927	7
4	1918	1917.72	0.1	194	192.974	5
5	229	228.49	2	82	81.977	3
6	2344	2350.48	3	471	472.252	3
7	146	143.47	17	256	258.448	10
8	321	326.391	17	250	252.667	11
t9	1746	1744.68	0.7	1069	1070.049	1
10	4523	4526.29	0.7	5430	5409.967	4

Note: Portion of the overall parcels; unit = m².

TABLE 12.2

The Statistical Results of the Corrected Values of Digitized Vertices

		15 ≤ v < 20 cm	10 ≤ v < 15 cm	5 < v < 10 cm	1 < v < 5 cm	v < 1 cm
Correction statistics in Example 1	v_x	3	7	96	297	259
	v_y	3	21	83	273	282
Correction statistics in Example 2	v_x	8	22	102	248	218
	v_y	8	20	68	250	252

Note: Unit = number of points.

in the range 15 cm and 20 cm are the fewest in terms of point number, those in the range 5 cm and 15 cm are relatively less, and corrective values of less than 5 cm are the greatest in terms of point number.

12.4.2 ESTIMATION OF VARIANCE COMPONENTS WITH RECTANGULAR CONSTRAINTS

Both registered areas and vertex coordinates in Examples 1 and 2 are now considered as observations with error. Variance component estimation with rectangular constraints are applied to constitute the conditional equations for the area adjustment.

TABLE 12.3
The Statistical Results of the Corrective Values

No. of Iteration		$10 \leq v < 15$ cm	$5 < v < 10$ cm	$1 < v < 5$ cm	$v < 1$ cm
Statistics after 1st iteration	v_x	3	19	254	386
(Example 1)	v_y	7	23	195	437
Statistics after 5th iteration	v_x	4	9	147	502
(Example 1)	v_y	5	17	123	517
Statistics after 1st iteration	v_x	11	44	240	303
(Example 2)	v_y	7	46	197	348
Statistics after 5th iteration	v_x	7	31	180	380
(Example 2)	v_y	4	34	155	405

Note: With registered areas taken as observations; unit = number of points.

Table 12.3 lists the statistical results of the adjustment as the corrective values of the area. The corrective values below 1 cm refer to a very large proportion, whereas values above 10 cm refer to a smaller proportion. As compared with the results in Table 12.2, the distribution of the corrective values, such as the change in the corrective value, can also be noticed. The main impression is that the number of corrective values in the range of 0–5 cm has increased, and the number of corrective value within the range of 5–15 cm has decreased, while the number in the range larger than 15 cm is 0. These results indicate that the overall adjustment accuracy has been improved.

Table 12.4 shows a portion of the area corrective values and errors of examples 1 and 2. It is noticed that for several iterations, the area corrective values increase, whereas the errors within the area observations decrease compared with the results of the fifth, third, and first iteration.

12.4.3 AREA ADJUSTMENT WITH SCALE PARAMETER AND RECTANGULAR CONSTRAINTS

Table 12.5 lists the results of the area adjustment calculations with the area scale parameter and rectangular constraints. The corrective values of digitized vertices and errors are further reduced compared with those in Table 12.3. The area scale parameters, in fact, correct the systematic errors in the land parcel data. The overall error reduces after the systematic errors are corrected. The systematic error due to scale effect should not be neglected in this example. It can be seen from the distribution of the statistical results in Table 12.5 that the majority of coordinate corrective values is below 1 cm and no point is located within the range above 10 cm. This means that, after using the scale factor in the area adjustment, the results improve.

Table 12.6 shows the iterative calculation results of area scale parameters in example 1. It is noticed that the errors of the area scale parameters are around 0.005. This indicates that systematic scale errors are introduced during parcel digitization. This may be due to transformations between different coordinate systems. Scale factors should not be neglected in the presence of a large system error on scale.

TABLE 12.4

Area Corrective Values and Errors of Examples 1 and 2

Example	No. of Area	Corrective Value and Error after 1st Iteration		Corrective Value and Error after 3rd Iteration		Corrective Value and Error after 5th Iteration	
		v_s	m_s	v_s	m_s	v_s	m_s
1	1	0.443	0.796	0.566	0.627	0.579	0.590
	2	0.652	1.439	0.993	1.097	1.030	1.028
	3	−0.018	0.575	−0.031	0.431	−0.033	0.404
	4	0.341	2.941	1.478	2.243	1.628	2.102
	5	1.474	1.514	2.351	1.186	2.450	1.115
	6	−0.619	1.369	−0.956	1.063	−0.995	0.997
	7	5.401	4.125	8.395	3.203	8.739	3.005
	8	−4.931	2.929	−6.425	2.295	−6.573	2.157
	9	−0.948	1.001	−1.274	0.782	−1.308	0.735
	10	−0.864	0.769	−1.158	0.593	−1.187	0.556
2	1	−0.999	3.191	−1.526	2.151	−1.530	2.143
	2	−0.823	2.244	−1.207	1.489	−1.211	1.483
	3	−11.584	5.482	−15.732	3.664	−15.759	3.650
	4	−0.624	1.630	−0.655	1.079	−0.654	1.075
	5	−0.375	1.181	−0.281	0.864	−0.281	0.862
	6	0.476	2.675	0.783	1.801	0.785	1.794
	7	1.379	1.822	1.966	1.226	1.970	1.221
	8	1.101	1.796	1.592	1.181	1.595	1.176
	9	0.396	2.773	0.675	1.850	0.676	1.843
	10	−12.866	5.913	−17.468	3.997	−17.490	3.982

Note: With registered areas as observations; unit = meter.

TABLE 12.5

The Statistical Results of the Corrective Values (with Scale Parameters; Unit: Number of Points)

No. of Iteration		$v < 10$ cm	$5 < v < 10$ cm	$1 < v < 5$ cm	$v < 1$ cm
Statistics after 1st iteration	v_x	0	8	211	443
(Example 1)	v_y	0	11	180	471
Statistics after 5th iteration	v_x	0	2	30	630
(Example 1)	v_y	0	0	34	628
Statistics after 1st iteration	v_x	2	14	218	364
(Example 2)	v_y	2	22	189	385
Statistics after 4th iteration	v_x	0	5	50	543
(Example 2)	v_y	0	1	48	549

TABLE 12.6
Results of Scale Parameters Computation

No. of Iteration	Scale Parameters $\hat{\lambda}$	Variance of Scale Parameter $\sigma_{\hat{\lambda}}^2$
1	0.999174	3.532e-007
2	0.998322	2.255e-007
3	0.997470	2.306e-007
4	0.996619	2.306e-007
5	0.995767	2.306e-007

TABLE 12.7
Area Corrective Values and Errors of Examples 1 and 2

Example	No. of Area	Corrective Value and Error after 1st Iteration		Corrective Value and Error after 3rd Iteration		Corrective Value and Error after 5th Iteration	
		v_s	m_s	v_s	m_s	v_s	m_s
1	1	0.534	1.048	0.730	0.322	0.730	0.320
	2	0.866	1.899	1.440	0.566	1.441	0.564
	3	0.006	0.755	−0.003	0.212	−0.003	0.211
	4	2.484	4.142	5.885	1.558	5.889	1.553
	5	1.824	2.001	3.207	0.621	3.208	0.619
	6	−0.473	1.802	−0.886	0.539	−0.887	0.536
	7	6.249	5.457	11.931	1.886	11.937	1.880
	8	−4.010	3.933	−6.552	1.366	−6.554	1.362
	9	−0.838	1.316	−1.333	0.395	−1.334	0.394
	10	−0.782	1.012	−1.235	0.298	−1.236	0.297
2	1	−0.114	3.261	−1.199	1.345	−1.199	1.344
	2	−0.249	2.309	−0.989	0.925	−0.989	0.924
	3	−9.295	5.712	−15.920	2.601	−15.921	2.600
	4	−0.656	1.660	−0.835	0.644	−0.835	0.644
	5	−0.317	1.194	−0.127	0.545	−0.127	0.545
	6	0.834	2.709	1.391	1.092	1.391	1.092
	7	1.743	1.859	2.778	0.744	2.778	0.744
	8	1.548	1.886	2.602	0.738	2.602	0.738
	9	1.451	2.849	2.118	1.236	2.118	1.235
	10	−8.810	6.597	−12.845	3.837	−12.846	3.836

Table 12.7 lists the corrective values and corresponding errors of a portion of the area observations of examples 1 and 2, when scale parameters have been considered in the adjustment. The area corrective values are larger than those in Table 12.43, and the area errors are reduced. This means that, after the addition of area scale parameter to the area adjustment process, the accuracy of the adjusted areas is higher than it would be without considering the scale parameter in the adjustment.

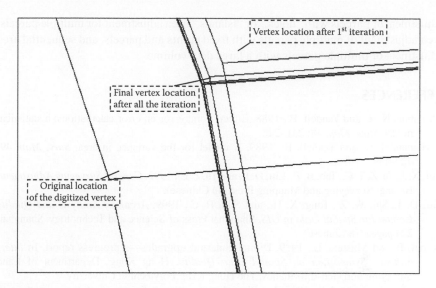

FIGURE 12.6 The position movement of a vertex in a multiple iteration of the adjustment computation.

Figure 12.6 shows the convergence situation of a parcel vertex during an iterative computation. This figure is enlarged for better visualization; the distance between the first and final iterations of the vertex looks large, but the actual distance between them is only about 2 cm.

12.5 SUMMARY

Inconsistency arising from multidata sources is a critical issue affecting the applications of GIS. A data quality control approach for solving the data inconsistency in object-based spatial data, based on the least square adjustment theory, has been proposed in the above chapter. Inconsistencies in cadastral data in object-based spatial data are taken as an example. Inconsistencies of land parcel area data were found to be due to the two different parcel area data sources of: (a) the registered area and (b) the area value calculated from the digitized vertex's coordinates.

A systematic approach to address land parcel area adjustments covering three error processing methods has been introduced. The first method lets the registered area of the parcel be the true value and the digitized vertex's coordinates be observations with random error for the establishment of the adjustment conditional equations. The second method lets both the registered area and the digitized vertex's coordinates be observations with a random error. The weights of these two observation types are estimated using a variance components method. The third method adds the scale parameter of the parcel to the adjustment process for handling systematic error due to scale effect. The weights of the observations are estimated based on variance components from conditional adjustments with parameters.

A quality control strategy for cadastral data has been provided for systematically handling the following five different cases: area adjustment for a single parcel, area adjustment of parcels with "holes," or "islands," area adjustment for multiple parcels, area adjustment of multiple parcels with fixed points and parcels, and sequential area adjustment of multiple parcels with a large data volume.

REFERENCES

Chrisman, N. R. and Yandell, B., 1988. Effects point error on error calculations: a statistical model. *Surv. Map.* 48: 241–246.

Chrisman, N. R. and Yandell, B., 1989. A model for the variance in area. *Surv. Map.* 49: 427–439.

Cui, X. J., Yu, Z. T. C., Tao, B. Z., Liu, D. J., and Yu, Z. L., 1992. *General Surveying Adjustment.* Beijing: Surveying and Mapping Press (in Chinese).

Liu, D. J., Shi, W. Z., Tong, X. H., and Sun, H. C., 1999. *Accuracy Analysis and Quality Control for Spatial Data in GIS.* Shanghai Press of Science and Technology. Shanghai, 224 pages (in Chinese).

Merrit, R. and Masters, E., 1999. Digital cadastral upgrades—a progress report. In *International Symposium on Spatial Data Quality.* Hong Kong: Department of Land Surveying and Geo-Informatics, The Hong Kong Polytechnic University.

Najeh, T. and Burkhard, S., 1995. A methodology to create a digital cadastral overlay through upgrading digitized cadastral data. *Surv. Land Inform. Syst.* 55(1): 3–12.

Shi, W. Z. and Tong, X. H., 2005. Improve accuracy of area objects in GIS based on the Helmert variance components estimation method (unpublished manuscript).

Tong, X. H., Shi, W. Z., and Liu, D. J., 2005, A least squares-based method for adjusting the boundaries of area objects. *Photogram. Eng. Rem. Sens.*, 71(2): 189–195.

13 Quality Control for Field-Based Spatial Data

Quality control for field-based spatial data is addressed in this chapter as follows. Satellite images are taken as examples of field-based spatial data, with quality control realized through geometric rectification of those images. The quality of field-based spatial data (satellite images) is controlled by firstly adopting point-based geometric rectification methods according to the point-based transformation models. Secondly, the line-based transformation model is proposed. Thirdly, both point-based and line-based methods are applied to both satellite image rectifications and spatial modeling.

13.1 INTRODUCTION

Satellites can provide remotely sensed images of the Earth's surface. With the development of space technologies, higher spatial, spectral, and temporal resolution satellite images have become available. Errors, however, still exist in satellite images due to the effect of terrain variation, the status of satellite sensors during imaging, and other factors. Therefore, a need exists in practical applications for the correction of the obtained satellite images to provide images that are more geometrically correct.

Two categories of image rectification methods have been distinguished: (a) rigorous physical models and (b) nonrigorous empirical models. A physical model reflects the physical reality of satellite imaging geometry, and may also include the relationship defined in map projection. An empirical model builds a relation between satellite image space and object space by using a transformation or polynomial model.

A physical model can reflect the real conditions of the sensor, the platform, and its image acquisition geometry. These are used for the geometric rectification of the satellite images. The collinear equation model for example is a physical model for frame camera-based images, describing the projection relationship between two-dimensional image space and three-dimensional object space.

In the absence of the sensor model, and precise ephemeris data and satellite orbit information, an empirical model can be applied as an alternative. It does not require satellite platform and sensor information.

In point-based image rectifications, both physical and empirical models have been developed for line scanner images, such as for the SPOT, MOMS-02, and IRS-1C/D satellites. El-Manadili and Novak (1996) used direct linear transformation (DLT) for the geometric rectification of SPOT imagery. Okamoto et al. (1998) proposed affine transformation models for SPOT image rectification. Gugan (1986) suggested an orbital parameter model based on the collinear equations and expanded by two orbital parameters to model the satellite movements along the path and the Earth's rotation. Tao and Hu (2001) applied rational function models for high resolution satellite image rectification and modeling. Fraser et al. (2002) proved the metric integrity of the IKONOS imaging system is comparable to that of medium resolution push-broom satellite sensors based on experimental studies.

Satellite images, from both frame and linear array scanners, can also be rectified on the basis of linear features. Related research has been reported, such as Mulawa and Mikhail (1988); Kanok (1995); Tommaselli and Tozzi (1996); Dare and Dowman (2001); Habib et al. (2003). Two-dimensional and three-dimensional transformation models, which can be applied to satellite image rectification, are introduced below.

13.2   TWO-DIMENSIONAL TRANSFORMATION MODELS

Two-dimensional transformation models are mainly used to build a geometric relationship between two-dimensional image space and two-dimensional object space. Examples of two-dimensional models are a generic two-dimensional polynomial transformation model, a conformal transformation model, and a projective transformation model.

13.2.1   THE GENERIC TWO-DIMENSIONAL POLYNOMIAL TRANSFORMATION MODEL

The generic form of two-dimensional polynomial model is given as:

$$x = a_0 + a_1 X + a_2 Y + a_3 X^2 + a_4 Y^2 + a_5 XY + \ldots + a_k X^n, \tag{13.1}$$

$$y = b_0 + b_1 X + b_2 Y + b_3 X^2 + b_4 Y^2 + b_5 XY + \ldots + b_k X^n, \tag{13.2}$$

where (a_i, b_i) $(i = 1, 2, \ldots, k)$ are model coefficients, n is the order of the polynomial, (x, y) is the pair of coordinates in the original image space, and (X, Y) is the pair of coordinates in object space (or in the rectified image space). For an image-to-image rectification process, (x, y) is the coordinate pair in the original image space before rectification and (X, Y) are those in the rectified image space.

The above polynomial model can be summarized as:

$$x = \left(\sum_{i=0}^{n} \right) \left(\sum_{j=0}^{i} \right) a_k \times X^{i-j} \times Y^j, \tag{13.3}$$

$$y = \left(\sum_{i=0}^{n} \right) \left(\sum_{j=0}^{i} \right) b_k \times X^{i-j} \times Y^j, \tag{13.4}$$

where the subscript is defined as

$$k = \frac{i(i+1)}{2} + j,$$

and n is the order of the polynomial, with $n = 1, 2, 3, ..., m$.

13.2.2 THE CONFORMAL TRANSFORMATION MODEL

An example of a polynomial model is the four-parameter similarity transformation model: the conformal transformation model. An object shape can be retained after a conformal transformation, which includes scaling, rotation, and translation. The conformal transformation model can be used to model relationships between two two-dimensional planes, such as two two-dimensional images according to the conformal transformation principle.

The conformal transformation model can be written as

$$x = S \cdot \cos\theta \cdot X - S \cdot \sin\theta \cdot Y + T_x, \tag{13.5}$$

$$y = S \cdot \sin\theta \cdot X + S \cdot \cos\theta \cdot Y + T_y, \tag{13.6}$$

where S is the scale factor, θ is a rotation angle, and T_x and T_y are translations in X and Y directions. Let $S \cdot \cos\theta = a$, $S \cdot \sin\theta = b$, $T_x = c$, $T_y = d$. Then, the Equations 13.5 and 13.6 can be written as

$$x = aX - bY + c, \tag{13.7}$$

$$y = bX + aY + d. \tag{13.8}$$

Least square adjustment is usually applied to solve the coefficients in Equations 13.7 and 13.8, and the polynomial equations thus obtained are then applied for an image rectification. The conformal transformation methods, for obtaining the coordinates of the points in the object space, include the following three steps: (a) the constitution of the equations by substituting the coordinates of ground control points in the image and the object space, (b) the solution of the coefficients by least square adjustment, and (c) the application of the conformal transformation model for each point of the image space.

13.2.3 THE TWO-DIMENSIONAL AFFINE TRANSFORMATION MODEL

A special case of the two-dimensional polynomial transformation is the two-dimensional affine transformation model, which is a first-order polynomial transformation involving only six coefficients:

$$x = a_0 + a_1 X + a_2 Y, \tag{13.9}$$

$$y = b_0 + b_1 X + b_2 Y. \tag{13.10}$$

13.2.4 THE PROJECTIVE TRANSFORMATION MODEL

The eight-parameter projective transformation model is defined by a ratio between two first-order polynomials, as follows:

$$x = (a_1 X + a_2 Y + a_3) / (a_4 X + a_5 Y + 1), \tag{13.11}$$

$$y = (a_6 X + a_7 Y + a_8) / (a_4 X + a_5 Y + 1). \tag{13.12}$$

The eight-parameter projective model can be used to model the relationship between two two-dimensional surfaces. In this application of image rectification, the two surfaces can be (a) the original image plane and (b) the projected image surface on the ground. The two-dimensional affine transformation model and the projective transformation model can be solved by a similar method used to solve the confirmation transformation model.

13.3 THREE-DIMENSIONAL IMAGE TO OBJECT SPACE TRANSFORMATION MODELS

Fully rigorous mathematical models for building the relationship between three-dimensional image space and object space has been developed on the basis of sensor calibration, satellite orbit, and satellite attitude information. In the absence of the necessary camera model and precise ephemeris data, however, sensor independent nonrigorous models provide alternative solutions. A three-dimensional transformation model aims to build a relationship between the two-dimensional image space and the three-dimensional object space on the ground. The nonrigorous models: the generic three-dimensional polynomial model, the three-dimensional affine transformation model, and the rational functions model, are introduced as follows.

13.3.1 GENERIC THREE-DIMENSIONAL POLYNOMIAL MODEL

A three-dimensional polynomial transformation model defines the transformation relationship between a point of a coordinate pair (x, y) in a two-dimensional space and corresponding point of coordinates (X, Y, Z) in a three-dimensional space. The following is the generic form of a three-dimensional polynomial transformation model:

$$x = \sum_{i=0}^{m} \sum_{j=0}^{n} \sum_{k=0}^{p} a_{ijk} X^i Y^j Z^k, \tag{13.13}$$

$$y = \sum_{i=0}^{m} \sum_{j=0}^{n} \sum_{k=0}^{p} b_{ijk} X^i Y^j Z^k. \tag{13.14}$$

The three-dimensional polynomial model serves as a basic model for other related transformation models, such as the eight-parameter affine model, rational function model, and others.

In three-dimensional modeling of satellite images, the three-dimensional polynomial transformation model can be used to establish a relationship between ground-level two-dimensional image space and three-dimensional object space. For building a three-dimensional model, a stereo pair of satellite images is used and the least square adjustment method is used to resolve the parameters of external elements. The coordinates of ground points can be computed by a spatial intersection.

13.3.2 Affine Transformation Model

A commonly used three-dimensional polynomial transformation model is the affine transformation model, which is

$$x = a_1X + a_2Y + a_3Z + a_4,\qquad(13.15)$$

$$y = b_1X + b_2Y + b_3Z + b_4.\qquad(13.16)$$

This model defines a relationship between a two-dimensional space and a three-dimensional space based on affine transformation. The application of this model for satellite image rectification and modeling is a transformation from the two-dimensional image space to the three-dimensional object space on the ground.

Least square adjustment can be applied to solve the coefficients. The relationship between point coordinates on stereo images and the corresponding point on the ground establishes the conditional equations. Each ground control point (GCP) forms two affine conditional equations. Space resection algorithms for generating stereo models and space intersection for computing ground coordinates are used for the corresponding calculations.

The affine transformation model is a robust method for image orientation and triangulation, both for line-scanner satellite images, such as for SPOT and others (Fraser et al., 2002). Compared with the rational function model below and other rigorous models, it has the advantage of not requiring detailed information about the satellite sensor model or the satellite ephemeris data during imaging.

13.3.3 Rational Function Model

The rational function model (RFM), also known as the rational polynomial camera model (RPC), has been developed to replace the physical sensor model. The RFM consists of a ratio of two polynomial functions. The RFM defines a relationship between three-dimensional ground coordinates and the corresponding image coordinates as in the following:

$$Row_n = \frac{P_1(X_n, Y_n, Z_n)}{Q_1(X_n, Y_n, Z_n)}.\qquad(13.17)$$

$$Column_n = \frac{P_2(X_n, Y_n, Z_n)}{Q_2(X_n, Y_n, Z_n)},$$

(13.18)

where Row_n and $Column_n$ are the normalized row and column pixel indices in the image space and X_n, Y_n, Z_n are the normalized coordinate point values in ground object space. P and Q are two polynomial functions.

Currently, Space Imaging and Digital Globe support their Ikonos and QuickBird images by the rational function model. Rational function models are also implemented in various remote sensing and digital photogrammetry systems.

A necessary addition to the point-based transformation models for either two-dimensional or three-dimensional transformations introduced above are the line-based transformation models. Such models require less ground control points for the transformation and are introduced and described below.

13.4   LINE-BASED TRANSFORMATION MODEL

The line-based transformation model (LBTM) has been developed (Shaker and Shi, 2003) based on the relationship between the *unit vector* components of a line in the image space and the unit vector components of the conjugate line in the object space. The conjugate relationship between *points* in image space and object space in the early models (such as affine transformation model) are replaced by a conjugate relationship between unit vectors in the image and object spaces.

13.4.1   RELATIONSHIP DEFINED BY UNIT VECTOR

In the image space, two points $p_1(x_1, y_1)$ and $p_2(x_2, y_2)$ form a vector $\bar{v}$ (see Figure 13.1). The unit vector is defined as

$$\bar{v} = \begin{bmatrix} v_x \\ v_y \\ 0 \end{bmatrix},$$

(13.19)

where

$$v_x = \frac{x_2 - x_1}{\sqrt{(x_2 - x_1)^2 + (y_2 - y_1)^2}},$$

(13.20)

$$v_y = \frac{y_2 - y_1}{\sqrt{(x_2 - x_1)^2 + (y_2 - y_1)^2}}.$$

(13.21)

If $P_1(X_1, Y_1, Z_1)$ and $P_2(X_2, Y_2, Z_2)$ are two points located on the conjugate line segment in the object space, the unit vector $\bar{V}$ in the object space is then

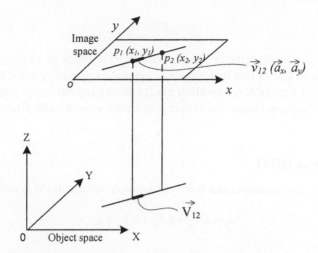

FIGURE.13.1 Relationship between a unit line vector in the image and object spaces. (From Shaker, A., 2004. The Line Based Transformation Model for High Resolution Satellite Imagery Rectification and Terrain Modeling. Ph.D. thesis, Department of Land Surveying and Geo-Informatics, Hong Kong Polytechnic University. With permission.)

$$\vec{V} = \begin{bmatrix} V_X \\ V_Y \\ V_Z \end{bmatrix} \tag{13.22}$$

where

$$V_X = \frac{(X_2 - X_1)}{\sqrt{(X_2 - X_1)^2 + (Y_2 - Y_1)^2 + (Z_2 - Z_1)^2}} \tag{13.23}$$

$$V_Y = \frac{(Y_2 - Y_1)}{\sqrt{(X_2 - X_1)^2 + (Y_2 - Y_1)^2 + (Z_2 - Z_1)^2}} \tag{13.24}$$

$$V_Z = \frac{(Z_2 - Z_1)}{\sqrt{(X_2 - X_1)^2 + (Y_2 - Y_1)^2 + (Z_2 - Z_1)^2}} \tag{13.25}$$

Note that the line $P_1 P_2$ is the conjugate line of $p_1 p_2$. However, P_1 in the object space is not the conjugate point of p_1 in the image space, and P_2 is not the conjugate point of p_2 (Figure 13.1).

The relationship between a unit vector in the object space and its conjugate unit vector in the image space can be defined by the following transformation:

$$\vec{v} = \lambda R \vec{V} + T, \tag{13.26}$$

where $\vec{v}$ is a vector of a line in image space; $\vec{V}$ is a vector of a conjugate line in the object space; λ is a scale factor (a diagonal matrix providing scale transformation in X, Y, and Z axis); R is a rotation matrix describing a rotation sequence in X, Y, and Z axis between image and object coordinate systems; and T is a translation vector.

13.4.2 Affine LBTM

Based on the above principle, the three-dimensional affine LBTM is defined as

$$v_x = a_1 V_X + a_2 V_Y + a_3 V_Z + a_4, \tag{13.27}$$

$$v_y = b_1 V_X + b_2 V_Y + b_3 V_Z + b_4, \tag{13.28}$$

where the elements of the unit vector $(v_x, v_y)^T$ in image space are given by Equations 13.20 and 13.21, with the elements of the unit vector (V_X, V_Y, V_Z) in the object space defined by Equations 13.23, 13.24, and 13.25. By comparing Equations 13.27 and 13.28 with 13.15 and 13.16, it can be seen that the elements of the unit vectors $(V_X, V_Y, V_Z, v_x, v_y)$ in the three-dimensional affine LBTM are used to replace the point coordinates (X, Y, X, x, y) in the three-dimensional affine transformation model, as described by Equations 13.15 and 13.16.

Similarly, the two-dimensional affine LBTM can be defined as

$$v_x = a_1 V_X + a_2 V_Y + a_3, \tag{13.29}$$

$$v_y = b_1 V_X + b_2 V_Y + b_3, \tag{13.30}$$

and the two-dimensional conformal LBTM as

$$v_x = a_1 V_X + a_2 V_Y + a_3, \tag{13.31}$$

$$v_y = a_2 V_X - a_1 V_Y + a_4. \tag{13.32}$$

Similar to the collinear equation, the two-dimensional affine LBTM can be used for spatial resection and spatial intersection for solving the ground coordinates. In addition, the two-dimensional affine LBTM and two-dimensional conformal LBTM can be used for image rectification or image-to-image matching.

In order to simplify the relationship between image and object coordinate systems, the following three assumptions were made in the development of the LBTM: (a) the satellite sensor moves linearly and stably in space; (b) the sensor orientation angles are constant; and (c) the satellite flight path is almost straight.

13.5   EXPERIMENTAL STUDIES ON POINT-BASED MODELS

An experimental study is described below. The objectives of this study are to test the performance of the two-dimensional transformation models for rectifying high-resolution satellite images with and without projecting GCPs onto a compensation plane. In addition, the performance of the three-dimensional transformation model for determining three-dimensional ground coordinates from stereoscopic images, regarding the terrain variation and the number of GCPs, is tested.

13.5.1   DATA SET

Two IKONOS images, covering the central part of Hong Kong, were used in this study. Image 1 (Figure 13.2a) covered an area of 11.60 km × 10.28 km over a portion of Hong Kong Island and the Kowloon district, and Image 2 (Figure 13.2b) covered an area of 6.62 km × 10.18 km of the same area. There was an overlapping area (approximately 2.5 km × 10 km) between Image 1 and Image 2. These two images thus form a stereoscopic image pair with the base-to-height ratio of the stereo images equaling 0.87. The maximum ground height difference in the test area was approximately 450 meter.

There were 38 ground control points (GCPs) identified for Image 1, and 18 for Image 2. The GCPs were measured by differential GPS techniques with two Trimble 4000 SSI systems. The accuracy of these measured points is of the order of 5 cm in X and Y directions and 10 cm in the Z direction, respectively. Several control points were landmarks and others were intersections between sidewalks and roads.

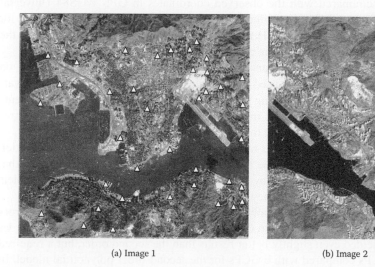

(a) Image 1   (b) Image 2

FIGURE 13.2   (See color figure following page 38.) Test images with ground control points.

13.5.2   Experiments Using the Two-Dimensional Transformation Models

This experimental study aims to test the performance of the two-dimensional transformation models in rectifying high-resolution satellite images, taking the two-dimensional polynomials and projective models as the examples.

Two solutions were tested: (a) use of the compensation plane (case 1) and (b) not using the compensation plane (case 2). The details of the experimental study are given in Shaker and Shi (2003).

The transformation to a compensation plane (or plane of control) is actually a transformation in easting coordinates (ΔX) and northing coordinates (ΔY) based on the following formulae:

$$\Delta X = \Delta Z \sin \alpha / \tan \varepsilon, \tag{13.33}$$

$$\Delta Y = -\Delta Z \cos \alpha / \tan \varepsilon, \tag{13.34}$$

where (α) is the azimuth angle, (ε) is the elevation angle of the satellite, and ΔZ is the height difference, with respect to the compensation plane. An elevation of 200 m (the average control point elevation) was adopted for the compensation plane, considering the highest elevation difference of 450 m in the area.

According to Space Imaging technical specifications for an IKONOS Geo product, its planimetric accuracy is 50 m (circular error) and 23.6 m associated root-mean-square error (RMSE) (Space Imaging, 2001), without taking the terrain displacement effects into the account.

To establish a baseline on the inherent accuracy of the IKONOS image for the Hong Kong test field, the geo-referenced coordinates of the control points were digitized and compared with the observed coordinates in GPS WGS84 Universal Transverse Mercator (UTM) coordinate system. The absolute planimetric errors for all points were between 1 m and 111 m in the Y direction and from 3 m to 32 m in the X direction, depending on the point elevations.

The experiments on two-dimensional transformation models were applied on Image 1. The number of GCPs used to compute the transformation parameters varied from 6 to 18 with the remaining points used as checkpoints. Table 13.1 presents the resulting RMSE discrepancies from the three transformation models for the two cases: with and without projecting the compensation plane.

In Case 1, the total RMS errors ranged from 5.83 m to 8.34 m in the X direction and from 13.47 m to 38.27 m in the Y direction. From these results, it is noticed that the accuracy of the rectification by these two-dimensional transformation models cannot reach an acceptable accuracy level for mapping.

In Case 2, the second-order polynomial produced the result with RMS errors of 0.46 m to 0.29 m and 0.49 m to 0.46 m in X and Y directions. The fourth-order polynomial yielded even slightly better results than the second order, but it required at least 16 GCPs compared with 6 GCPs for the second-order polynomial model. In general, the rectified accuracy from the two-dimensional transformation model can reach an acceptable level of accuracy: the RMS error discrepancy is less than one pixel in both the X and Y directions.

TABLE 13.1

RMS Error Values at the Checkpoints via Two-Dimensional Rectification Models

Model Name	Number of GCPs	Number of Check Points	Case 1 (Before Projecting Coordinates to a Compensation Plane) RMS Errors (m)		Case 2 (After Projecting Coordinates to a Compensation Plane) RMS Errors (m)	
			X	Y	X	Y
Second-order polynomial	6–18	32–20	6.77–5.83	34.98–29.85	0.46–0.29	0.49–0.46
Fourth-order polynomial	16–18	22–20	7.14–6.79	38.27–35.39	0.38–0.34	0.44–0.43
Projective model	6–18	32–20	8.34–8.21	34.88–13.47	0.82–0.59	0.61–0.49

From the above results, it can be seen that no significant improvements in terms of the total RMS errors were achieved when the number of the GCPs was increased from 6 to 18. When the number of GCPs reaches a certain level (such as 5 GCPs), the quality rather than quantity of GCPs is more important for building two-dimensional image rectification transformation models. Furthermore, the second-order polynomial produces a better image rectification than the projective model, when control coordinates are projected onto a compensation plane.

13.5.3 EXPERIMENTS ON THE THREE-DIMENSIONAL TRANSFORMATION MODEL

The three-dimensional affine transformation model was taken as an example for testing the performance of the three-dimensional transformation model to determine ground coordinates where spatial resection and spatial intersection were performed, using a least-squares adjustment.

There were 18 GCPs identified on the stereo model, as shown in Figure 13.3. Multiple sets of 4, 6, 8, 10, and 12 well-distributed GCPs were tested for the resection computation. Other GCPs were used as checkpoints. The RMS errors of the three-dimensional checkpoints are presented in Figure 13.4.

The total RMS errors in the X, Y, and Z directions decrease considerably with an increase in the number of GCPs. The error in the Z direction is about twice as high as errors in the X, Y directions.

If 4 well-distributed GCPs were used to generate the affine transformation model and 14 GCPs were used as checkpoints, RMS errors would be obtained equal to 1.38 m, 1.98 m, and 3.20 m in the X, Y, and Z directions, respectively. RMS error results improved significantly to 0.58 m and 0.63 m in the X, Y directions and 0.98 m in the Z direction when 12 GCPs were used. An almost inverse linear relationship apparently, exists between the RMS errors and the number of ground control points.

FIGURE 13.3   (See color figure following page 38.) The overlapped area of IKONOS image with GCPs for building stereo model.

The experimental results demonstrate that two-dimensional rectification accuracy can be improved significantly by projecting the GCP coordinates to a compensation plane. The second-order polynomial model presents the best two-dimensional transformation model results. The model can reach an accuracy of 1 m for Ikonos images.

The affine model results show that an accuracy of 1 m for a three-dimensional ground point is achievable for stereo images and without any further need for satellite sensor model and ephemeris data information. Furthermore, increasing the number of GCPs significantly improves the accuracy of the results when the affine model is applied to an area with different terrain types.

13.6   AN EXPERIMENTAL ON LINE-BASED TRANSFORMATION MODEL

The proposed line-based transformation model is widely applicable for image rectification, spatial triangulation and resection, and image-to-image matching. In an

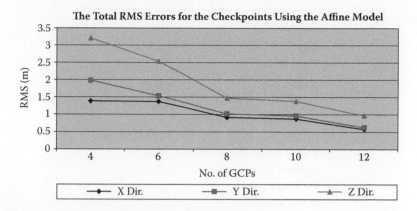

FIGURE 13.4 (See color figure following page 38.) Shows the accuracy of the checkpoints from the three-dimensional affine transformation model. (From Shaker, A., and Shi, W. Z, 2003. *Proceedings of Asian Conference on Remote Sensing* [ACRS 2003] and International Symposium on Remote Sensing, Korea. On CD. With permission.)

experimental study now described, the line-based transformation model (LBTM) for image-to-image registration using line features as control elements (Shi and Shaker, 2006) is tested. In contrast to the point-based transformation models, line segments on linear features can be used to recover the model transformation coefficients. Based on the recovered coefficients, the whole image is rectified or modeled, using the related transformation models.

13.6.1 DATA SETS

Four data sets include three IKONOS images for Hong Kong (China), Melbourne (Australia), and Zagazeg (Egypt), and one Quickbird image for the province of Fayoum in Egypt). A set of checkpoints for verifying the LBTM has been obtained for each data set, using GPS. GPS coordinates were provided for the WGS84 Datum and UTM projection. The accuracy of the point observed by GPS is less than 5 cm in both the X and Y directions. The accuracy of the points on the image is about half the size of a pixel. Details of the characteristics of the Hong Kong data set are presented in Table 13.2.

Each of the four data sets consists of a master image and reference image. The master image is the original image from the vendors, whereas the reference image is the image from a polynomial transformation model, after removing the relief displacement, if any, by projecting the ground coordinates to a compensation plane. In all cases, the accuracy level of the rectified images was approximately equal to one pixel.

Several line segments representing linear features have been digitized on the master and the reference images and are used as ground control lines (GCLs) for the recovery of the coefficients of the LBTM. Figure 13.5 shows the distribution of the line segments of the GCLs and the checkpoints for the Hong Kong date sets.

TABLE 13.2
Characteristics of the Hong Kong
Data Set

Specification	Hong Kong
Satellite	Ikonos
Mode (off nadir)	Pan
Coverage Area (km)	11.64×10.28
Ground Sample Distance (GSD)	0.91
Elevation Range (m)	500
Coordinate System:	
Master Image	
Reference Image	
WGS 84 UTM	
WGS 84 UTM	
Number of GCLs	54
Number of Checkpoints	38

13.6.2 RESULTS AND ANALYSIS

The two-dimensional affine and the conformal LBTM for a test of the model for image registration were applied. The following operational procedure was applied for image registration: (a) Digitize the GCLs in the master and the reference images, (b) calculate the unit vector components of the GCLs, (c) determine the coefficients of the LBTM using the GCLs, (d) correct the calculated translation coefficients by the aid of one point, and finally (e) substitute the coefficients calculated from the LBTM forms to the corresponding point-based forms to register the whole image. Tests have been conducted using the real data sets, and the results are summarized and presented in Figure 13.6 in terms of RMS errors of the independent checkpoints in the X and Y directions.

It is noticed from the results that the LBTM performed well in all experiments. The results obtained using the two-dimensional conformal LBTM are more consistent than those resulting from the application of the two-dimensional affine LBTM. The checking points RMS errors for the image rectified by two-dimensional conformed LBTM are consistently around 1 m.

A higher GCL accuracy gives the better results. Generally, the accuracy of the results does not change significantly when the GCL count exceeds a certain number. Such numbers range from 12 to 16 GCLs. The results of the nadir registration or near-to-nadir images are slightly better than those of off-nadir images.

The experiments show that both forms of the LBTM are applicable for image-to-image registration of high-resolution satellite imagery. In particular, a higher accuracy (better than two pixels) can be achieved using a moderate number of GCLs for building the LBTM.

FIGURE 13.5 **(See color figure following page 38.)** Configurations of the ground control lines (GCLs) and checkpoints over the Hong Kong data set.

13.7   SUMMARY AND COMMENT

Coordinate transformations from image space to object space can be realized by two approaches: (a) point-based models and (b) line-based models. These two approaches can be either rigorous physical models or nonrigorous empirical models. In the absence of the satellite orbit and other related information, the empirical models provide a main stream solution, where polynomials or rational functions are used as examples.

If not enough control points are available, the line-based transformation model (LBTM) provides an alternative solution. The model is based on the unit vector and forms the relationship between image space and ground object space. Both point-based and line-based transformation models can be applied for image rectification, spatial resection, and intersection, being key techniques for modeling a three-dimensional surface based on satellite images.

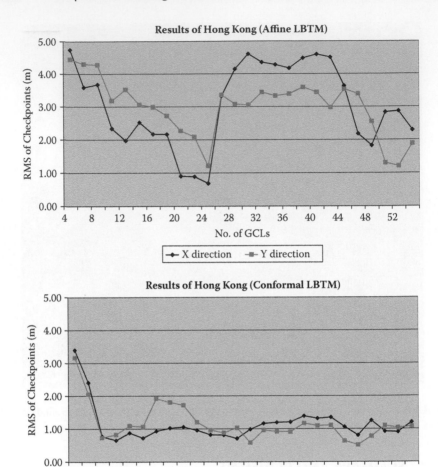

FIGURE 13.6   **(See color figure following page 38.)** Results of image rectification of different data sets. (From Shi, W. Z. and Shaker, A., 2006. *Int. J. Rem. Sens.* 27(14): 3001–3012. With permission.)

REFERENCES

Dare, P. and Dowman, I., 2001. An improved model for automatic feature-based registration of SAR and SPOT images. *ISPRS J. Photogram. Rem. Sens.* 56(1): 13–28.

El-Manadili, Y. and Novak, K., 1996. Precision rectification of SPOT imagery using the direct linear transformation model. *Photogram. Eng. Rem. Sens.* 62: 67–72.

Fraser, C. S., Hanley, H., and Yamakawa, T., 2002. Three-dimensional geopositioning accuracy of IKONOS imagery. *Photogram. Rec.* 17(99): 465–479.

Gugan, D., 1986. Practical aspects of topographic mapping from SPOT imagery. *Photogram. Rec.* 12(69): 349–355.

Habib, A. F., Lin, H. T., and Morgan, M. F., 2003. Line-based modified iterated Hough trans- form for autonomous single-photo resection. *Photogram. Eng. Rem. Sens.* 69(12): 1351–1357.

Kanok, W., 1995. Exploitation of Linear Features for Object Reconstruction in digital Photogrammetric Systems. Ph.D. thesis, Purdue University.

Mulawa, D. C. and Mikhail, E. M., 1988. Photogrammetric treatment of linear features. *Proceedings of the XVIth Congress of ISPRS*, Commission III, 27(part B10), ISPRS, Elsevier, Amsterdam, The Netherlands, pp. 383–393.

Okamoto, A., Fraser, C. S., Hattori, S., Hasegawa, H., and Ono, T., 1998. An alternative approach to the triangulation of SPOT imagery. *Int. Arch. Photogram. Rem. Sens.* 32(4): 457–462, Stuttgart.

Shaker, A. and Shi, W. Z., 2003. Line-Based Transformation Model (LBTM) for high-resolution satellite imagery rectification. *Proceedings of Asian Conference on Remote Sensing (ACRS 2003) and International Symposium on Remote Sensing*, Korea. On CD.

Shaker, A., 2004. The Line Based Transformation Model for High Resolution Satellite Imagery Rectification and Terrain Modeling. Ph.D. thesis, Department of Land Surveying and Geo-Informatics, Hong Kong Polytechnic University.

Shi, W. Z. and Shaker, A., 2006. The line-based transformation model (LBTM) for image to image registration of high-resolution satellite image data. *Int. J. Rem. Sens.* 27(14): 3001–3012.

Shi, W. Z. and Shaker, A., 2003. Analysis of terrain elevation effects on IKONOS imagery rec- tification accuracy by using non-rigorous models. *Photogram. Eng. Rem. Sens.* 69(12): 1359–1366.

Space Imaging, 2001. Company web site. http://www.spaceimaging.com (accessed: September 2001).

Tao, C. V. and Hu, Y., 2001. A comprehensive study of the rational function model for photo- grammetric processing. *Photogram. Eng. Rem. Sens.*, 67(12): 1347–1357.

Tommaselli, A. M. G. and Tozzi, C. L., 1996. A recursive approach to space resection using straight lines. *Photogram. Eng. Rem. Sens.* 62(1): 57–66.

14 Improved Interpolation Methods for Digital Elevation Model

This chapter addresses the methods for improving the quality of digital elevation models (DEM). Of particular interest is the presentation of two improved interpolation methods for DEM generation. Unlike the existing interpolation methods, the improved methods include terrain change in the model design. In view of the fact that a terrain may be composed of both high and low frequency elevation changes, a hybrid interpolation model, which contains both linear and nonlinear functions, is proposed. In recognition of the possible different elevation change frequencies in different orientations, a bidirectional interpolation model is also presented.

14.1 INTRODUCTION

A DEM is commonly used to represent the earth terrain. Traditionally, it is generated by applying an interpolation method to sample data captured from the terrain surface. The quality of a generated DEM is thus determined by (a) the accuracy of the original sample data, and (b) the applied interpolation method. The number of samples and the spatial distribution of the samples are also factors that affect the accuracy of the generated DEM.

In DEM interpolation, the elevation at an unknown point can be estimated based on an interpolation function applied to a number of sample points. A polynomial function is commonly used for this purpose. Of the available polynomial interpolation functions, the bilinear and the bicubic interpolation algorithms are frequently used. Improvement in the interpolation methods may lead to improvements in the accuracy of the generated DEM, and consequently improvement in the quality of any terrain analysis based on the DEM.

The two new interpolation methods, the hybrid and the bidirectional interpolation methods, introduced in this chapter and described below are considered to be improvements on the existing single function or single direction interpolation methods for DEM generation.

14.2 HYBRID INTERPOLATION METHOD

A natural terrain is usually composed of terrain elements considered as either possessing (a) lower frequency of change in elevation or (b) higher change frequency. Linear interpolation methods are considered appropriate for modeling a terrain with lower frequency change in elevation, whereas nonlinear models better approximate a terrain with higher frequency elevation changes. In fact, the nature of the interpolation can be viewed from a filtering perspective (Shenoy and Parks, 1992), and the image filtering techniques can be transferred for DEM interpolation. For terrain with lower frequency of elevation change, i.e., equivalent to an image with low frequency signals, a linear interpolation method such as the bilinear interpolation method is more desirable than a nonlinear interpolation method. Nonlinear interpolation is more suitable for representing terrain with higher frequency of elevation change, i.e., equivalent to an image with a rich higher frequency signal. Bilinear and bicubic interpolations are adopted as linear and nonlinear interpolation methods, respectively, in the following studies.

As indicated above, a digital terrain can be regarded as a composition of two types of elementary terrains: (a) elementary terrain with high frequency change in elevation (equivalent to high frequency signal in image) and (b) elementary terrain with low frequency change in elevation (equivalent to low frequency signal in image), respectively. According to the properties of the linear and the nonlinear interpolation methods, an integration of both linear and nonlinear interpolation methods are able to provide a better representation of a natural or real-world terrain, than that obtained by a single type interpolation method. Such an integrated model may describe both the general outline of a terrain and also detailed terrain with frequent change in elevation. A hybrid interpolation method based on this contention has been proposed for generating DEM (Shi and Tian, 2006). This method is described in detail, below.

14.2.1 Mathematical Expression of Hybrid Interpolation

A hybrid interpolation is derived as follows. Firstly, bilinear interpolation is given by

$$f_1(x, y) = a_1 x + a_2 y + a_3 xy + a_4. \tag{14.1}$$

Secondly, the 10-term bicubic interpolation, a typical nonlinear model, is given by

$$f_2(x, y) = b_1 + b_2 x + b_3 y + b_4 x^2 + b_5 xy + b_6 y^2 + b_7 x^3 + b_8 x^2 y + b_9 xy^2 + b_{10} y^3. \tag{14.2}$$

Based on these two interpolations, a hybrid interpolation is defined for $0 \leq \rho \leq 1$ as:

$$I = \rho A + (1 - \rho) B, \tag{14.3}$$

where I, A, and B represent the hybrid interpolator, the bilinear interpolator $f_1(x, y)$, and the bicubic interpolator $f_2(x, y)$. Then, ρ is a coefficient associated with the proportion of low-frequency information contained in a terrain. When $\rho = 0$, the hybrid interpolation degenerates to the bicubic interpolation. When $\rho = 1$, the hybrid interpolation is then the bilinear interpolation. By putting Equations 14.1 and 14.2 into Equation 14.3, the following is given:

$$f^\rho(x,y) = \rho(a_1 x + a_2 y + a_3 xy + a_4) \tag{14.4}$$

$$+ (1-\rho)(b_1 + b_2 x + b_3 y + b_4 x^2 + b_5 xy + b_6 y^2 + b_7 x^3 + b_8 x^2 y + b_9 xy^2 + b_{10} y^3)$$

$$= c_1 + c_2 x + c_3 y + c_4 xy + c_5 x^2 + c_6 y^2 + c_7 x^3 + c_8 x^2 y + c_9 xy^2 + c_{10} y^3,$$

where f^ρ denotes the hybrid interpolator, and the coefficients c_i, for $i = 1, 2, \ldots, 10$ are expressed as

$$c_1 = \rho a_4 + (1-\rho)b_1 \qquad\qquad c_2 = \rho a_1 + (1-\rho)b_2$$

$$c_3 = \rho a_2 + (1-\rho)b_3 \qquad\qquad c_4 = \rho a_3 + (1-\rho)b_5$$

$$c_5 = (1-\rho)b_4 \qquad\qquad\qquad c_6 = (1-\rho)b_6$$

$$c_7 = (1-\rho)b_7 \qquad\qquad\qquad c_8 = (1-\rho)b_8$$

$$c_9 = (1-\rho)b_9 \qquad\qquad\qquad c_{10} = (1-\rho)b_{10}$$

Equation 14.4 shows that the hybrid interpolation has a similar form as the bicubic interpolation but with different coefficients. In terms of computational load, the hybrid model is comparable with the bicubic interpolation model.

The bilinear and the bicubic interpolations are performed to realize the hybrid model. Adding the weights ρ and $1 - \rho$ to their results gives the hybrid interpolation result. The coefficient ρ of the linear combination plays an important role in the hybrid model.

14.2.2 THEORETICAL INTERPRETATION OF ρ

An original terrain data set, f, and a resample of that data set, f_0, which is a simplification of the original data set, is discussed as follows:

An interpolation method is applied on the resampled data set. The interpolation is evaluated by comparing an interpolation in the resampled data with the original terrain data set; the closer the two, the better.

Let, as before, f_1, f_2, and f^ρ denote the results of the bilinear, the bicubic and the hybrid interpolations. The root-mean-square error (RMSE) of f^ρ is used for evaluation. It is defined by

$$RMSE = \sqrt{\frac{\displaystyle\sum_{i=1}^{n}\sum_{j=1}^{m}(f_{i,j} - f_{i,j}^\rho)^2}{n \times m}}, \tag{14.5}$$

where $f_{i,j}$ and $f_{i,j}^\rho$ denote the elements of f and f^ρ at (i, j). Replacing $f_{i,j}^\rho$ with f_1 and f_2, we can also obtain the RMSEs of the bilinear and the bicubic interpolations.

Let $\Delta f = |f - f^{\rho}|$. Error in f^{ρ} is then expressed as

$$
\begin{aligned}
\left\| \Delta f \right\|^2 &= \left\langle f - f^{\rho}, f - f^{\rho} \right\rangle \\
&= \left\langle f, f \right\rangle + \left\langle f^{\rho}, f^{\rho} \right\rangle - 2 \left\langle f, f^{\rho} \right\rangle,
\end{aligned}
\tag{14.6}
$$

where $<\bullet, \bullet>$ is the inner product of vectors in mathematics. Equation 14.5 is equivalent to the square root of Equation 14.6, except the constant term. It implies

$$
\frac{\partial \left\| \Delta f \right\|^2}{\partial \rho} = \frac{\partial \left\langle f^{\rho}, f^{\rho} \right\rangle}{\partial \rho} - 2 \frac{\partial \left\langle f, f^{\rho} \right\rangle}{\partial \rho}.
\tag{14.7}
$$

By considering $f^{\rho} = \rho f_1 + (1-\rho) f_2 = \rho A f_0 + (1-\rho) B f_0$, we have

$$
\begin{aligned}
\left\langle f^{\rho}, f^{\rho} \right\rangle &= \left\langle \rho A f_0 + (1-\rho) B f_0, \rho A f_0 + (1-\rho) B f_0 \right\rangle \\
&= \rho^2 \left\| A f_0 \right\|^2 + (1-\rho)^2 \left\| B f_0 \right\|^2 + 2\rho(1-\rho) \left\langle A f_0, B f_0 \right\rangle,
\end{aligned}
\tag{14.8}
$$

and

$$
\begin{aligned}
\left\langle f, f^{\rho} \right\rangle &= \left\langle f, \rho A f_0 + (1-\rho) B f_0 \right\rangle \\
&= \rho \left\langle f, A f_0 \right\rangle + (1-\rho) \left\langle f, B f_0 \right\rangle.
\end{aligned}
\tag{14.9}
$$

Differentiating these two equations gives

$$
\frac{\partial \left\langle f^{\rho}, f^{\rho} \right\rangle}{\partial \rho} = 2\rho \left\| A f_0 \right\|^2 - 2(1-\rho) \left\| B f_0 \right\|^2 + 2 \left\langle A f_0, B f_0 \right\rangle - 4\rho \left\langle A f_0, B f_0 \right\rangle,
\tag{14.10}
$$

and

$$
\frac{\partial \left\langle f, f^{\rho} \right\rangle}{\partial \rho} = \left\langle f, A f_0 \right\rangle - \left\langle f, B f_0 \right\rangle = \left\langle f, A f_0 - B f_0 \right\rangle.
\tag{14.11}
$$

Equation 14.7 is then rewritten as

$$
\frac{\partial \left\| \Delta f \right\|^2}{\partial \rho} = 2\rho \left\| A f_0 \right\|^2 - 2(1-\rho) \left\| B f_0 \right\|^2 + 2 \left\langle A f_0, B f_0 \right\rangle
\tag{14.12}
$$

$$
- 2\rho \left\langle A f_0, B f_0 \right\rangle - 2 \left\langle f, A f_0 - B f_0 \right\rangle.
$$

By setting

$$\frac{\partial \|\Delta f\|^2}{\partial \rho} = 0,$$

Equation 14.12 becomes

$$\rho \left(\|Af_0\|^2 + \|Bf_0\|^2 - 2\langle Af_0, Bf_0 \rangle \right) = \|Bf_0\|^2 + \langle f, Af_0 - Bf_0 \rangle - \langle Af_0, Bf_0 \rangle. \quad (14.13)$$

That is,

$$\rho = \frac{\langle Bf_0 - Af_0, Bf_0 - f \rangle}{\|Af_0 - Bf_0\|^2}. \quad (14.14)$$

This equation reveals that the parameter ρ is smaller if the difference between Bf_0 and f is smaller. Here, Bf_0 contains a high-frequency component of the original data set f. The smaller difference means that the original data are mainly high frequency with a small part having low frequency. This reconfirms that the parameter ρ of the hybrid model is closely related to the proportion of low-frequency components of the original data.

14.2.3 EXPERIMENTAL RESULTS

Two experiments have been conducted to validate the hybrid model and to study the problem of parameter determination. The validation was carried out using the following steps.

Step 1: Provide the original data set of a terrain.
Step 2: Resample the original data with a given sampling ratio.
Step 3: Perform the bilinear, the bicubic, and the hybrid interpolation on the resample data.
Step 4: Compare the interpolated results with the original data set for the above three interpolations.

14.2.3.1 Data Set A

The original data set A for a terrain is shown in Figure 14.1A: (a) The sampling ratio for resampling is equal to 16:1, leading to the resampled data set shown in Figure 14.1A (b). Results from the three interpolations methods are shown in Figures 14.1A: the bilinear interpolation result (c), the bicubic interpolation result (d), and the hybrid method, with $\rho = 0.2$ (e).

It can be seen from Figure 14.1A, that the result of the bilinear interpolation (14.1A [c]) is so smooth that many detailed features of the original terrain

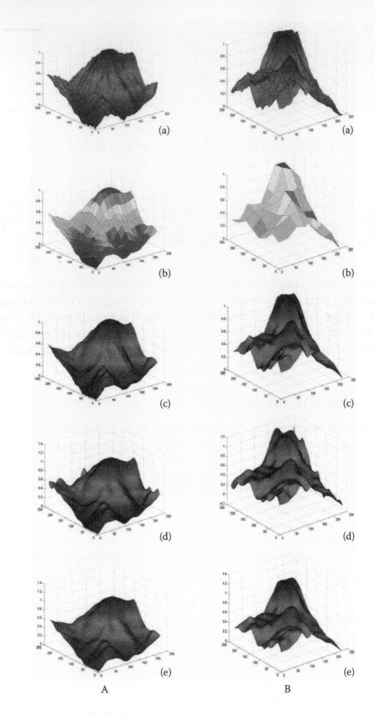

A   B

FIGURE 14.1 (See color figure following page 38.) The experimental results following the application of interpolation methods for the data sets A and B. Where A-(a) is the original data set A, A-(b) is the resampled data set of data set A; (*continued on facing page*)

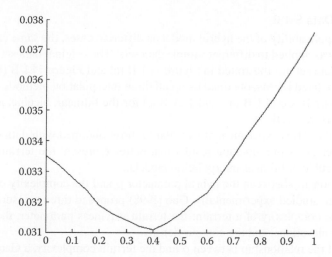

FIGURE 14.2   The RMSE versus the hybrid model parameter value for data set A. (From Shi, W. Z. and Y. Tian, 2006. *Int. J. Geog. Inform. Sci.*, 20(1): 53–67. With permission.)

(Figure 14.1A [a]) are lost. From the results of the bicubic interpolation (14.1A[d]) it can be seen that more detailed features are preserved. However, at the same time, false terrain is generated. The hybrid interpolation model gives a more realistic interpolation result, as shown in Figure 14.1A (e). The experimental results show that the hybrid interpolation model not only preserves the basic outline of the original data set, but also preserves the detailed terrain image without any false undulations.

Questions can be raised regarding the relationship between the accuracy of the interpolated DEM with the choice of parameter ρ. By applying the same data set, the value of ρ can vary from 0 to 1, increasing its value by 0.05 each time; ρ can thus have a series of different values. The corresponding RMSE for each value of the ρcoefficient is computed and the result is illustrated in Figure 14.2.

It is noted that the RMSE of the bicubic interpolation (ρ = 0) equals 0.0335, whereas that of the bilinear interpolation (ρ = 1) is 0.0376. The RMSE value decreases when the hybrid parameter increases within the interval [0, 0.4], reaching its minimum of 0.0312 at ρ = 0.4. The RMSE increases monotonically as the hybrid parameter increases within the interval [0.4, 1]. When ρ = 0.7, the RMSE is equal to 0.0335, which is similar to the case when ρ = 0. Obviously, if ρ takes values within the open interval (0, 0.7), the RMSE of the hybrid model is always smaller than the RMSE of the bilinear interpolation model and that of the bicubic interpolation model. This means that when ρ lies between (0, 0.7), the hybrid model is more effective than either the bilinear or bicubic interpolation models.

FIGURE 14.1   (*see facing page*) A-(c) is the result of bilinear interpolation; A-(d) is the result of bicubic interpolation; and A-(e) is the result following the hybrid interpolation for data set A, with ρ = 0.2. B-(a) is the original data set B; B-(b) the resampled data set of B; B-(c) the result of bilinear interpolation; B-(d) is the result of bicubic interpolation; and B-(e) is the result following the hybrid interpolation for data set B.

14.2.3.2   Data Set B

To test the applicability of the hybrid model on different cases, the same experimental method was applied to different sample data sets. The original data set B and the resampled data set are illustrated in Figure 14.1 B (a) and Figure 14.1 B (b), respectively. Interpolated DEMs, obtained using all three interpolation methods are shown in Figures 14.1 B (c), 14.1 B (d), and 14.1 B (e) for the bilinear, bicubic, and hybrid interpolations, respectively.

The results of this experiment are similar to those obtained with data set A. The hybrid model gives interpolation results that better represent the terrain than can either the bicubic or bilinear interpolation models.

The relationship between the hybrid parameter ρ and the complexity of a terrain has also been studied experimentally. Gao (1998) proposed three measures for representing the complexity of a terrain: the terrain waviness parameter, the standard deviation of elevations, and the contour density.

To model the relationship between ρ and the terrain complexity, a Gauss surface function was used to generate terrains with various complexities (Liu, 2002). The Gauss surface is expressed as:

$$z = A\left[1-\left(\frac{x}{m}\right)^2\right]e^{-\left(\frac{x}{m}\right)^2-\left(\frac{y}{n}+1\right)^2} - B\left[0.2\left(\frac{x}{m}\right)-\left(\frac{x}{m}\right)^3-\left(\frac{y}{n}\right)^5\right]e^{-\left(\frac{x}{m}\right)^2-\left(\frac{y}{n}\right)^2}$$

$$-Ce^{-\left(\left(\frac{x}{m}\right)+1\right)^2-\left(\frac{y}{n}\right)^2}.$$

(14.15)

If the optimal hybrid parameter is regarded as a function of the terrain complexity, this function, according to the experimental results, tends to be a monotonic increasing function. Therefore, the hybrid parameter can be determined based on the terrain complexity level.

In a practical application, experimental regions that cover the various complex levels of the terrain are first chosen, and the optimized hybrid parameter value at each complexity level can then be determined. These optimized values of the hybrid parameter are then applied at various complexity levels to the whole terrain area in a hybrid interpolation process. The optimal hybrid parameter is computed for each of the complex levels of the terrain. The average of all optimal hybrid parameter values serves as the optimal hybrid parameter for the whole study area.

14.3   BIDIRECTIONAL INTERPOLATION METHOD

Natural terrains can be of anisotropy, which means that the frequency of elevation change may be different in different directions. Different interpolation models incorporating the characteristics of terrain change in different directions should then be applied. This is the theoretical foundation for proposing the directional interpolation methods for DEM.

Jensen and Anastassiou (1990) proposed an interpolation method that preserved the edges of an image. This method detects the edges and conducts interpolations along the orientation of each edge. Algazi et al. (1991) developed a directional interpolation technique to enhance the perception of the interpolated image. This work, however, focused on interpolating edge regions in one direction only, and only a simple averaging (Bayrakeri and Merseneau, 1995). Kwon et al. (2003) studied directional interpolation for images; five distinct directions were predefined, and different interpolation methods were used in each of these directions. In general, the above methods aim at interpolating an image based on the one common understanding: that pixel values change more frequently in the direction across the edge but change more steadily along the edge direction.

For digital terrain modeling, research on directional interpolation algorithms is limited. In Sections 14.3.1 to 14.3.3 below, a bidirectional interpolation method for DEM models is proposed and presented. In the new interpolation method design, terrain changes in two different directions allowances are made, and correspondingly different interpolation strategies are applied for the different directions.

14.3.1 Mathematical Formula for Bidirectional Interpolation

Bidirectional interpolation conducts terrain interpolation along two perpendicular directions (i.e., the x and y directions) by applying two different interpolation models. According to the different terrain complexity levels in the x and the y directions, different interpolation (linear and nonlinear) models are applied. The problem of a two-dimensional interpolation is then decomposed into a problem of three one-dimensional interpolations. The commonly used cubic interpolation is taken as the nonlinear method, as follows:

Linear interpolation and a cubic interpolation in one-dimension are given by

$$z = a_1 + a_2 x \tag{14.16}$$

and

$$z = a_1 + a_2 x + a_3 x^2 + a_4 x^3, \tag{14.17}$$

where a_i ($i = 1, 2, 3, 4$) are interpolation coefficients.

The parameters of Equation 14.17 are estimated, based on two known points together with their derivatives. The value of each derivative is estimated by using finite difference approximations such as the forward difference, the backward difference, and the central difference.

Given locations of points A, B, C, and D, as in Figure 14.3, the elevation of O can be derived from the following bidirectional interpolation:

Step 1: Estimate the slopes of $A^1 D^1$ and $B^1 C^1$ (denoted by $G_{A^1 D^1}$ and $G_{B^1 C^1}$) with a threshold ε. If $G_{A^1 D^1} \geq \varepsilon$, the one-dimensional cubic method is then employed to achieve interpolated result EE^1. Otherwise, if $G_{A^1 D^1} < \varepsilon$, one-dimensional linear interpolation is applied to achieve the interpolated result EE^1.

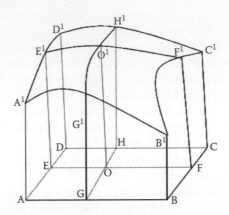

FIGURE 14.3   An example of the bidirectional interpolation.

Step 2: Follow Step 1 to obtain the interpolated result FF^1 at the point F.
Step 3: Repeat Step 1, but now interpolation is carried out along the EF direction, and the interpolated result OO^1 at the point O is obtained.

In a bilinear interpolation, linear interpolation models are used in both x and y directions, whereas nonlinear interpolation models are applied in both directions for a nonlinear (such as a bicubic) interpolation method, disregarding the difference of terrain change in the two directions. In a bidirectional interpolation, however, a linear interpolation model can be used for the interpolation in one direction, and a nonlinear interpolation model is employed for another direction, depending on the nature of terrain changes in two different directions. The bidirectional interpolation method can be described, mathematically, as follows.

In the x direction,

$$z = \begin{cases} a_1 + a_2 x, & \text{if } slope_x(P^*) \leq \varepsilon \\ a_1 + a_2 x + a_3 x^2 + a_4 x^3, & \text{if } slope_x(P^*) > \varepsilon \end{cases}. \tag{14.18}$$

In the y direction,

$$z = \begin{cases} a_1 + a_2 y, & \text{if } slope_y(P^*) \leq \varepsilon \\ a_1 + a_2 y + a_3 y^2 + a_4 y^3, & \text{if } slope_y(P^*) > \varepsilon \end{cases}, \tag{14.19}$$

where $slope_x(P^*)$ and $slope_y(P^*)$ are the slopes at point P^* along x and y directions, respectively.

14.3.2 SLOPE ESTIMATION

The classical slope estimation, such as forward difference, is illustrated in Figure 14.4. The slope at the fixed point P^* is

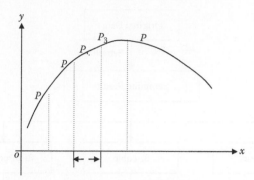

FIGURE 14.4   An example of slope estimation at a point.

$$Slope(P^*) \approx \left|\left(\left|O_3P_3\right| - \left|O_2P_2\right|\right)/h\right| \tag{14.20}$$

where $\left|O_3P_3\right|$ and $\left|O_2P_2\right|$ are the elevations of points P_3 and P_2, respectively, and h is an incremental interval equal to the difference between P_2 and P_3 in the x direction.

The elevation at a point is closely correlated to its neighboring points. The degree of the correlation decreases when the neighboring points move from that point. This classical slope estimation method uses two nearby points and ignores the slope change at farther points. To improve the method, it is suggested that the slope is estimated at P^* as follows:

$$Slope(P^*) \approx \frac{1}{2}\left|\left(\left|O_3P_3\right| - \left|O_2P_2\right|\right)/h\right| + \frac{1}{4}\left|\left(\left|O_2P_2\right| - \left|O_1P_1\right|\right)/h\right| + \frac{1}{4}\left|\left(\left|O_4P_4\right| - \left|O_3P_3\right|\right)/h\right|. \tag{14.21}$$

This method considers impacts of both the two closest points (P_2 and P_3) and the farthest points (P_1 and P_4) on slope estimation. The point level contributing to the estimation is governed by the weights. In the following experiment, the slope with Equation 14.21 is calculated.

14.3.3 Experimental Results

To assess the accuracy of bidirectional interpolation, a series of experiments was conducted. They are as follows. Bidirectional interpolation was applied to two real data sets. Two sampling ratios were set individually. In this way, the applicability of the proposed interpolation method to different terrains was tested. Two error measures, RMSE and maximum error (MAXE), were utilized to offer an objective description of errors in the interpolations.

The experimental procedure for testing the performance of the bidirectional interpolation method is outlined in Figure 14.5. Three steps are described as follows:

Step 1: Resample a given original data set f_0 with a certain sampling ratio and obtain a resampled rough grid data f_1, which is to be interpolated by each of the three different interpolation methods in Step 2.

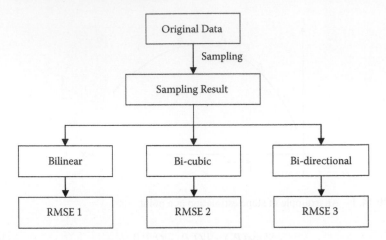

FIGURE 14.5   An experiment for testing performance of the bidirectional interpolation.

Step 2: Interpolate the rough grid data f_1 by using the bilinear, the bicubic and the bidirectional interpolations, and denote the results by f_2, f_3, and f_4, respectively.

Step 3: Calculate RMSE and MAXE for each of the interpolated results and compare their differences in terms of interpolated DEM accuracy.

The experimental results for two data sets (A and B) to examine the effectiveness of the bidirectional interpolation are given as follows.

14.3.3.1   Data Set A

The first set of experimental data is shown in Figure 14.6A (a), with the resampled data given in Figure 14.6A (b). In the latter case, the sampling ratio is set to be 4:1. The interpolation results are as follows: the bilinear interpolation result is given in 14.6A (c), the bicubic interpolation result in 14.6A (d), and the bidirectional interpolation result (with increment interval $\varepsilon = 0.0937$) in 14.6A (e).

No visual significant difference between the three interpolation methods appears in the interpolation results, shown in Figure 14.6A. However, if the interpolation results are enlarged fourfold, it can be seen that the bilinear interpolation result loses many details of the original terrain, whereas the bicubic interpolation result appears unnaturally smooth. The RMSE and the MAXE of the interpolations for data set A is summarized in Table 14.1.

From Table 14.1, it is noticed that the maximum error (0.0564) for the bidirectional interpolation is smaller than that for both the bilinear interpolation (0.627) and the bicubic interpolation (0.6013). The MAXE difference between the bilinear interpolation and the bidirectional interpolation equals 0.5707, whereas that between the bicubic and the bidirectional interpolation equals 0.5449. The bidirectional interpolation has the smallest RMSE and MAXE. This experimental result has demonstrated that the bidirectional interpolation method is more desirable.

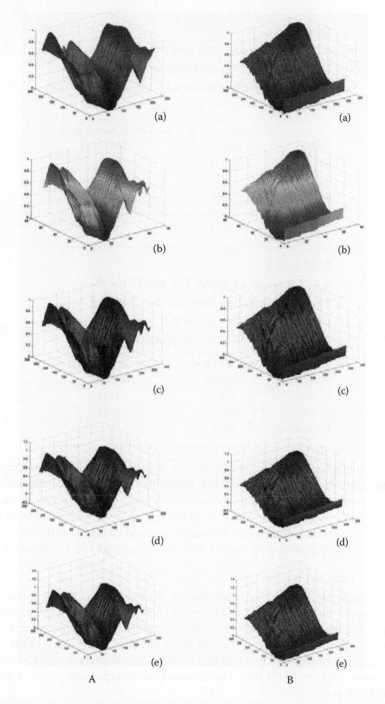

(a)   (a)

(b)   (b)

(c)   (c)

(d)   (d)

(e)   (e)

A   B

FIGURE 14.6 **(See color figure following page 38.)** Experimental results for bidirectional interpolations based on data sets A and B.

TABLE 14.1

A Comparison of RMSE and MAXE for the Three Interpolation Methods for Data Set A (With the Sampling Ratio 4:1)

	Index	
Method	Maximum Error (MAXE)	Root-Mean-Square Error (RMSE)
Bilinear	0.6271	0.0850
Bicubic	0.6013	0.0732
Bidirectional	0.0564	0.0397

TABLE 14.2

A Comparison of RMSE and MAXE for the Three Interpolation Methods for Data Set A (With the Sampling Ratio 8:1)

	Index	
Method	Maximum Error (MAXE)	Root-Mean-Square Error (RMSE)
Bilinear	0.8933	0.1436
Bicubic	0.8715	0.1384
Bidirectional	0.092	0.0574

Other interpolation results are obtained by repeating the above procedures, but with the sampling ratio set equal to 8:1 instead of 4:1, and the hybrid parameter equal to 0.01402 with an increment interval equal 1/10. Visually, the interpolation results are similar to those in Figure 14.6. The maximum error and the RMSE are tabulated in Table 14.2. This table also shows the maximum error and the RMSE in the bidirectional interpolation to be significantly smaller than that of the other two interpolations.

14.3.3.2   Data Set B

The third experiment uses the terrain data B. Both data set B and the corresponding resampled data are shown in Figures 14.6B (a) and 14.6B (b), respectively. By repeating the same operation procedure as for data A, the interpolated results for the bilinear and the bicubic interpolations are obtained (Figure 14.6B [c] and 14.6B [d]). The threshold ε is set equal to 0.0697 for the average slope of the resample terrain (Figure 14.6B [b]). The bidirectional interpolation result is given in Figure 14.6B (e).

In Figures 14.6B (c) to 14.6B (e), the interpolated results from the three algorithms are apparently similar. Similar to the first experimental results, by enlarging the

TABLE 14.3

A Comparison of RMSE and MAXE of the Three Interpolation Methods for Data Set B (With the Sampling Ratio 4:1)

	Index	
Method	Maximum Error (MAXE)	Root-Mean-Square Error (RMSE)
Bilinear	0.2344	0.0159
Bicubic	0.1738	0.0132
Bidirectional	0.0205	0.0108

TABLE 14.4

A Comparison of RMSE and MAXE of the Three Interpolation Methods for Data Set B (With the Sampling Ratio 8:1)

	Index	
Method	Maximum Error (MAXE)	Root-Mean-Square Error (RMSE)
Bilinear	0.4588	0.0327
Bicubic	0.3169	0.0289
Bidirectional	0.0310	0.0243

results fourfold, it is noticed that the bidirectional method generates a terrain that is closer to the original terrain than that of the other two methods. The maximum error and the RMSE are given in Table 14.3. It is noticeable that the maximum error and the RMSE of the bidirectional method are the smallest, whereas differences among the three interpolated results in terms of the maximum error and the RMSE are more obvious. For a sampling ratio set to 8:1, the maximum error and RMSE are given in Table 14.4. The two experiments with different sampling ratios reach the same conclusion: the bidirectional interpolation method has higher interpolation accuracy.

A comparison of the results shown in Table 14.1 with those shown in Table 14.2 for the data set A indicate that both RMSE and maximum error for the bidirectional interpolation method for the sampling ratio case to be 4:1 are smaller than that of the sampling ratio case, which is 8:1. This is true, too, for the results of the data set B listed in Table 14.3 and Table 14.4. In fact, this is also applicable to the bilinear and the bicubic interpolations. This means that the accuracy of interpolation is related to the sampling ratio. This conclusion is consistent with the results obtained by Tang (2000) and Toutin (2002).

14.4 SUMMARY

Described in the above chapter are two newly developed interpolation methods, (a) hybrid interpolation and (b) bidirection interpolation, designed to improve DEM interpolation accuracy. Both interpolation methods take the terrain change in elevation into the consideration in their design. A comparison of the two proposed interpolation methods with existing ones is also presented. The experimental results have indicated that the two newly proposed interpolation methods have a higher DEM interpolation accuracy.

REFERENCES

Bayrakeri, S. D. and Mersereau, R. M., 1995. A new method for directional image interpolation. *Proc. Int. Conf. Acoustics, Speech, and Sig. Proc.* 4: 2383–2386.

Gao, J., 1998. Impact of sampling intervals on the reliability of topographic variables mapped from grid DEMs at a mirco-scale. *Int. J. Geog. Inform. Sci.* 12: 875–890.

Jensen, K. and Anastassiou, D., 1990. Spatial resolution enhancement of images using nonlinear interpolation. *Proc. ICASSP* 1990, pp. 2045–2048.

Kwon, O., Sohn, K., and Lee, C., 2003. Deinterlacing using directional interpolation and motion compensation. *IEEE Trans. Cons. Electro.* 49: 198–203.

Liu, X. J., 2002. *On the Accuracy of the Algorithm for Interpreting Grid-based Digital Terrain Model*, Ph.D. dissertation, Wuhan University Press.

Shenoy, R. G. and Parks, T. W., 1992. An optimal recovery approach to interpolation. *IEEE Trans. Sig. Proc.* 40: 1987–1996.

Shi, W. Z. and Tian, Y., 2006. A hybrid interpolation method for the refinement of regular grid digital elevation model. *Int. J. Geog. Inform. Sci.*, 20(1): 53–67.

Tang, G. A., 2000. *A Research on the Accuracy of Digital Elevation Models*. Beijing: Science Press, Beijing.

Toutin, T., 2002. Impact of terrain slope and aspect on radargrammetric DEM accuracy. *ISPRS J. Photogram. Rem. Sens.* 57: 228–240.

Section VI

Presentation of Data Quality Information

15 Visualization of Uncertainties in Spatial Data

Valuable information has been gained from the rigorous procedures involved in the creation and analysis of uncertainty models, regarding the uncertainty present in spatial data, spatial models, and spatial analyses. Such uncertainty information, for its value to be realized, has to be transferred to the final users of spatial data to enable a better understanding of the nature of the data quality and hence its appropriate utilization. Section VI (Chapters 15 to 17) of this book provides the methods to present spatial data quality and uncertainty information.

This chapter presents a summary of uncertainty visualization methods, based mainly on the techniques in cartography, to enable the visualization of the nature of the uncertainty in spatial data. The uncertainty level is indicated by the cartographic variables. Effective visualization methods enable users not only to distinguish the level of uncertainty in spatial data, but also to understand spatial distribution of that uncertainty.

15.1 INTRODUCTION

As indicated in previous chapters, various uncertainties exist in spatial data and spatial analysis. Some of these uncertainties can be well modeled and some cannot; for example, uncertainty in the form of ambiguity in natural phenomena, and uncertainties governed by human perception and psychology so far cannot be well modeled. Interactive visualization is a form of uncertainty modeling and is particularly useful in cases where uncertainties in spatial data are too complex for common modeling methods. By using this visualization modeling method, the uncertain nature of spatial data and spatial analyses, such as distribution and uncertainty quantity, can be more clearly presented.

Visualization approaches have been developed for different purposes, audiences, and levels of interactivity. These approaches, which can be classified into static mode and dynamic mode (Goodchild et al., 1994), have application potential for visualizing uncertainties in spatial data and analyses. The static mode refers to the conventional cartographic display that presents static snapshots of uncertainty information in data surfaces and objects. Uncertainty is then represented by utilizing the graphical variables, which include size, value, texture, color, orientation and shape, and location in space (Bertin, 1983).

In the representation of uncertainty information, any figure can vary in one of the graphical variables. That figure can be expressed in the form of a geometric object, such as a column of membership values, and others. In the following sections, several static methods for displaying uncertainty information are presented. They include the error ellipse method, the arrow representation method, the gray-scale map method, the color map method, a three-dimensional representation method, and the symbol-based method.

As indicated above static cartography is applicable for the representation of uncertainties in spatial data. In this context, the manipulation of time is also conceived to be a cartographic variable. Temporal data have been treated in the past as an attribute to be mapped in the form of duration date or location change in relation to a specific time period (MacEachren, 1994). By integrating regional static maps over time a period, a dynamic and animated visualization, known as *dynamic cartography*, can be formed. Uncertainty in dynamic spatial analysis is thus represented. The methods whereby uncertainty information is displayed in an animated form are presented and described in the sections below.

Both static and dynamic cartographic representations are visual methods. However, these visual methods can possibly complicate or degrade maps embedded with uncertainty information. Fisher (1994a) proposed the additional use of sound to represent cartographic uncertainties in an attempt to alleviate such potential problems.

15.2   ERROR ELLIPSE APPROACH

In land surveying, geodesy, and engineering surveys, an ellipse is used to depict positional error of a point, two-dimensionally. This ellipse, known as the *error ellipse*, is established around the error point to indicate precision regions of different probabilities. The orientation of the error ellipse (relative to the Cartesian coordinate system) depends on the correlation between errors in *x* and *y* directions in that ellipse (Figure 15.1).

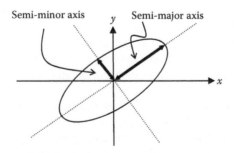

FIGURE 15.1   An example of error ellipse method.

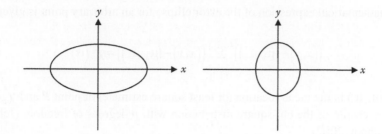

FIGURE 15.2   (a) At left, error ellipse with orientation parallel to the axes and (b) at right, error circle.

If errors in x and y directions are uncorrelated, the two semi-axes of the error ellipse will be parallel to the x and y axes (Figure 15.2a). If the two semi-axes have the same length, the error ellipse becomes an error circle (Figure 15.2b).

15.2.1   BASIC EQUATION

The mathematical basis of error ellipse is as follows: Suppose that point $P = (x, y)$ has a covariance matrix in the form of

$$\Sigma = \begin{pmatrix} \sigma_x^2 & \sigma_{xy} \\ \sigma_{xy} & \sigma_y^2 \end{pmatrix},$$ (15.1)

where σ_x^2 and σ_y^2 are the variances of x and y, respectively, and σ_{xy} is the covariance of x and y. The semi-manor and semi-minor axes of the error ellipse are equal to square roots of two solutions of

$$\lambda^2 - \left(\sigma_x^2 + \sigma_y^2\right)\lambda + \left(\sigma_x^2\sigma_y^2 - \sigma_{xy}^2\right) = 0.$$ (15.2)

Note that the orientation of this error ellipse is determined by computing an interior angle between the x axis and the semi-major axis:

$$\tan(2\theta) = \frac{2\sigma_{xy}}{\sigma_x^2 - \sigma_y^2}.$$ (15.3)

The probability of falling on or inside this error ellipse is 0.3935. This error ellipse is also called a *standard error ellipse*. The standard error ellipse can be enlarged to gain different probability values by multiplying a scale multiplier (Mikhail, 1976).

A mathematical expression of the error ellipse for an arbitrary point is given by

$$\left((x,y)-(\mu_x,\mu_y)\right)^T \Sigma^{-1}\left((x,y)-(\mu_x,\mu_y)\right) = \chi_p^2(\alpha), \tag{15.4}$$

where (μ_x,μ_y) is the mean location (or least square estimate) of point P and $\chi_p^2(\alpha)$ is the α percentile of the chi-square distribution with p degrees of freedom (Johnson and Wichren, 2002).

15.2.2   Confidence Region Model for Line Features

The confidence region model elaborated in Section 4.5 presents a solution using the confidence region to describe the positional uncertainty of a line. Figure 4.4 presents the confidence region model for two-dimensional point, line and polygon features.

15.2.3   An Analysis of the Error Ellipse

The error ellipse approach is a visual method to graphically represent positional uncertainty of a point feature in object-based data format in GIS. The magnitude of this uncertainty is characterized by the size of the error descriptor in the form of a geometrical object. The larger the error ellipse, the larger the positional uncertainty of the measurement point.

The error descriptor defined in the error ellipse approach is derived based on the assumption that the positional uncertainty follows a normal distribution. This assumption is feasible based on the central limit theorem.

The error ellipse approach can be applied for simultaneously displaying the positional uncertainty of many spatial features. This provides an overall picture of spatial feature uncertainty on a map. It also enables users to have a preliminary comparison of the spatial features in terms of their accuracy. Its disadvantage is that the map may be confusing when the spatial features and their error descriptors are shown simultaneously, as some spatial features might overlap with the error descriptors of other spatial features. This can be overcome by displaying error descriptors of one or several spatial features, individually, at any one time.

15.3   ARROW-BASED APPROACH

An uncertainty vector in two-dimensional space can be visualized by means of an arrow representation. The arrow is governed by two graphical variables: orientation and size. The orientation measures discrepancy from an assumed value in terms of direction, whereas size measures discrepancy according to the magnitude of error.

The following gives an example of uncertainty in image rectification, and the arrow-based approach is applied to visualize of the uncertainty in the rectification.

15.3.1 AN EXAMPLE OF IMAGE RECTIFICATION

The following steps are conducted to obtain a rectified image. First, several control points within the image area for registration are selected, and an image transformation is chosen for image rectification. Second, parameters of this transformation are estimated by substituting the control points into the transformation. Third, the transformation of the estimated parameters is applied, yielding a rectified image.

Image rectification accuracy is subject to the applied transformation model and the accuracy, number and spatial distribution of the control point. The commonly used transformations are introduced in Sections 13.1 and 13.2, such as the conformal transformation, the affine transformation, and the polynomial transformation. Here, the conformal transformation is taken as an example for illustrating the arrow-based uncertainty visualization method.

15.3.1.1 Conformal Transformation

Suppose n control points are selected and their coordinates on the ground are denoted by (x_i, y_i) for $i = 1, 2, \ldots, n$. Let (ξ_i, η_i) denote the ith control point on a map before rectification. Based on the conformal transformation, the following equation is obtained:

$$\begin{cases} x_i = T_x + \mu \cos(\alpha)\xi_i - \mu \sin(\alpha)\eta_i \\ y_i = T_y + \mu \sin(\alpha)\xi_i + \mu \cos(\alpha)\eta_i \end{cases}, \tag{15.5}$$

where T_x and T_y are translation parameters, μ is a scaling parameter, and α is a rotation parameter. By letting

$$\begin{cases} a = \mu \cos(\alpha) \\ b = \mu \sin(\alpha) \end{cases}, \tag{15.6}$$

Equation 15.5 simplifies to

$$\begin{bmatrix} x_i \\ y_i \end{bmatrix} = \begin{bmatrix} 1 & 0 & \xi_i & -\eta_i \\ 0 & 1 & \eta_i & \xi_i \end{bmatrix} \begin{bmatrix} T_x \\ T_y \\ a \\ b \end{bmatrix}. \tag{15.7}$$

Substituting all control points into Equation 15.7 provides a system of equations in matrix form

$$A\lambda = L + v, \tag{15.8}$$

where

$$
\lambda = \begin{bmatrix} T_x \\ T_y \\ a \\ b \end{bmatrix}, \quad A = \begin{bmatrix} 1 & 0 & \xi_1 & -\eta_1 \\ 0 & 1 & \eta_1 & \xi_1 \\ \cdots & \cdots & \cdots & \cdots \\ 1 & 0 & \xi_n & -\eta_n \\ 0 & 1 & \eta_n & \xi_n \end{bmatrix}, \quad L = \begin{bmatrix} x_1 \\ y_1 \\ \cdots \\ x_n \\ y_n \end{bmatrix} \text{ and } v = \begin{bmatrix} v_{x(1)} \\ v_{y(1)} \\ \cdots \\ v_{x(n)} \\ v_{y(n)} \end{bmatrix}.
$$

The column vector v represents the positional uncertainty introduced in the conformal transformation. Using least squares, the parameters are derived as

$$
\lambda = \left(A^T A\right)^{-1} A^T L, \tag{15.9}
$$

and their covariance matrix is (Wolf and Ghilani, 1997)

$$
Q_{\lambda\lambda} = \left(A^T A\right)^{-1} = \frac{1}{P} \begin{bmatrix} \sum_{i=1}^{n}\left(\xi_i^2 + \eta_i^2\right) & 0 & -\sum_{i=1}^{n}\xi_i & \sum_{i=1}^{n}\eta_i \\ 0 & \sum_{i=1}^{n}\left(\xi_i^2 + \eta_i^2\right) & -\sum_{i=1}^{n}\eta_i & -\sum_{i=1}^{n}\xi_i \\ -\sum_{i=1}^{n}\xi_i & -\sum_{i=1}^{n}\eta_i & n & 0 \\ \sum_{i=1}^{n}\eta_i & -\sum_{i=1}^{n}\xi_i & 0 & n \end{bmatrix}, \tag{15.10}
$$

where

$$
P = n\sum_{i=1}^{n}\left(\xi_i^2 + \eta_i^2\right) - \left[\left(\sum_{i=1}^{n}\xi_i^2\right)^2 + \left(\sum_{i=1}^{n}\eta_i^2\right)^2\right].
$$

Accuracy of the rectification is computed by the following root-mean-square error for all control points:

$$\begin{cases} S_x = \dfrac{1}{\sqrt{n-2}} \sqrt{\sum_{i=1}^{n} \left(v_{x(i)} \right)^2} \\ \\ S_y = \dfrac{1}{\sqrt{n-2}} \sqrt{\sum_{i=1}^{n} \left(v_{y(i)} \right)^2} \end{cases} . \tag{15.11}$$

For each control point (x_i, y_i), the positional error in the x and y directions is given by

$$\begin{cases} R_{x(i)} = v_{x(i)} \\ R_{y(i)} = v_{y(i)} \end{cases}, \tag{15.12}$$

and the radial error is given by

$$R_i = \sqrt{R_{x(i)}^2 + R_{y(i)}^2} . \tag{15.13}$$

The accuracy for an arbitrary point in the image is derived as follows. Let $(\tilde{x}, \tilde{y})$ denote its coordinate for an arbitrary point in the image before rectification. After the rectification process, this becomes

$$\begin{bmatrix} x \\ y \end{bmatrix} = \begin{bmatrix} 1 & 0 & \tilde{x} & -\tilde{y} \\ 0 & 1 & \tilde{y} & \tilde{x} \end{bmatrix} \lambda . \tag{15.14}$$

The covariance matrix is given by

$$\begin{bmatrix} q_x^2 & q_{xy} \\ q_{xy} & q_y^2 \end{bmatrix} = \begin{bmatrix} 1 & 0 & \tilde{x} & -\tilde{y} \\ 0 & 1 & \tilde{y} & \tilde{x} \end{bmatrix} Q_{\lambda\lambda} \begin{bmatrix} 1 & 0 & \tilde{x} & -\tilde{y} \\ 0 & 1 & \tilde{y} & \tilde{x} \end{bmatrix}^T$$

$$= \frac{1}{P} \begin{bmatrix} \sum_{i=1}^{n} (\tilde{x} - \xi_i)^2 + \sum_{i=1}^{n} (\tilde{y} - \eta_i)^2 & 0 \\ 0 & \sum_{i=1}^{n} (\tilde{x} - \xi_i)^2 + \sum_{i=1}^{n} (\tilde{y} - \eta_i)^2 \end{bmatrix} . \tag{15.15}$$

Since the control points on the rectified image contain positional uncertainty, the x and y variances of the arbitrary point on the rectified map are expressed as

$$
\begin{cases}
\sigma_x^2 = \dfrac{1}{(n-2)P} \sum_{i=1}^{n} \left(v_{x(i)}\right)^2 \left[\sum_{i=1}^{n} \left(\tilde{x}-\xi_i\right)^2 + \sum_{i=1}^{n} \left(\tilde{y}-\eta_i\right)^2 \right] \\[4mm]
\sigma_y^2 = \dfrac{1}{(n-2)P} \sum_{i=1}^{n} \left(v_{y(i)}\right)^2 \left[\sum_{i=1}^{n} \left(\tilde{x}-\xi_i\right)^2 + \sum_{i=1}^{n} \left(\tilde{y}-\eta_i\right)^2 \right]
\end{cases} . \tag{15.16}
$$

This equation considers that the error of the rectified point comes mainly from the error of the control points.

15.3.1.2 Inverse Distance Interpolation

Inverse distance interpolation, also known as inverse distance weighted interpolation, gives the interpolated values as weighted averages, with the weights determined in inverse proportion to the distances of the known points from an interpolated point. If the distance of a point to a control point decreases, the impact of that control point on the point to be interpolated is then larger.

If an interpolated point coincides with a grid node, the distance between that point and the grid node vanishes, and the point is then assigned a weight equal to one. Weights for all other grid nodes are zero. The smoothing parameter is a mechanism for buffering this behavior. When a nonzero smoothing parameter is assigned, there is no point with a weight of zero or one. The equation used for the inverse distance interpolation is (Liu et al., 2004)

$$
\hat{Z} = \frac{\displaystyle\sum_{i=1}^{n} \left(Z_i / h_{ij}^{\beta}\right)}{\displaystyle\sum_{i=1}^{n} \left(1 / h_{ij}^{\beta}\right)} , \tag{15.17}
$$

where $\hat{Z}$ is the interpolated value for a grid node, Z_i is the value of control point (x_i, y_i), β is the weighting power, and h_{ij} is the effective separation distance between the grid node and the control point.

The effective separation distance is given by

$$
h_{ij} = \sqrt{d_{ij}^2 + \delta^2} , \tag{15.18}
$$

where d_{ij} is the distance between the interpolated grid node and the control point (x_i, y_i), and δ is the smoothing parameter.

FIGURE 15.3   The rectification image and distribution of control points. (From Liu et al., 2004.)

Positional uncertainty for each grid node can then be visualized in an arrow map. This arrow map results by comparing points derived from Equation 15.7 with the corresponding interpolated point computed from Equation 15.17.

15.3.1.3   Visualization of Uncertainty in a Rectified Image

By taking uncertainty for control points as the datum mark, uncertainty for any point over the image can be interpolated. Uncertainty in the rectified image can thus be estimated and visualized. Here, an IKONOS image is used as an example. The image shows the Kowloon area in the Hong Kong Special Administrative Region (Figure 15.3). This figure also shows the marks for the control points for the image registration, which is performed based on conformal transformation. Control point uncertainty is then computed, based on Equations 15.11 and 15.12, and summarized in Table 15.1.

Using the control point shown in Figure 15.3, the rectified IKONOS image is generated. Inverse distance interpolation is carried out, in which the weighting power and smoothing parameter are set equal to one. An arrow map for visualizing uncertainty of the rectified image is given in Figure 15.4.

15.3.2   ANALYSIS OF THE ARROW-BASED APPROACH

In the arrow-based approach, the size of the arrow represents the error quantity of the point. The arrow size can be used to compare errors of different points within the same data set. However, it cannot be used to distinguish the positional error of different data sets. Therefore, the arrow on each map can only represent uncertainty of homogeneous data.

TABLE 15.1

Error of Control Points After the Image Registration

Point_ID (i)	Error in the x Direction ($R_{x(i)}$)	Error in the y Direction ($R_{y(i)}$)	Radial Error (R_i)
5	6.05	−2.53	6.56
8	−13.13	2.56	13.37
11	8.47	−0.01	8.47
13	8.92	−1.02	8.98
16	−8.27	−1.85	8.48
17	−3.72	0.34	3.74
18	−4.15	−4.30	5.97
19	−5.15	12.97	13.96
20	11.50	−5.15	12.60
21	8.20	−3.08	8.76
27	−11.07	1.20	11.13
28	4.70	−0.43	4.72
29	−2.36	1.30	2.69

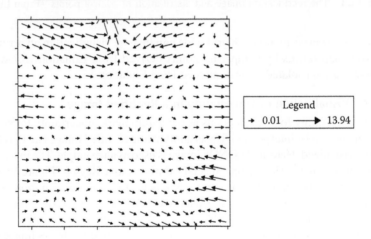

FIGURE 15.4   An arrow map for visualizing uncertainty in image rectification. (From Liu et al., 2004.)

The orientation variation also shows the discrepancy from the assumed location in terms of directions. Therefore, the arrow-based approach is applicable to describe positional uncertainty in an image at grid level or in a point map at feature level.

15.4   GRAY-SCALE MAP

A gray-scale map describes a data class subject to uncertainty, but without a straightforward measure to represent that uncertainty. This data class may be expressed

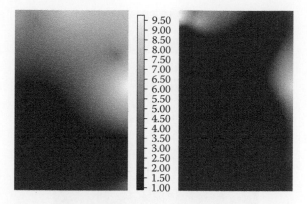

FIGURE 15.5 Two gray-scale maps for spatial uncertainty distribution.

in either the field-based data model or the object-based data model. In the field-based data model, variations within grid cells are lost. The object-based data model (e.g., polygon) partitions a map into a number of polygons of arbitrary shape and homogenous class. In such a case, variations within polygons are lost.

Figure 15.5 gives two gray-scale maps that describe the image rectification uncertainty distribution. The location of the darkest gray value has the lowest uncertainty. The left map has less control points and thus higher uncertainty. The spatial distribution of the uncertainty is also illustrated by the gray-scale map. In contrast, the right map contains more control points and thus has less uncertainty.

15.4.1 MAXIMUM LIKELIHOOD CLASSIFICATION

In a maximum likelihood classification, a pixel is classified as a class in which the pixel has the maximum probability, based on the spectral information. With the maximum likelihood classification, each pixel in Figure 15.6a is allocated to a class in which it has the highest probability, as shown in Figure 15.6b.

Section 5.1.4 of this book defines relative accuracy (U_R), which is used to describe the probable misclassification between two classes (C_{li} and C_{lj}) for a pixel (Z_T). The value of the parameter U_R is within the range [0, 1]. The larger the parameter value, the easier the discrimination between the C_{li} and C_{lj} classes for the pixel Z_T. That is to say, there is less chance that the pixel Z_T is misclassified between classes C_{li} and C_{lj}. For example, if $P(C_{l1}/Z_T)$ is 85% and $P(C_{l2}/Z_T)$ is 5%, then U_R is 0.80. Therefore, it is most likely that pixel Z_T is classified as class C_{l1}, rather than class C_{l2}. In another example, $P(C_{l1}/Z_T) = 44\%$ and $P(C_{l2}/Z_T) = 40\%$, then $U_R = 0.04$. In this case, it is not sure whether pixel Z_T should be classified as class C_{l1} or C_{l2}. Misclassification may easily happen on such occasions.

15.4.2 SINGLE CLASS MAPPING

Single class mapping, probability values for k classes generate k separate maps. Grid cells in each separate map show the probability values for the corresponding class.

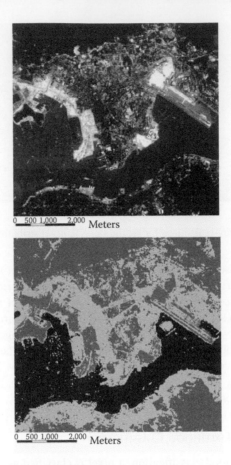

0 500 1,000   2,000 Meters

0 500 1,000   2,000 Meters

FIGURE 15.6   (See color figure following page 38.) (a) At top, Landsat image of Hong Kong's Victoria Harbor taken in 1996, and (b) at bottom, the result of the maximum likelihood classification method. The classification of maximum membership functions is as follows: vegetation (blue), low-reflection urban area (green), high-reflection urban area (red), and water (black).

The probability values can be retained in gray-scale maps (as shown in Figure 15.7). In this way, more details concerning the uncertainty of the original image are represented.

In the gray-scale approach, uncertainty in raster data is quantified by probability values, and these probability values are represented by intensity levels. The dark end of the scale corresponds to the higher certainty of affiliation to the class.

15.5   COLOR MAP

In the context of spatial data analysis color is considered as a visual variable. In cartography, color has been implemented to represent map information qualitatively and quantitatively. Many color models exist for raster graphics in computer graphics.

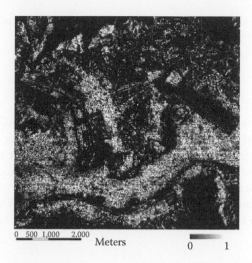

FIGURE 15.7 The probability for the classification result in Figure 15.6b.

These include the RGB (red, green, blue) used with color cathode-ray tube monitors, the YIQ model used in the broadcast TV color system, and the CMY (cyan, magenta, yellow) used for certain color printing devices (Foley et al., 1994). The standard color model in spatial data analysis is the RGB model (Brown and Feringa, 1999). This standard color model may fall short, however, for more than three classes, and even for the three cases if the sum of the class memberships is not normalized to one (Goodchild et al., 1994). The other color models, however, do not have a direct relationship with intuitive color notions of hue, saturation, and intensity (HSI). Therefore, Jiang et al. (1995) developed HSI for the visualization of uncertainty.

15.5.1 RGB Color Model

The RGB model can be represented with the unit cube defined on R, G, and B axes (Figure 15.8). Each color point within this cube is expressed in triple form of

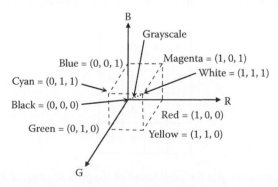

FIGURE 15.8 The RGB color model where grays are on the dotted main diagonal.

(R, G, B) where the element values of the triple are assigned in the range from 0 to 1. That is, a color C in this model is expressed as

$$C = r \cdot R + g \cdot G + b \cdot B, \tag{15.19}$$

where $R = (1,0,0)$, $G = (0,1,0)$ and $B = (0,0,1)$.

In a classified image, each class is associated with a point in RGB space, and each grid cell is assigned with a color that is determined by mapping its class membership vector to an intermediate point in the color space by linear interpolation. This concept was adopted in a pixel mixture technique and a color mixture technique for visualizing multiple memberships.

Three equal parameters in Equation 15.19 yield a gray color. Thus, the main diagonal of the cube in Figure 15.8 represents the gray levels from black of (0, 0, 0) to white of (1, 1, 1).

15.5.1.1 Pixel Mixture

The pixel mixture (PM) technique was introduced by de Gruijter et al. (1997). For each cell in an image, a grid of 3×3, 4×4, or $N \times N$ colored pixels is generated where N is a natural number. The colors in these generated pixels are assigned according to the probability proportional to the membership grade in the class. This probability can be interpreted as a visual impression of both the possible classes and their mixture. A cell, for example, has membership values equal to 0.25, 0.30, and 0.45 for classes A, B, and C, respectively. Its generated grid is given in Figure 15.9 where the proportion of the number of pixels for each class to the total number of pixels is equal to its corresponding probability value. The proportion for class A is equal to $4/16 = 0.25$.

If these three classes are represented by different colors, multiple memberships for each cell on the image can thus be visualized.

15.5.1.2 Color Mixture

The color mixture technique was proposed by Hengl et al. (2002). This technique is applicable for most classes. Suppose that the three classes in the above example

A	B	C	C
B	A	B	A
B	C	A	C
C	C	B	C

FIGURE 15.9 A generated grid for a cell based on the pixel mixture. (a) Probability distribution for water class, (b) probability distribution for low-reflection land use, (c) probability distribution for high-reflection land use, and (d) probability distribution for vegetation.

of the pixel mixture technique are represented by colors in RGB space. The color at each cell in the image is defined as an average intensity of RGB bands: for a cell i,

$$
\begin{cases}
R_i = \dfrac{\displaystyle\sum_{c=1}^{k} \mu_{i,c} R_c}{\displaystyle\sum_{i=1}^{n} \mu_{i,c}} \\[3em]
G_i = \dfrac{\displaystyle\sum_{c=1}^{k} \mu_{i,c} G_c}{\displaystyle\sum_{i=1}^{n} \mu_{i,c}}, \\[3em]
B_i = \dfrac{\displaystyle\sum_{c=1}^{k} \mu_{i,c} B_c}{\displaystyle\sum_{i=1}^{n} \mu_{i,c}}
\end{cases}
\tag{15.20}
$$

where the R_i, G_i, and B_i are the new derived mixed colors, the R_c, G_c, and B_c are the digital values (ranging from 0 to 255) for selected class colors, and k is the number of classes.

Figure 15.10 shows one example of the color mixture approach for the classification result given in Figure 15.6a and the maximum likelihood classification result Figure 15.6b. Probability distribution maps are shown in Figure 15.10 The probability value for each class, being inversely proportional to the probability value, is given the gray value.

Hengl et al. (2004) stated that a problem in the RGB model might occur when the color representing one class is similar to a mixture of the colors representing the other two classes. In such a case, the result of cell classification for the one class appears the same color as for the two classes. Therefore, Hengl et al. (2004) proposed a fuzzy-metric color legend, based on the HSI color model, for interpreting the cell's color generated from the color mixture. This legend is actually defined by converting the RGB model to the HSI model.

15.5.2 HSI Color Model

The HSI color model is user-oriented with color parameters being the hue (H), saturation (S), and intensity (I). It is defined in a hexcone in a three-dimensional space (see Figure 15.11). The top of this hexcone is a hexagon the boundary of which represents

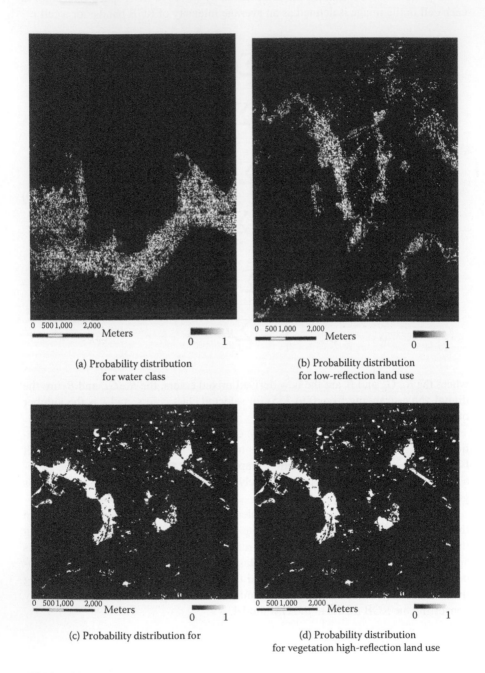

(a) Probability distribution
for water class

(b) Probability distribution
for low-reflection land use

(c) Probability distribution for

(d) Probability distribution
for vegetation high-reflection land use

FIGURE 15.10   Multiple probability maps for the Figure 15.6b.

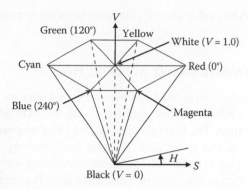

FIGURE 15.11   The HSI color model.

the various hues. The horizontal and the vertical axes in the hexagon represent saturation and intensity, respectively.

Many different software techniques exist for conversion between this color model and the RGB model. The following is the conversion used to define the legend of the color mixture technique by Hengl et al. (2004):

$$
\begin{cases}
H = \dfrac{360}{2\pi} \tan^{-1}\left(\dfrac{\sqrt{3}}{2}\left[G-B\right],\ R-\dfrac{G+B}{2} \right) \cdot \dfrac{240}{360} \\[4mm]
S = \sqrt{R^2+G^2+B^2-RG-RB-GB} \cdot \dfrac{240}{255} \qquad \alpha, \\[4mm]
I = \dfrac{R+G+B}{3} \cdot \dfrac{240}{255}
\end{cases}
\tag{15.21}
$$

where the input R, G, and B range from 0 to 255, whereas the output H, S, and I range from 0 to 240.

15.5.3   ANALYSIS OF THE COLOR APPROACH

The use of color is a popular description method for the perceptible difference in an image. Color in uncertainty visualization of classified images is used to represent multiple membership values for each class at each grid cell. Two common solutions to the visualization of the multiple probability values are the pixel mixture and the color mixture techniques. With the two techniques, each class is associated with an RGB space point or HSI space point; each grid cell is also colored in the corresponding color space.

Two points should be noted on the color approach. Firstly, in an image for soil classification, different soil types might show similar colors. In such situations, it might be difficult to distinguish one soil type from the other. Secondly, a color-blind person might misinterpret a color map. Therefore, to represent the uncertainty in the image effectively, color can be combined with other visual variables (such as texture).

15.6   OTHER SYMBOL-BASED APPROACHES

Symbols are the graphic marks used to visualize uncertainty on a map. The above four sections have introduced four applications of the cartographic symbol for uncertainty visualization. The first two applications focus on positional uncertainty visualization in vector and raster data, whereas the latter two relate to attribute uncertainty visualization in raster data. Attribute uncertainty visualization in vector data is briefly described in this section. This uncertainty is presented in a quantitative map. However, what needs to be presented is the uncertainty quantity at different locations.

Among the graphical variables given in Bertin (1983), size is the dimensional quality in the quantitative map. Figure 15.12 shows a polygon map, each polygon in the map has different values of attribute uncertainty. The attribute uncertainty quantity is represented by circles (symbols) inside individual polygons. By varying the size of the circle symbol, the polygon that contains a larger value of the attribute uncertainty, or that which contains the smaller, can easily be distinguished.

When the location is intrinsic to the symbol itself, in symbol-based uncertainty visualization approaches, that symbol can be defined, based on shape, size, color, orientation, texture, and location. During the selection of symbols used for representing uncertainty, it is necessary to consider the map scale. The symbol must have the facility to be alterable, according to scale.

15.7   THREE-DIMENSIONAL APPROACH

The visualization tools, described above, focus on displaying two-dimensional data uncertainty information. This information is treated as an attribute of the data set. An alternative to the two-dimensional approaches is to use a three-dimensional technique. In doing so, the uncertainty information is treated as the "z-value" of the data set. In addition, a contour map (used to display lines of constant scalar value for the data set distributed over a surface) shows the range and variation of the data value

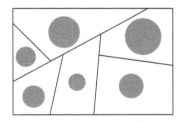

FIGURE 15.12   An example of attribute uncertainty visualization based on the symbol of circle size.

	1.60
	1.40
	1.20
	1.00
	0.80
	0.60
	0.40
	0.20
	0.00
	0.20
	0.40
	0.60
	0.80
	1.00
	1.20
	1.40
	1.60
	1.80
	2.00
	2.20
	2.40
	2.60

FIGURE 15.13   A contour map used to depict the uncertainty distribution for a region: the higher the contour value, the larger the uncertainty of the surface.

uncertainty information, over the region of space (Figure 15.13). A three-dimensional approach can be applied for visualizing uncertainty both for vector and raster data.

15.7.1   THE 3D APPROACH FOR VECTOR DATA

For vector data, two main types of uncertainty measures exist: (a) geometric measure, and (b) scalar quantity. For example, the error ellipse is a geometric measure, which gives a confidence region for two-dimensional points so that this region contains the true location of the point with a predefined probability. In a three-dimensional case, the error ellipse model evolves into an error ellipsoid model. This error ellipsoid model aims to describe positional uncertainty for a three-dimensional point by means of an ellipsoid (Figure 15.14).

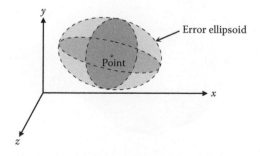

FIGURE 15.14   An error ellipsoid surrounding a point in a three-dimensional case.

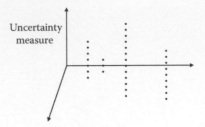

FIGURE 15.15   A point map with the uncertainty information in a three-dimensional space.

The second measure type is the scalar quantity. A scalar quantity example is a buffer uncertainty measure—the normalized discrepant area. The covariance matrix of a point can be simplified as a scalar quantity by the root-mean-square error. Since these uncertainty measures are distributed over a map, they can be visualized as vertical bars rising from the surface. Figure 15.15 gives an example of point map uncertainty visualization, the higher the vertical bar, the higher the uncertainty in that location.

15.7.2 THE 3D APPROACH FOR RASTER DATA

An uncertainty measure for raster data and its associated analysis is usually expressed in the form of single quantity. Raster data are distributed over an image, as are their uncertainty data. The uncertainty values can be interpolated and subsequently form a smooth surface.

By following Equation 15.17, uncertainty for any point over the rectified image can be estimated. This uncertainty is assessed by a radial measure based on the root square sum of the x- and y-directional variances of that point.

Three-dimensional uncertainty visualization for all points over the rectified image is displayed in Figure 15.16. The lighter the "elevation" at a location on the rectified image, the larger the uncertainty in that location.

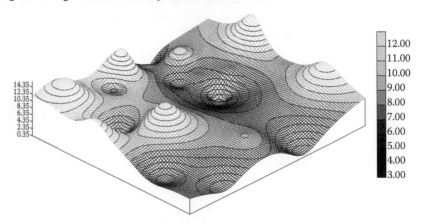

FIGURE 15.16   (See color figure following page 38.) A three-dimensional uncertainty visualization in the rectified image. (From Liu et al., 2004.)

15.7.3   An Analysis of Three-Dimensional Approach

A three-dimensional approach is appropriate for the uncertainty representation of both vector and raster data in three-dimensional space. However, this approach is more complex than the two-dimensional approach, due to display devices being normally two-dimensional.

To transform three-dimensional uncertainty representation into two-dimensional viewing, several points should be taken into consideration. Firstly, a projection transforming 3D objects onto a 2D projection plane has to be determined. Secondly, the viewer's eye position and the location of the viewing plane must be specified. Varying any or both of these parameters yields various representations of the perspective image. Thirdly, the contents of the projection of the 3D uncertainty map are transformed into a display viewpoint. Further information about such transformations can be found in Foley et al. (1994).

15.8   GRAPHIC ANIMATION

Uncertainty can also be represented by animation. It is possible to show a series of individual realizations in quick succession to depict a change caused by uncertainty. These individual realizations are mainly given in Monte Carlo simulation according to the source map uncertainty data.

15.8.1   The Animation Approach for Vector Map

Location and attribute in a vector map are two main information aspects. To visualize location and attribute uncertainties by animation, two randomizations are required. First, the spatial feature locations in such a map are simulated based on positional uncertainty distributions of the spatial features. The second step is aspatial randomization. Attribute uncertainty for each spatial feature is simulated based on that attribute's statistical accuracy measure. After performing these two steps, one realization is achieved. Animation for uncertainty visualization in the source map can be realized by running a series of these realizations.

15.8.2   Raster Image Animation Approach

In raster format, where each cell of an image represents its attribute value, the attribute uncertainty impact is visualized by means of animation on the source map. Animation can be performed by continuously selecting cells and reevaluating the cell value to give a constant change display. This is known as *random animation* (Fisher, 1994b, 1996).

Serial animation as an alternative (Blenkinsop et al., 2000) produces α-cuts of fuzzy membership images. By varying α, users can obtain an image showing those pixels with membership above or lower than this alpha value, or an image showing those pixels with a small range of fuzzy membership.

In the process of stringing a series of realizations, the combination of random animation and serial animation are likely to give a disjointed result as there will be many gaps between the realizations. Sequential animation overcomes this problem (Ehlschlaeger et al., 1997). It involves not only the stringing process but also the process of generating intermediate images between the realizations. The intermediate images within the transition between any two successive realizations are generated using an interpolation method. Playing the realizations and the intermediate images progressively gives a sequential animation. This animation method ensures that the realization transition looks smooth and natural (MacEachren and DiBiase, 1991).

15.8.3 An Analysis of the Animation Approach

Animation shows realizations cartographically, based on different uncertainty levels. The total time length occupied by the animated map display can reflect that map's accuracy. If the animated map does not change for a long period of time, the original map is then likely to be considered highly accurate. Hence, the speed of display change connects to an interpretation of "time period lengths" and the subsequent relationship to map accuracy. Therefore, a pertinent legend would be of help and perhaps even guide understanding of the relationship between display speed and map accuracy. If maps are displayed with too-long intervals between each display, users could fail to notice display changes and wrongly conclude that the original map is of higher accuracy. Additionally, if the map display change is too rapid, users will be unable to focus on and study the impact of elements of uncertainty. Further investigation of the implementation of animation in relation to uncertainty visualization is both a necessary and interesting challenge.

15.9   SUMMARY AND COMMENT

Methods for visualizing uncertainty in spatial information have been summarized in this chapter. These methods aim to make it possible to visualize the uncertainties in spatial data and analyses. Hence, users can have a better understanding of the nature of these features and a subsequent correct interpretation of the uncertainties. Evaluation results have shown that a combination of several visualization techniques is a better approach than the application of a single visualization method for uncertainty visualization.

Two solutions can be applied to implement the presented visualization methods in this chapter. In the first solution, uncertainty is treated as an additional piece of information—that is, additional to the original spatial data. In the second solution, a new graphic window is needed in the GIS system design to show the incorporation of the uncertainty information in the original data. However, in the latter solution, the data and associated uncertainty information cannot be visualized separately.

REFERENCES

Bertin, J., 1983. *Semiology of Graphics*. Madison, WI: University of Wisconsin Press.
Blenkinsop, S., Fisher, P. F., Bastin, L., and Wood, J., 2000. Evaluating the perception of uncertainty in alternative visualization strategies. *Cartographica*. 37: 1–14.

Brown, A. and Feringa, W., 1999. *A Color Handbook for GIS Users and Cartographers.* Enschede: International Institute for Geo-information Science and Earth Observation, The Netherlands.

De Gruijter, J. J., Walvoort, D. J. J., and van Gaans, P. F. M., 1997. Continuous soil maps—a fuzzy set approach to bridge the gap between aggregation levels of process and distribution models. *Geoderma.* 77: 169–195.

Ehlschlaeger, C. R., Shortridge, A. M., and Goodchild, M. F., 1997. Visualizing spatial data uncertainty using animation. *Comput. Geosci.* 23: 387–395.

Fisher, P. F., 1994a. Hearing the reliability in classified remotely sensed images. *Cart. Geog. Inform. Syst.* 21: 31–36.

Fisher, P. F., 1994b. Visualization of the reliability in classified remotely sensed images. *Photogram. Eng. Rem. Sen.* 60: 905–910.

Fisher, P. F., 1996. Animation of reliability in computer-generated dot maps and elevation models. *Cart. Geog. Inform. Syst.* 23: 196–205.

Foley, J. D., van Dam, A., Feiner, S. K., Hughes, J. F., and Phillips, R. L., 1994. *Introduction to Computer Graphics.* Reading, MA: Addison-Wesley.

Goodchild, M., Buttenfield, B., and Wood, J., 1994. Introduction to visualizing data validity. In *Visualization in Geographical Information Systems*, Hearnshaw, H. M. and Unwin, D. J. (eds.), Chichester, England: John Wiley & Sons, pp. 141–149.

Hengl, T., Walvoort, D. J. J., and Brown, A., 2002. Pixel and color mixture: GIS techniques for visualization of fuzziness and uncertainty of natural resource inventories. In *Proceedings o the 5th International Symposium on Spatial Accuracy Assessment in Natural Resources and Environmental Sciences (Accuracy2002)*, Melbourne: University of Melbourne.

Hengl, T., Walvoort, D. J. J., Brown, A., and Rossiter, D. G., 2004. A double continuous approach to visualization and analysis of categorical maps. *Int. J. Geog. Inform. Sci.* 18: 183–202.

Jiang, B., Brown, A., and Ormeling, F., 1995. Visualization of uncertainty with fuzzy color system. In *Proceedings of the 17th ICA Conference: Cartography Crossing Borders*, Barcelona: International Cartographic Association, pp. 507–508.

Johnson, R. A. and Wichern, D. W., 2002. *Applied Multivariate Statistical Analysis.* Upper Saddle River, NJ: Prentice Hall.

Liu, C., Shi, W. Z., and Zhu, S. L., 2004. Spatial visualization of image rectification accuracy based on spatial interpolation. *Journal of Remote Sensing* 8(2): 142–147.

Mikhail, E. M., 1976. *Observations and Least Squares.* New York: IEP.

MacEachren, A. M. and DiBiase, D., 1991. Animated maps of aggregate data: conceptual and practical problems. *Cart. Geog. Inform. Syst.* 18: 221–229.

MacEachren, A., 1994. Time as a cartographic variable. In *Visualization in Geographical Information Systems*, Hearnshaw, H. M. and Unwin, D. J. (eds.), Chichester, England: John Wiley & Sons, pp. 115–130.

Wolf, P. R. and Ghilani, C. D., 1997. Adjustment computations: Statistics and least squares in surveying and GIS. New York: John Wiley & Sons.

16 Metadata on Spatial Data Quality

Metadata are data that give further information about specific data sets. Over time, various metadata standards have been developed and the widely accepted standards are reviewed and analyzed in this chapter. Particular focus is on metadata associated with spatial data quality information. Spatial data quality metadata fall into the following classes: completeness, logical consistency, positional accuracy, thematic accuracy, and temporal accuracy. Also introduced in this chapter is a newly proposed object-oriented metadata organization method, which has been designed to provide uncertainty metadata at an object level. The development of an improved error metadata management system for solving problem on data quality metadata management in the real world is described.

16.1 INTRODUCTION

Spatial metadata, as its name implies, provides further information relating to the original spatial data. The purpose of metadata is to enable the documentation of described spatial data. Metadata, in general, cover content, quality, source organizations, data format, data currency, spatial references, and other related characteristics on the spatial data.

Metadata has an important role in GISci, in that information so provided can be used to assist data users in effectively archiving and searching related data. Metadata can also be used to provide information for spatial data sharing. Based on metadata information, users can decide whether a data set is suitable for a specific application. The Federal Geographic Data Committee (FGDC) has summarized three major uses of metadata (FGDC, 2000). They are:

- To maintain an organization's investment in data. This can avoid any waste or misuse of data in normal circumstances, during personnel change or owing to the passage of time.
- To provide information about data catalogues and clearinghouses. This can aid an organization find the relevant data partners to share a data collection and maintenance, and customers to purchase data.
- To provide information for data transfer. This can enable an organization to receive and interpret the data, and to incorporate data already held.

A number of organizations have developed standards for storing and maintaining metadata. These organizations are mainly divided into two groups: national government organizations and independent standard bodies. The former group is responsible for approving standards for use by their constituencies, such as the FGDC in the United States. The independent standard bodies are usually representatives from government agencies, professional organizations, and private companies. Their duty is to build and promote standards. The European Committee for Standardization (CEN) and the International Organization for Standardization (ISO) are examples of such organizations.

16.1.1 THE FEDERAL GEOGRAPHIC DATA COMMITTEE METADATA STANDARD

The FGDC developed and approved the Content Standard for Digital Geospatial Metadata (CSDGM) version 1 in 1992 and 1994. This is the first metadata standard in the geospatial community. According to the Executive Order 12906, Coordinating Geographic Data Acquisition and Access: The National Spatial Data Infrastructure, all federal agencies in the United States have used this standard to document metadata since 1995. The second version of the CSDGM was endorsed in 1998 (FGDC, 2000). Apart from some amendments to the wording of the first revision, the second version includes free-text descriptions and guidelines for the development of profiles and user-defined metadata entities and elements.

The FGDC standard concentrates on metadata content; however, it does not cover practical implementation methods to promote standards. Metadata standards are defined by seven main sections and three supporting sections. The main sections aim at describing the major aspects about geospatial data. These are:

- Identification—basic information about the data set, including descriptions of the data set, data status, the state of and maintenance information on the data set, the geographic attribute domain of the data set, access constraints, etc.
- Data quality—general quality assessment information regarding the data set
- Spatial data organization—information about the mechanism used to represent spatial information in the data set
- Spatial reference—the description of the reference frame for coordinates in the data set
- Entity and attribute—information concerning content of the data set, including the entity types and their attributes, and the domains from which attribute values are assigned
- Distribution—information about the distributor of and options for obtaining the data set
- Metadata reference—information on the status of the metadata information, and the responsible party

The FGDC supporting sections define the citation, data time period, and contact information within the main sections.

Each section contains three parts: (1) section definition, which also covers the name and (2) production rules, which provides descriptions of the section in terms of lower-level component elements. The list of component elements includes the name and definition of each component element in the section. (3) A list of compound elements.

Sections, compound elements, and data elements in this standard, are either "mandatory," "mandatory if applicable," or "optional." "Mandatory" elements must be provided in a data set, whereas "mandatory if applicable" elements must be provided if the data set contains information defined by the element. The provision of "optional" metadata elements are left to the discretion of the data provider.

16.1.2 THE ANZLIC STANDARD

The Australia/New Zealand Land Information Council (ANZLIC) has built a metadata standard, mainly for describing characteristics of spatial data sets maintained in the public and private sectors in Australia and New Zealand. The ANZLIC's standard was developed in 1996, based originally on the FGDC's standard and kept consistent with the Australian/New Zealand Standard on Spatial Data Transfer AS/NZS 4270. The second version of the ANZLIC was issued in 2001 (ANZLIC, 2001). This version provides a minimum set of elements in metadata and is simpler to the FGDC standard. Data managers may require gaining supplementary information about the data for specific purposes.

The ANZLIC standard version 2 contains 41 core elements. These core elements are grouped into 10 categories:

- Data set—information about the content of the data
- Custodian—information on the agency responsible for data collection and maintenance
- Description—information concerning the geographic area it covers
- Data currency—the time frame of the data described, including the earliest and latest dates at which the phenomena in the data set actually occurred
- Data set status—information about the status of the process of creation of the data set, and maintenance and update frequency
- Access—information about data format and constraints on the use of the data set
- Data quality—information on data quality including lineage, positional accuracy, attribute accuracy, logical consistency and completeness
- Contact information—description of address details for the contact person who is responsible for delivery of the data set to other users
- Metadata date—information about the currency of the directory entry
- Additional metadata—additional information that supports documentation of the data set

The following properties are given for each core element: category, name, definition, obligation (mandatory, optional, conditional), maximum occurrence, allowable content, format rules, field type, and field length.

16.1.3 THE CEN/TC287 STANDARD

This metadata standard was endorsed by the European Committee for Standardization and has been widely adopted in Europe (CEN, 1998). It covers the following topics:

- Identification;
- Spatial data organization;
- Spatial reference;
- Data quality;
- Content;
- Distribution;
- Metadata reference.

As compared with the FGDC's standard, the CEN content standards are more flexible in terms of the number of metadata elements.

16.1.4 THE ISO METADATA STANDARD

Technical Committee ISO/TC211, Geographic Information/Geomatics, endorsed the ISO 19115 metadata standard for geographic information in May 2003. The metadata standard was developed based on metadata standards of different organizations, including the FGDC, the ANZLIC, the CEN/TC287, and other metadata standards. Therefore, this international standard is more comprehensive than the existing metadata standards.

The structure of ISO 19115 is similar to that of the FGDC standard. In the ISO standard, metadata for geospatial data are presented in 11 Unified Modeling Language (UML) packages (similar to the section defined in the FGDC's standard). These UML packages are (ISO 19115, 2003):

- Identification information—information to uniquely identify the data, including citation of the resource, an abstract, the purpose, credit, the status, and points of contact
- Constraint information—information about the restrictions placed in data
- Data quality information—description of a general assessment of the quality of the data set, and information concerning the sources and production processes used in producing the data set
- Maintenance information—information about the scope and frequency of updating data
- Spatial representation information—information about the mechanisms used to represent spatial information in the data set
- Reference system information—the description of the spatial and temporal reference system(s) used in the data set
- Content information—information identifying the feature catalogue used or information describing the content of a coverage data set

- Portrayal catalogue information—information identifying the portrayal catalogue used
- Distribution information—information about the distribution of a resource
- Metadata extension information—information about user-specified extensions
- Application schema information—information about the application schema used to build the data set

Each UML package contains one or more entities. Entities are composed of elements used to identify the discrete units of metadata. Here, the entities and the elements can be referred to the compound elements and the data elements in the FGDC standard. Both standards define metadata and identification information sections as mandatory parts, whereas other sections are optional. However, the minimum set of metadata required to serve the full range of metadata applications should contain the following core:

- Data set title (mandatory)
- Data set reference date (mandatory)
- Data set responsible party (optional)
- Geographic location of the data set (conditional)
- Data set language (mandatory)
- Data set character set (conditional)
- Data set topic category (mandatory)
- Spatial resolution of the data set (optional)
- Abstract describing the data set (mandatory)
- Distribution format (optional)
- Additional extent information for the data set (optional)
- Spatial representation type (optional)
- Reference system (optional)
- Lineage (optional)
- On-line resource (optional)
- Metadata file identifier (optional)
- Metadata standard name (optional)
- Metadata standard version (optional)
- Metadata language (conditional)
- Metadata character set (conditional)
- Metadata point of contact (mandatory)
- Metadata data stamp (mandatory)

The ISO 19115 is the most comprehensive metadata spatial data standard. It not only integrates components of the existing metadata standards, but also provides conditional details for a data set, such as special coverage of raster and image information. It provides a data classification metadata scheme, hierarchical levels, and multilingual support for free text metadata elements.

The ISO standard is an international standard. Many metadata standard organizations for geospatial data are reviewing their individual standards with the objective of creating an ISO-19115-compliant profile. For instance, the FGDC endorsed remote sensing metadata extensions for the CSDGM. These extensions are used to describe geospatial data obtained from remote sensing. The current FGDC standard accompanying the remote sensing metadata extensions can be converted to comply with the ISO standard.

16.2   QUALITY INDICATORS IN METADATA

A data set can be viewed as a data set series, to which the data set belongs, as the data set itself, or as a smaller data grouping that shares common characteristics such as belonging to the same feature type or attribute value within the data set. The smaller data groupings may be as small as a single feature or a single attribute value. The quality of all smaller data groupings belonging to a data set is often the same; the data set quality can then reflect the quality of each smaller data grouping. If one of these smaller data groups has a different quality, it is then necessary to report the quality of the data set for each of the smaller data groupings.

The ISO standard provides more conditional elements to document information about data quality than the existing metadata standards. The data specified by a data quality scope may be a data set series, the data set itself, or a smaller data grouping that has the same feature type or attribute value. ISO 19115 explicitly provides for the recording of quality information as metadata for data set series, the data set, and the smaller data groupings with the exception of the single feature or attribute value. Single feature quality information (or attribute value), which differs from its parent type, may be treated as an attribute of the corresponding feature (or the attribute value) and recorded in the GIS database. The ISO standard, different from the existing standards that mainly document data quality as metadata for an individual data set, allows the description of the metadata quality information for the data specified the quality contents.

16.2.1   Data Quality Classes and Subclasses

In the ISO standard, data quality for the data specified by the data quality contents consists of two classes: lineage and quality information reports. The quality information report is constituted of five subclasses covering completeness, thematic accuracy, logical consistency, temporal accuracy, and positional accuracy. Figure 16.1 shows the classes and subclasses of data quality.

16.2.1.1   Lineage

The lineage class is mandatory if the quality information class report is not provided. It provides nonquantitative quality information about the lineage of the specific data set. The information includes a general explanation of the data creator's knowledge about the lineage of the specific data, a description of events in the life of the specific data set, or information about the source data used in creating the specific data set.

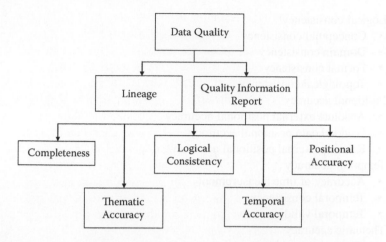

FIGURE 16.1 Data quality classes and subclasses in ISO 19115. (From FGDC [Federal Geographic Data Committee], 1994. Content Standards for Digital Geospatial Meta data. Washington, D.C.: Federal Geographic Data Committee. With permission.)

16.2.1.2 Quality Information Report

The quality information report provides quantitative quality information for the specific data set. This report is mandatory if the lineage class is not provided. The quality information is documented, according to the five subclasses of the data quality report. Each subclass includes the following items:

- Name of the test applied to the data
- Code identifying a registered standard procedure
- Description of the measure
- Type of method used to evaluate quality of the data set
- Description of the evaluation method
- Reference to the procedure information
- Date or range of dates on which a data quality measure was applied
- Value (or set of values) obtained from applying a data quality measure or the outcome of evaluating the obtained value (or set of values) against a specified acceptable quality level.

The quality information report contains references to temporal information that the data creator may require over a time period for the generation of a data quality report. Different evaluation methods may, also, be applied for each subclass of the quality information report. Hence, the quality information report is repeatable.

According to the ISO 19115 (2003), entities of each subclass of the data quality report are listed as follows:

- Completeness
  - Commission
  - Omission

- Logical consistency
  - Conceptual consistency
  - Domain consistency
  - Format consistency
  - Topological consistency
- Positional accuracy
  - Absolute external positional accuracy
  - Gridded data positional accuracy
  - Relative internal positional accuracy
- Temporal accuracy
  - Accuracy of time measurement
  - Temporal consistency
  - Temporal validity
- Thematic accuracy
  - Thematic classification
  - Non-quantitative attribute accuracy
  - Quantitative attribute accuracy

16.2.2 COMPLETENESS

Completeness in the data quality report is measured in terms of commission and omission errors. In terms of the report scope, *commission* refers to excess data present in the data set, while *omission* refers to data absent from the data set. Consider a data set specified to represent all buildings in a region, with heights greater than a specified value. The term *commission* specifies to what degree the buildings presented in the data set have been misidentified in terms of the specification requirement. The collection of all buildings meeting the requirement is called the *universe of discourse*. The term *omission* specifies to what degree buildings with heights taller than the specified value in the region are missing from the data set. The *commission* and *omission*, as defined here, concentrate on entity object completeness.

The alternative to data completeness is attribute completeness (Brassel et al., 1995). Attribute commission relates to excess attributes throughout the data set as described by a data quality scope. Attribute omission specifies to what degree the attribute values are missing throughout the data set. For example, a data set contains buildings in a specific region with the attributes of the number of stories, the height, and the address of each building in the data set. Commission specifies excess attributes that do not belong to the number of stories, the height, or the address of each building on a local level, or excess attributes throughout the whole data set on a global level. Omission relates to missing attributes for some buildings presented in the data set on a local level, or missing attributes throughout the whole data set.

Entity object completeness can be assessed when the data, specified by the scope, belong to the data set series, the data set, or the data smaller groupings which share the same feature type. Attribute completeness can be assessed when the data specified by the scope is the data set series, the data set, or the smaller data groupings that share the same attribute value. Completeness can be measured either in terms of a Boolean value, an integer, or an estimated percentage value.

The Boolean result can be gained by comparing the number of data items against the number of items in the universe of discourse, which, as indicated above, is the perfect data set that corresponds to the product specification. If the number of excess items in the data set exceeds the acceptable level of commission, this commission test will fail and give the value "false"; otherwise, the Boolean result for this evaluation test will be "true." If the number of absent items in the data set is larger than the acceptable level of omission, this omission test will fail and report "false"; otherwise, the omission test will pass and give the result as true.

The second method is again based on a comparison of the number of items in the data set and the number of items in the universe of discourse. The excess data number and the absent data number in the data set are reported as commission and omission, respectively.

The excess item number in the data set can be further divided by the total number of items in the universal abstract and multiplied by 100. In a similar manner, omission can be measured by dividing the number of missing items in the data set into the number of items in the universal abstract and multiplying the result by 100. This approach gives accuracy measures of commission and omission in percentage form.

16.2.3 LOGICAL CONSISTENCY

Logical consistency is the degree of adherence to the logical data structure rules and spatial data attribute rules. ISO 19115 classifies logical consistency into conceptual consistency, domain consistency, format consistency, and topological consistency. Conceptual consistency deals with rules of the conceptual schema. Domain consistency is the degree of adherence of values to the value domain. Format consistency considers the degree to which data are stored in accordance with the physical data structure. Topological consistency deals with the correctness of the explicitly encoded topological characteristics of a data set.

Each portion of the logical consistency subclass can be described in terms of Boolean variables, number, and percentage.

16.2.3.1 Measures for Conceptual Consistency

Conceptual consistency aims to report any violation of the conceptual schema of features. It can be measured as follows.

The number of features and relationships between features, which violate the conceptual data schema, is counted. If this number is not greater than the acceptable violation level in the data set, a Boolean variable will be assigned as true. If the number is greater than the acceptable level of violation in the data set, the Boolean variable will be assigned as false. This Boolean variable is one measure of conceptual consistency.

Conceptual consistency can also be assessed by counting the feature numbers and the relationships between features that violate the data set conceptual schema, or by dividing the result into the number of features and relationships between data set features and multiplying the result by 100. In these ways, number and percentage accuracy descriptions are provided.

16.2.3.2   Measures for Domain Consistency

Domain consistency deals with item attributes (features in vector format or pixels in raster format) within a data quality scope that does not lie within a certain domain.

The Boolean variable for describing domain consistency is determined based on the comparison of item attributes within the data as specified by the scope against acceptable attribute domains. The number of items with attributes outside the acceptable attribute domain is counted. If this number is larger than the acceptable level of items with attribute violations, the Boolean variable will have the value "false"; otherwise, the Boolean variable will have the value "true."

The number of items within the scope, the attributes that are outside the acceptable attribute domain, can also be reported in metadata for that data specified by the data quality scope. This number may then be divided by the number of items with attribute violations into the total number of items in the data set, as specified by the data quality scope; the result is multiplied by 100. This result expresses the accuracy measure in percentage terms.

16.2.3.3   Measures for Format Consistency

Format consistency is the degree of adherence to the format of a specified field. Similar to the above consistencies, a Boolean variable can be given to a number or percentage to depict the format consistency.

The Boolean variable is derived by comparing the record structure for all items with the specified field definitions and structures. Those items having format violations are counted. If the number of these items is not larger than the acceptable level of the items violating the specified format, the Boolean variable will be assigned as true; otherwise, the Boolean variable will be assigned as false.

The number of those items within the data set scope that violate the specified format is counted and recorded in metadata. Or the percentage is defined by dividing the number of the items violating the specified format into the total number of items within the data set scope and multiplying the result by 100.

16.2.3.4   Measures for Topological Consistency

Topological consistency comprises topological constraints for a cell complex or for a feature type.

For each cell complex or feature type, the number of items within the data set scope having topological inconsistencies is counted. This number is then compared with an acceptable level of the number of items having topological violations. If the number is larger than the acceptable level, the Boolean variable will have the value "false"; otherwise, the Boolean variable will have the value: true.

A number measure is defined by counting the number of items within the data set scope that have topological inconsistencies. The percentage is defined by dividing the number of items within the scope with topological inconsistencies into the number of items within the data set scope, and multiplying the result by 100.

16.2.4 Positional Accuracy

Positional accuracy deals with accuracy of the position of a feature or the position of a cell (or a group of cells). Chapter 4 provides a number of models for describing positional GISci spatial data quality. Positional accuracy can be divided into absolute or external accuracy, relative or internal accuracy, or position accuracy of gridded data. The first two forms of accuracy relate to a vector feature in an object-based spatial data model. The position accuracy of gridded data is for a raster cell in a field-based spatial data model.

16.2.4.1 Absolute or External Accuracy

Absolute or external accuracy reflects the closeness of reported coordinate values of objects to values accepted as or being true. This closeness is valued, based on a comparison between absolute location of a feature and its location in the universe of discourse (or the coordinate accepted as true). In a vector format, a feature may be a point, a line, or a polygon.

16.2.4.1.1 Point Measurement

A horizontal, vertical, and radial positional point accuracy can be derived from a statistical test analysis, and the results can be described by error indicators such as root-mean-square error (RMSE) and standard derivation (SD). Based on results of the statistical test analysis, an error ellipse, for example, can be created to visualize the positional accuracy for that point in a graph.

The RMSE statistic has been widely implemented in surveying for describing accuracy encompassing both random and systematic errors. The RMSE statistic for positional uncertainty in a point is defined as follows:

$$\text{RMSE} = \sqrt{\frac{1}{n}\sum_{i=1}^{n}\left[\left(x_i - x\right)^2 + \left(y_i - y\right)^2\right]}, \tag{16.1}$$

where n is the total number of point in the data set; (x_i, y_i) is the absolute coordinate value of the ith point in the data set; and (x, y) is the absolute coordinate value of the point in the universe of discourse. Furthermore, RMSE_x and RMSE_y for x- and y-direction positional accuracies for the point are given, respectively, by

$$\begin{cases} \text{RMSE}_x = \sqrt{\dfrac{1}{n}\sum_{i=1}^{n}\left(x_i - x\right)^2} \\[4mm] \text{RMSE}_y = \sqrt{\dfrac{1}{n}\sum_{i=1}^{n}\left(y_i - y\right)^2} \end{cases}. \tag{16.2}$$

The SD statistics for x- and y-direction, and radial positional accuracies for the point are expressed as

$$\left\{ \begin{array}{l} SD_x = \sqrt{\dfrac{1}{n-1}\displaystyle\sum_{i=1}^{n}\left(x_i - x\right)^2} \\[3ex] SD_y = \sqrt{\dfrac{1}{n-1}\displaystyle\sum_{i=1}^{n}\left(y_i - y\right)^2} \\[3ex] SD = \sqrt{\dfrac{1}{n-1}\displaystyle\sum_{i=1}^{n}\left[\left(x_i - x\right)^2 + \left(y_i - y\right)^2\right]} \end{array} \right. \qquad (16.3)$$

Equations 16.1–16.3 provide a positional accuracy measure for a single point feature. These two statistics can be further extended to a single data grouping within the data set itself, or a data set series to which the data set belongs. The sum of the error distance between absolute coordinate values of all (or sample) points within the data set and those in the universe of discourse can be calculated, and the result divided by the total number of all (or sample) points within the data set in the RMSE method, or the total number of all (or sample) points within the data set, minus 1 in the SD method. In this way, positional accuracy for points within the data set is obtained.

An alternative way deals, individually, with positional accuracy for the point type data set and gives an accuracy interval for describing positional accuracy for all points within the data set. Positional accuracy for each point (or each sample point) within the data set is assessed. The minimum and maximum values of the positional accuracies of all points (or all sample points) within the data set are then computed. These two values form an accuracy interval for describing positional accuracy for points within the data set.

According to ISO 19115, metadata can be represented in various forms, such as Boolean variables, numbers, tables, binary images, matrices, free texts, and others. The various forms of metadata enable the depiction of positional accuracy for point or all points within the metadata data set in terms of error ellipse. The error ellipse model is derived from standard deviation, and outlines a feasible region in which the true location of the point is most probably located. Error ellipse model details and the corresponding visualization are given in Section 15.1. The map to which the data set belongs and in which error ellipses for all points are embedded can thus be stored. Positional accuracy information for a related data set series and the data set itself are by this means documented.

16.2.4.1.2   Line Measurement

Three possible ways enable the assessment of positional accuracy for a single line. They are based on (a) characteristics of the line, (b) error distances between sample line absolute positions in the data set and also those in the universe of discourse, and (c) the confidence region model introduced in Chapter 4.

Line positional uncertainty can affect the line characteristics. The characteristics may include length or orientation. The length is the sum of distances between line nodes, while orientation can be interpreted as an aspect of fronting to the east. A characteristic difference between an absolute position of a data set line and that in the universe of discourse is used to indicate the line positional accuracy. If there is more than one observation for the data set line, the RMSE of the characteristics difference will be computed to indicate line positional accuracy:

$$\left\{ \begin{array}{l} \mathrm{RMSE}(l) = \sqrt{\dfrac{1}{n}\sum_{i=1}^{n}(l_i - l)^2} \\[2em] \mathrm{RMSE}(o) = \sqrt{\dfrac{1}{n}\sum_{i=1}^{n}(o_i - o)^2} \end{array} \right., \tag{16.4}$$

where l and l_i represent the line length in the universe of discourse and that in the data set, respectively, o and o_i represent the orientation of the line in the universe of discourse and that in the data set, respectively, and n is the total number of line observations in the data set.

Assessing line positional accuracy, based on a characteristic of that line, is an indirect method that aims to study impact of the positional accuracy of the characteristic. A direct approach to positional accuracy assessment can be derived from a distance between an absolute location of a line in the data set and that in the universe of discourse. A distance between two lines can be determined by a discrepant area (Shi et al., 2003). The discrepant area is the area of a region bound by the two lines (see Figure 16.2). In case there are multiple sample line absolute locations in the data set, the RMSE of the discrepant area can be computed.

The confidence region model for a line described in Section 4.5 can also be used to report positional accuracy of the line in metadata. A confidence region around an absolute line location in the data set is a feasible region, inside which the absolute line location in the universe of discourse is located with a specific probability. This is a graphical representation of line positional accuracy. Furthermore, dividing this confidence region area by the total line length in the universe of discourse is one possible line positional accuracy measure. If values of the specific probability for confidence regions around two data set lines are identical, the line with the larger value of this positional accuracy measure will contain a larger value of positional

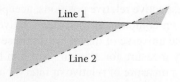

FIGURE 16.2 A discrepancy between line 1 and line 2.

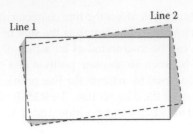

FIGURE 16.3   A discrepancy between polygons 1 and 2.

uncertainty. Therefore, the confidence region model not only provides a graphical representation of the positional accuracy of each line within the data set, but it further provides a numerical indicator of line positional accuracy.

To assess data set positional uncertainty, a data set series to which the data set belongs or a smaller data grouping within the data set, an accuracy interval for the positional uncertainty should be documented in metadata. The lower bound of this accuracy interval is the minimum value of a positional error measure of a line, and the upper bound is the maximum value of the positional error measure.

16.2.4.1.3   Measurement of a Polygon

Similar to the positional accuracy measurement of a line, positional accuracy of a polygon can be measured based on characteristics of the polygon, and a distance between an absolute location of the polygon in the database and that in the universe of discourse. Characteristics of the polygon may include area, perimeter, and center of gravity. The RMSE for each characteristic can be computed to describe the polygon positional accuracy. In addition to this indirect approach, a distance between absolute locations of the polygon in the database and in the universe of discourse can be computed, based on the discrepant area (Shi et al., 2003). This discrepant area between two polygons is derived from a region bound by the polygons (see Figure 16.3).

The polygon positional accuracy measure can be extended to describe positional accuracy for a data set series to which the data set belongs, the data set itself, or a smaller data grouping within the data set. An accuracy interval, on a global level, contains the minimum and maximum of a polygon positional accuracy measure as the lower and upper bounds of the accuracy interval, respectively.

16.2.4.2   Relative or Internal Accuracy

Relative or internal accuracy corresponds to the closeness of the relative positions of data set features to their respective relative positions, accepted as or being true. This closeness is valued based on comparisons between relative positions of features in the data set and those in the universe of discourse (or those accepted as true).

The positional accuracy measure for a feature given in the above subsection can also be applied to provide a measure of relative or interval accuracy by replacing the absolute coordinate with the relative coordinate. For example, the RMSE of the error distance for a point in the relative or interval accuracy is defined by

$$\text{RMSE} = \sqrt{\frac{1}{n}\sum_{i=1}^{n}\left[\left(\tilde{d}_i - \tilde{d}\right)^2\right]}, \tag{16.5}$$

where n is the total number of sample relative distance values of the lines in the data set; $(\tilde{d}_i)$ is the ith sample relative distance value of the point in the data set; and $(\tilde{d})$ is the relative distance value of the line in the universe of discourse.

16.2.4.3 Gridded Data Position Accuracy

Gridded data position accuracy is the closeness of gridded data position values to values accepted as or being true. This accuracy can be determined by comparing the absolute position of a gridded point in a data set with that in the universe of discourse (or that accepted as true) on a local level. The RMSE of the difference between these two absolute positions is then computed as a measure of this position accuracy of the gridded point. On a global level, the RMSE of differences between absolute positions of all (or sample) gridded points in the data set with those in the universe of discourse, is estimated.

In addition, the area (or the center of mass) of a cell group in the data set and that in the universe of discourse can be calculated. The RMSE of the area difference (or the difference in terms of the center of mass) is then computed to describe positional accuracy of this cell group. This approach is applicable for a data set series to which the data set belongs, the data set itself, or a smaller grouping of data within the data set.

16.2.5 Temporal Accuracy

Temporal accuracy is the accuracy of temporal information in a GIS database. It comprises three subelements, including accuracy of a time measurement, temporal consistency, and time validity.

16.2.5.1 Accuracy of a Time Measurement

Accuracy of a time measurement deals with the correctness of reported time measurement. Data within the data set, as specified by a specific scope, may contain temporal information, such as data occurrence time. To assess temporal information accuracy, for example, the difference between each data occurrence time in the data set and that in the universe of discourse is measured. The RMSE of this time measurement is then computed from the occurrence time differences.

16.2.5.2 Temporal Consistency

Temporal consistency corresponds to the correctness of ordered events or sequences. It can be detected by checking each historical event to ensure that the event is correctly ordered against the rest of events. The temporal consistency in terms of the number of items, or the percentage of those items wrongly ordered, can then be reported in the metadata. If an acceptable level of the number of items having temporal inconsistency is given, by comparing the number of items ordered wrongly with the acceptable level, a Boolean metadata result is reported.

16.2.5.3   Temporal Validity

Temporal validity is data validity of with respect to time. It can be determined by checking each data item with the data set, as specified in the data set scope, to assure that it was captured on the date as specified in the lineage. The number of items failing the check are either then counted and reported in metadata, or the result is divided by the total number of items in the data set and then multiplied by 100 and presented in metadata in percentage form. In addition, if an acceptable level of the number of items with temporal invalidity is given, a Boolean result will be obtained and reported in metadata.

16.2.6   THEMATIC ACCURACY

Thematic accuracy, also referred to as attribute accuracy, deals with the correctness of feature classifications and their relationships, the correctness of nonquantitative attributes, and the accuracy of quantitative attributes.

16.2.6.1   Classification Correctness

Classification incorrectness occurs when the wrong class is assigned to points or features, or when assigned classes are based on different data capture techniques (or from different observers) are inconsistent.

Many techniques have been developed to assess image classification accuracy, as discussed in Chapter 5. These techniques involve image sampling locations and comparisons of the class assigned to each location in a reference image. The results are then tabulated in the form of an error (or misclassification) matrix. For example, Table 16.1 shows an example of the error matrix for two classes. The columns define the classes in the true data, while the rows define the classes in the data set. The error matrix can be directly documented in metadata in global level matrix form.

According to the error matrix, two overall accuracy measures—the PCC and the kappa coefficient—are defined as

$$PCC = \frac{1}{c_{..}} \sum_{i=1}^{n} c_{ii},$$ (16.6)

TABLE 16.1

An Example of the Error Matrix for Two Classes

		Data Set Class		
		A	**B**	**Count**
True class	**A**	c_{11}	c_{12}	$c_{11} + c_{12}$
	B	c_{21}	c_{22}	$c_{21} + c_{22}$
		$c_{11} + c_{21}$	$c_{12} + c_{22}$	$c_{11} + c_{12} + c_{21} + c_{22}$

and

$$\kappa = \frac{1}{c_{..} - \sum_{i=1}^{n}[c_{i.}c_{.i}/c_{..}]}\left[\sum_{i=1}^{n}c_{ii} - \sum_{i=1}^{n}[c_{i.}c_{.i}/c_{..}]\right], \qquad (16.7)$$

where n is the number of classes, i and j represent the rows and the columns, respectively; c_{ij} is the table entries of the error matrix,

$$c_{..} = \sum_{j=1}^{n}\sum_{i=1}^{n}c_{ij}, \; c_{.i} = \sum_{j=1}^{n}c_{ji} \text{ and } c_{i.} = \sum_{j=1}^{n}c_{ij}.$$

Both measures can be documented in metadata in the form of number.

One additional overall accuracy measure is derived by comparing the assigned class against true class in the universe of discourse. Dividing the number of correctly classified items by the total number of items in the data set and multiplying the result by 100 can gain a percentage value for the overall classification accuracy.

The above accuracy measures, derived from the error matrix, are global. Each cell probability vector provides a thematic accuracy measure on a local level. Further probability vector details and the parameters derived from probability vectors are given in Section 5.2.4.

16.2.6.2 Nonquantitative Attribute Accuracy

A nonquantitative attribute is an attribute that is not represented in the form of ordinal data. It may include geographic names, addresses, land use types, and others. Accuracy of the nonquantitative attribute can be assessed by examining whether all (or sample) items within the data set have attribute values differing from those in the universe of discourse. The number of items with incorrect attribute values, or the percentage of these items in the data set, can be metadata documented.

16.2.6.3 Quantitative Attribute Accuracy

A quantitative attribute refers to an attribute represented in the form of ordinal data. The attribute accuracy can be described statistically, such as by using RMSE measure. RMSE can be computed from differences between an attribute value in the data set and that in the universe of discourse. Furthermore, the number of items with the difference exceeding the specification limit can be counted and the metadata documented. This count can also be presented in percentage form.

16.3 AN OBJECT-ORIENTED METADATA SYSTEM

An error metadata management system (EMMS) (Gan and Shi, 2002) was designed to provide quality metadata information for spatial objects at object level. The system was developed based on an object-oriented modeling method. It updates any change in data quality and temporal information of the individual spatial objects during database maintenance. Feature metadata created in the EMMS can be used in the automatic creation of CSDGM compliant metadata.

The spatial data, that is the 1:20,000 topographic data set of the study area in Hong Kong, originated from many data sources. Accuracy of each data source is not identical; it is subject to the data capture method by which that data source was obtained, to the characteristics of the data source (e.g., whether the data is vague), and to human perception of the data source, which causes different degrees of generalization error.

16.3.1 DATABASE DESIGN

Data life cycles comprise three stages: birth (initialization, growth); processing, including displacement, line thinning, and other generalization processes; and death (obsolescence). EMMS was designed to update data automatically whenever any geometrical change to the data set occurred so that dynamic metadata can be obtained on a feature level.

EMMS organizes quality and temporal information (including spatial data quality elements, temporal information, aggregation error information, generalization effects and information of data source, data capture methods, and data capture parameters) into five tables, two of which are supporting tables (Figure 16.4). In addition, user-defined tables can also be linked to a feature through the unique "MSLINK" number. A brief explanation of each metadata data field is given in Table 16.2.

EMMS has been developed to automatically manage feature-based metadata for any data manipulation. However, data producers should take the responsibility for validating the quality information of each feature within the data set. Figure 16.5 shows an example of system interface for checking quality information.

EMMS can be used to automatically manage metadata for a data set on a local level. It can input, store, and update the lineage, and quality and temporal information for each data item within the data set, subject to data manipulation. Data manipulation includes data initialization, generalization processing, and obsolescence. The lineage, quality, and temporal information relating to the data are captured automatically with minimal operator input. It can be retrieved by users with an SQL command. The users are enabled to determine the fitness of use of a specific data item for their own application, with query functionality.

The EMMS can work with CSDGM-compliant metadata, as the quality and temporal information captured by the system is defined according to the CSDGM standard. Both the feature-based metadata and the CSDGM compliant metadata can be stored in XML formation for Web-based queries.

In addition to the generalization process handled in the EMMS, change may be caused by spatial analyses and other spatial processes. Therefore, there is a need to further develop the system regarding managing data quality and the automatic documentation GIS of quality information.

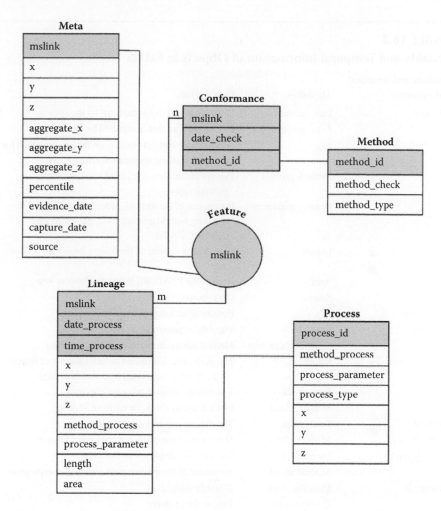

FIGURE 16.4 The EMMS database structure. (From Gan, E. and Shi, W. Z., 2002. In *Spatial Data Quality*, Shi, W. Z., Goodchild, M. F., and Fisher, P. F. (eds.). London: Taylor and Francis. With permission.)

16.4 SUMMARY AND COMMENT

Several existing worldwide metadata standards have been summarized in this chapter. Most are CSDGM-compliant. ISO 19115 was the first international metadata standard endorsed in 2003. The current metadata standards have been revised to ensure compliance with the ISO international metadata standard. The metadata standards provide the benefit of internal data sharing, such as within a company and externally throughout the world.

Spatial data quality information is an important part of spatial metadata. Spatial data quality metadata covers completeness, logical consistency, positional accuracy,

TABLE 16.2
Quality and Temporal Information of Objects in EMMS

Quality and Temporal Information	Attributes	Description
Lineage	Date_process	Processing date that is captured by the system
	Time_process	Processing time that is captured by the system
	x, y, z	Signed uncertainty in position on the process taken for calculation of the aggregation errors
	Method_process	Process undertaken (e.g., head-up digitizing, generalization)
	Process_parameter	Parameters of the method_process (for example, in scanning with MapSetter 4000, parameters are OFFSET, GAIN)
	Length	Length or parameter of the feature before the process step
	Area	Area of the FEATURE before the process step
	Source	Data source
Positional_ accuracy	x, y, z	Positional accuracy of individual features
	Date_check	The date of positional accuracy checking
	Method_check	Method use to check the positional accuracy
Attribute accuracy	Percentile	The percentile ascertained to be a category of feature (e.g., feature categorized by remote sensing)
	Date_check	The date of attribute accuracy checking
	Method_check	Method use to check the attribute accuracy
Logical_ consistency	Date_check	The date of consistency checking
	Method_check	Statement of compliance among other features
Completeness	Date_check	The date of completeness checking
	Method_check	Statement of compliance with technical specification
Temporal	Evidence_date	Date of evidence
	Capture_date	Date of data capture
Others	Aggregate_x, aggregate_y, aggregate_z	Aggregation errors from graphical manipulations on features.

Source: Gan, E. and Shi, W. Z., 2002. Error Metadata Management System. In *Spatial Data Quality*, Shi, W. Z., Goodchild, M. F., and Fisher, P. F. (eds.). London: Taylor and Francis. With permission.

thematic accuracy, and temporal accuracy. The existing metadata standards provide a spatial data quality framework for these aspects. The details given in this chapter—together with the corresponding measurement methods described in the previous chapters of this book, such as positional uncertainty measurement methods in Chapter 4—provide contributions to further the successful measurement of the different aspects of data quality elements.

An object-oriented metadata organization method has been proposed to document metadata to a feature or object level. Correspondingly, the EMMS has been

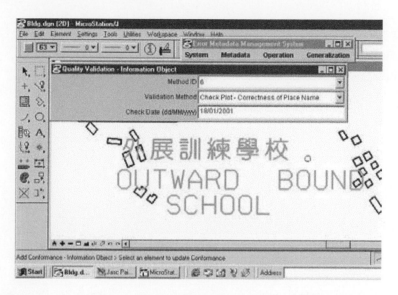

FIGURE 16.5 An example of system interface for checking quality information using EMMS. (From Gan, E. and Shi, W. Z., 2002. In *Spatial Data Quality*, Shi, W. Z., Goodchild, M. F., and Fisher, P. F. (eds.). London: Taylor and Francis. With permission.)

designed, based on the proposed method, for solving problem on data quality metadata management in the real world.

REFERENCES

ANZLIC (Australia/New Zealand Land Information Council), 2001. ANZLIC Metadata Guidelines: Core metadata elements for geographic data in Australia and New Zealand, version 2. http://www.anzlic.org.au/asdi/metagrp.htm.

Brassel, K., Bucher, F., Stephan, E. M., and Vckovski, A., 1995. Completeness. In *Elements of Spatial Data Quality*, Guptill, S. C. and Morrison, J. L. (eds.). Oxford: Elsevier.

CEN (European Committee for Standardization), 1998. Geographic Information—Fundamentals Overview. http://forum.afnor.fr/servlet/ServletForum?form_name=cForumFich&file_name=Z13C%2FPUBLIC%2FDOC%2F13425.PDF&login=invite&password=invite.

FGDC (Federal Geographic Data Committee), 1994. Content Standards for Digital Geospatial Meta data. Washington, D.C.: Federal Geographic Data Committee.

FGDC (Federal Geographic Data Committee), 2000. Content Standard for Digital Geospatial Metadata Workbook Version 2.0. Washington, D.C.: Federal Geographic Data Committee.

Gan, E. and Shi, W. Z., 2002. Error Metadata Management System. In *Spatial Data Quality*, Shi, W. Z., Goodchild, M. F., and Fisher, P. F. (eds.). London: Taylor and Francis.

Shi, W. Z., Cheung, C. K., and Zhu, C. Q., 2003. Modeling error propagation in vector-based buffer analysis. *Int. J. Geog. Inform. Sci.* 17(3): 251–271.

17 The Web Service-Based Spatial Data Quality Information System

This chapter addresses dissemination of quality information on spatial data and spatial analyses based on one of the latest Web technologies, the Web service technology. A new system designed by the author, the Web Service-Based Spatial Data Quality Information System (DQIS), has been developed for delivering two types of quality information: (a) numerical quality information on point, line, and area spatial objects and buffer and overlay spatial analyses, and (b) visual representation of quality information. With this system, quality information on spatial data and analyses will be widely accessible to the users of spatial data through the Web. The system is now described in this chapter.

17.1 INTRODUCTION

Spatial data quality information is not only essential for spatial data-based GIS applications, but is also a basic element enabling the interoperation and integration of multiple geographic data. Web service technology is a new technology enabling the distribution information through the Web, and can thus be used to distribute spatial data quality information. Web technology organizations such as World Wide Web Consortium (W3C) and international standard organizations, such as International Organization for Standardization, Technical Committee 211 Geographic Information/ Geomatics (ISO/TC211) and Open Geospatial Consortium (OGC) are working on geographic information interoperability. Both types of organizations have proposed solutions to capacitate GIS and Web Service technology integration. W3C focuses on Web service technological and implementation aspects, whereas OGC focuses on Web service concepts, with an emphasis on Web service geographic information status and functions. That is to say, OGC concentrates on the Web service framework and concepts, whereas W3C concentrates on Web service implementation.

The Web service offers networking components equipped with basic functions to enable internet distribution, computing, and integration. It has already become an integrated platform for heterogeneous systems, as it has the features of open communication standard and emphasis on interoperability. Web service technology can be applied to directly connect and link various application programs, services, and devices, and thus, by using a consistent mode, can realize the data (including data quality information) exchange among different platforms.

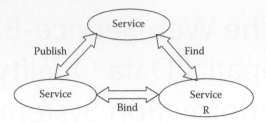

FIGURE 17.1   The conceptual model of Web Service.

TABLE 17.1
A List of Protocols Used by Web Service

Protocol Layer	Protocol
Workflow	WFSL
Service-discovering, integration	UDDI
Service-defining	WSDL
Information layer	SOAP
Transmission layer	HTTP, FTP, SMTP
Internet	IPV4, IPV6

The Web service describes a service-oriented and components-based application architecture. Standardized service providers can register and publish their services in the registration center, while service users can find and use these services. The Web service conceptual model is illustrated in Figure 17.1.

The Web service has the following characteristics: encapsulation, loose coupling, standardized protocols and norms, highly integration capability, and common data format. The service is composed of the series of protocols listed in Table 17.1.

To realize the interoperability of spatial data, it is necessary to have a common protocol. Each application service, when utilized, is interrelated via interfaces. Therefore, interoperability between spatial data is not the only issue of the data model or an application programming interface (API).

17.2   DESIGN OF THE SYSTEM

To provide users with quality information about spatial data and spatial analyses through the Web, the DQIS has been newly designed by the author. The system is composed of two parts: a server and a client. The server provides functional services for program providers based on the requirements of the Web service, whereas the requirements of the client are service functions based on specific operations.

17.2.1   DATA TRANSFORMATION SERVICE

DQIS is designed to comply with the data sharing standards of Open Geospatial Consortium (OGC). The system transfers data with various formats (such as E00 of

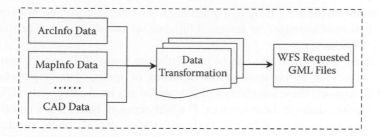

FIGURE 17.2 Data transformation service.

ArcInfo, Shapefile of ArcView, Mif of MapInfo, and others) into a system requested format as shown in Figure 17.2.

OGC and ISO/TC211 have proposed geographic information service standards based on XML, including the following three data-sharing standards: Web Map Service (WMS), Web Feature Service (WFS), and Web Coverage Service (WCS). These three standards are introduced briefly below.

Web Map Service: the aim is to generate maps based on the visualization of geographical data. Three operations are defined in WMS. The first operation, *GetCapabilities*, returns the metadata described in XML at the service level. The second operation, *GetMap*, returns a map image with respect to the requested parameters. Its spatial parameters and parameter values are explicitly defined. The returning map images can be in the format of GIF, JPEG, PNG, or SVG. The third operation, *GetFeatureInfo*, returns specific map information. This latter operation is optional.

Web Feature Service: the aim is to return GML codes at the feature level, and it also offers functional operations, such as adding, modifying, or removing features. Therefore, it is concerned with a relatively deeper level of service as compared to WMS. The standardized WFS define the following five operations. The first operation is the *GetCapabilities* operation, which returns the XML documents that describe the capability of WFS. The second operation, *DescribeFeatureType*, returns the WFS-described feature type schemas. The third operation, *GetFeature*, returns the acquired feature GML documents. Clients can specify which attributes of the obtained features are needed and can also use spatial or nonspatial queries with the specific conditions. The fourth, operation *transaction*, provides routine request services. One routine work is for creation, modification, and geographic feature removal. The fifth operation, *LockFeature*, can handle locking requests for single feature or for the multi-features events. It can thus guarantee that the system can support sequential events.

Web Coverage Service (WCS): the aim is to provide a spatial image-oriented service. WCS supports electronic retrieval of geospatial data as "coverages," and provides access to potentially detailed and rich sets of geospatial information. WCS is composed of three operations, *GetCapabilities*, *GetCoverage*, and *DescribeCoverageType*. *GetCapabilities* returns the description XML documents of service and the data set. *GetCoverage* is executed after *GetCapabilities*, confirming which queries can be executed and which data can be acquired. It returns the values or attributes of geographic locations, bundled in a well-known coverage format. *DescribeCoverageType* returns a full description of a coverage provided by concrete WCS server.

The above three standards can be used either as spatial data service standards in Web service or as implementation standards of spatial data interoperability. When GIS supports these interfaces and is deployed at a local server, other GISs can acquire required data via these services. The Web services-based DQIS described in this chapter supports these WFS standards, with the aid of GML as the standardized format for data transformation in the network. The advantage of adopting the WFS standards as the system format is that much commonly used GIS software, such as ArcIMS, ArcGIS, and MapInfo, support the WMS, WFS, and WCS standards of OGC. The DQIS can directly call the WFS of these GISs. However, the drawback is that GML is in a text format and requests large data volume, resulting in a relatively slow data transmission speed.

17.2.2   SERVICE ON DATA QUALITY INFORMATION

As indicated above, DQIS has been designed to provide quality information on spatial data and spatial analyses by using Web service-based technology. The service provides two types of quality information: (1) spatial data quality information and (2) spatial analyses quality information. Spatial data includes points, lines, and area objects, while spatial analyses cover buffer and overlay analyses.

After the service data are transmitted toward the system, data in GML format can be provided in the server for GIS function services. Function services are realized by Web service technology, and Simple Object Access Protocol (SOAP) supported protocol. Data transmitting standards comply with OGC WMS and WFS standards. As Web service technologies possess the network component characteristics, related GIS function services can be encapsulated within Web services. The components can be easily integrated into a system. The application of Web service technology for managing and disseminating GIS data quality information is described below, and the design of the DQIS is illustrated in Figure 17.3.

Currently, two solutions exist to implement the Web service: (a) Java language based on the J2EE standard and (b) a development platform based on Microsoft .Net software. Java language possesses complete and comprehensive Web Service tools, and the Java virtual machine is installed at the client side. However, .Net software for the development platform is strong in Web Service encapsulation, and requires the installation of a NET framework environment at the client side. The DQIS, described in this chapter, is developed in C# language and on Microsoft .NET platform.

17.2.3   THE CLIENT DESIGN

The client provides data services and function services related to quality information of spatial data and spatial analysis. These services are provided in response to user requests from the client. A user makes a request from the client side, and the client then sends that user's request to the server, which then executes the corresponding Web service-related request. Finally, the generated resultant document is returned to

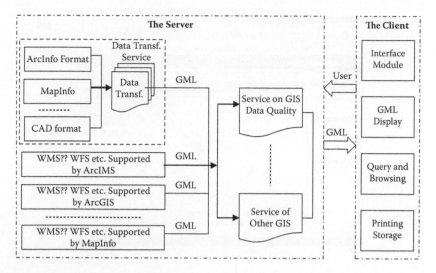

FIGURE 17.3 The design of the DQIS.

the client side in GML. Graphical visualization results, relating to the original request, can be made by the client through an interpretation of the GML documents.

The following tasks are performed at the client side:

- User requests are explained through a user interface, and the user's request is processed by calling for the corresponding Web service in the server.
- The obtained GML document at the client end is interpreted and visualized after the Web server at the server side produces the response GML document.
- A simple zoom-in, zoom-out, panning, and other browsing functions at the client end are provided.

17.3 FUNCTIONS OF THE DATA QUALITY INFORMATION SYSTEM

The data quality information function service includes (a) providing uncertainty information for spatial objects covering that information for point, line and area objects; and (b) providing uncertainty information for spatial analyses—buffer and overlay.

17.3.1 QUALITY INFORMATION OF POINT OBJECTS

Points are fundamental elements to describe spatial features in GIS. A point is generally described by its two-dimensional coordinates (x_1, x_2) or three-dimensional coordinates (x_1, x_2, x_3). Point quality can be described by its positional error, with an error indicator.

Random variable error can be indicated through its variance and covariance matrices. The accuracy of random vector of a point coordinate depends on the accuracy of the measured value and the relationship between these measured values. With the variance and covariance matrices, a point error can be quantified and visualized (e.g., using an error ellipse or the error circle model, as shown in Figure 17.4). The point error is expressed by related error ellipses. Details of object

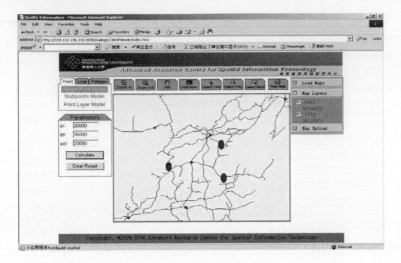

FIGURE 17.4   **(See color figure following page 38.)** Visualization of quality of points with error ellipses model.

positional uncertainty modeling, including that for point, line, and area objects in GISci are given in Chapter 4.

17.3.1.1   Definition of Interfaces

a. The *PntErrEllipse* Interface: The error ellipse of a point is acquired based on the inputted point (PntX, PntY) and its error parameters. The Web service returns the calculated error ellipse in XML documentation form.

Name of Parameter	Data Type	Description of Function
PntX	String	Input X coordinate of a point
PntY	String	Input Y coordinate of a point
g, gy, gxy	String	Input error parameter of a point

b. The *PntErrEllipses* Interface: According to the coordinate string, strPntXY, of various inputted points and their error parameters, the error ellipses of those points are obtained and returned to the client as XML documents.

Name of Parameter	Data Type	Description of Function
StrPntXY	String	Input X, Y coordinate strings of various points
gx, gy, gxy	String	Input error parameters of points

c. The *PntErrLayer* Interface: With the inputted points and point error parameters of a layer, the error ellipses can be obtained for the layer and are returned to the client as an XML document.

Name of Parameter	Data Type	Description of Function
URL	String	Input the information of an XML document of a point layer
LayerName	String	Input the name of a point layer
gx, gy, gxy	String	Input error parameter of a point layer

17.3.2 Quality Information of Line Objects

Line uncertainty is related to component point uncertainties, and to the relationship between the point errors. A number of line error models are presented in Chapter 4. These include the confidence region model, error band models, and maximum error model. Line quality information can be generated and visualized based on these models. As an example, line quality based on the error band model is visualized in Figure 17.5.

a. The *LineErrBand* Interface: The XML error band document is acquired in accordance with the input information of the line and related error parameters. The commutated XML document is then returned to the client.

Name of Parameter	Data Type	Description of Function
URL	String	Input the XML document of a line layer
LayerName	String	Input the name of a line in the layer
O_ID	String	Input the designated O_ID of a line
gx1, gx1x2, gx2, gx1y1, gx1y2, gx2y1, gx2y2, gy1, gy1y2, gy2	String	Input the error parameters of a line

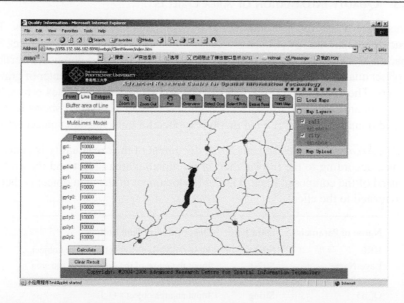

FIGURE 17.5 (See color figure following page 38.) Visualization of line quality with an error band model.

b. The *LineErrBands* Interface: The XML line error band document is acquired in accordance with the inputted multilines and the error parameter information. The XML computed multiline error band document is then returned to the client.

Name of Parameter	Data Type	Description of Function
URL	String	Input the XML documents of line layers
LayerName	String	Input the names of line layers
StrO_IDs	String	Input the designated O_ID strings of lines with the separator of ",": O_ID1, O_ID2, O_ID3 ..., O_IDn
gx1, gx1x2, gx2, gx1y1, gx1y2, gx2y1, gx2y2, gy1, gy1y2, gy2	String	Input error parameters of lines

c. The *LineErrLayer* interface: The XML document is acquired by inputted line layer and the error parameter information. The computed XML document results; the lines' error band is then returned to the client.

Name of Parameter	Data Type	Description of Function
URL	String	Input the XML document of a line layer
LayerName	String	Input the name of a line layer
gx1, gx1x2, gx2, gx1y1, gx1y2, gx2y1, gx2y2, gy1, gy1y2, gy2	String	Input the error parameter of lines

17.3.3   QUALITY INFORMATION OF AREA OBJECTS

An area object uncertainty is determined by the composed lines and vertices. Two approaches are introduced in Chapter 4 for modeling uncertainty of an area object. The approaches are (a) based on the composed lines, and (b) based on composed vertices. Area object quality can be indicated by parameter error such as area, center of gravity, and others. The quality of an area object can also be indicated by vertex errors. The quality of an area object is visualized in Figure 17.6 as an example. It shows how the composed boundary line error band union is applied for modeling area object error.

a. The *AreaErrObj* interface: an XML document of the area quality is generated according to the information of an inputted area object and the error band of the composed lines. The XML document of the area object is then returned to the client.

Name of Parameter	Data Type	Description of Function
URL	String	Input an XML document of an area object
LayerName	String	Input the name of the layer where the area object belongs to
O_ID	String	Input an area object's O_ID
Region	String	Input the region area object

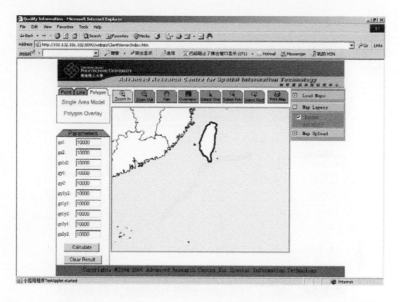

FIGURE 17.6 **(See color figure following page 38.)** Quality information of an area object.

b. The *AreaErrObjs* interface: According to the information of inputted multi-area objects and the area object regions, an XML document of area objects can be generated and an XML document of the area objects is returned to the client.

Name of Parameter	Data Type	Description of Function
URL	String	Input an XML document of the area objects
LayerName	String	Input the name of the layer that the area objects belong to
Str_O_IDs	String	Input the designated O_ID strings of lines with the separator of ",": O_ID1, O_ID2, O_ID3...O_IDn
Regions	String	Input the region of the area objects

17.3.4 QUALITY INFORMATION OF BUFFER SPATIAL ANALYSIS

It is reported in Chapter 10 that uncertainty of a buffer spatial analysis is quantified by errors of commission and omission and a normalized discrepant area. Details of these models are given in the chapter. An example of buffer analysis visualization is given in Figure 17.7.

a. The *BufferObj* interface: According to the inputted information of a line object and the radius of a buffer, an XML document of the line's buffer can be obtained. The XML document with the calculated error band of the object is then returned to the client.

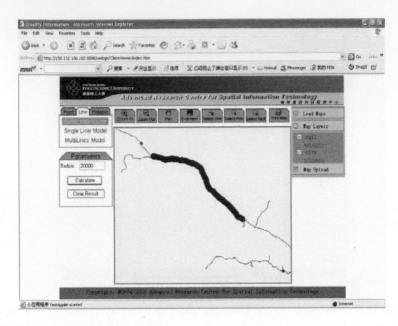

FIGURE 17.7   (See color figure following page 38.) Visualization of quality information of a buffer spatial analysis.

Name of Parameter	Data Type	Description of Function
URL	String	Input an XML document of an object layer
LayerName	String	Input the name of a the object layer
O_ID	String	Input the designated O_ID of the object
BufRadius	String	Input the radius of a buffer

b. The *BufferObjs* interface: According to the information of multiline objects and the radius of buffers as input, an XML document of the line' buffers are acquired. The XML document with the calculated multi-line buffer error bands are then returned to the client.

Name of Parameter	Data Type	Description of Function
URL	String	Input an XML document of line objects' layers
LayerName	String	Input names of line object layers
Str_O_IDs	String	Input the designated O_ID strings of line objects with the separator of ";": O_ID1, O_ID2, O_ID3…, O_IDn
BufRadius	String	Input the radius of buffers

c. The *BufferObjVar* interface: According to the information of inputted line objects, the width of buffers and a variance-covariance matrix, buffer values of commission errors and omission errors can be computed by the lines and returned to multipolygonal redundant errors and omission errors of calculation-resulted line buffers.

Name of Parameter	Data Type	Description of Function
URL	String	Input an XML document of layer with line objects
LayerName	String	Input the name of the layer
O_id	String	Input a line object's ID
BufRadius	String	Input the radius of a buffer
StrCovPoly	String	Input a variance-covariance matrix of a line
VovPolyRow	String	Input the row number of a variance-covariance matrix of a line
CovPolyCol	String	Input the column number of a variance-covariance matrix of a line
SimulNum	String	Number of simulation computation

17.3.5 Quality Information on Overlay Spatial Analysis

Overlay spatial analysis uncertainty propagation is described in Chapter 9. Two approaches are introduced: (a) an analytical approach and (b) a simulation-based approach. Figure 17.8 gives a visual description of overlay analysis for providing quality information in the analysis.

a. The *OverlayDeviation* interface: The maximum and minimum of both a standardized deviation and positional distance error is calculated, according to the variance-covariance matrix of the polygon component vertices. The XML document with the maximum and minimum of the standardized deviation and distance positional error is then returned to the client.

Name of Parameter	Data Type	Description of Function
ErrNum	Long	Input the number of variance-covariance matrix of points
ErrValues	String	Input the values of concrete variance-covariance matrix and separated with the separator, ";"

b. The *Intersect PntVar* interface: A variance-covariance matrix of an intersected point by two line segments is calculated and a variance-covariance

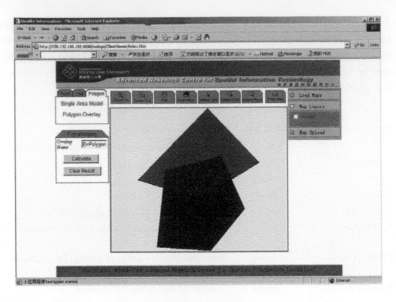

FIGURE 17.8   **(See color figure following page 38.)** A visual description of overlay for providing quality information in the analysis.

matrix of the two line segment calculated intersected point—as a 2 by 2 matrix is returned to the client.

Name of Parameter	Data Type	Description of Function
PntA1, pntA2	String	Start point and end point of the first line segment
PntB1, pntB2	String	Start point and end point of the second line segment
PntVar	String	Variance-covariance matrix [8x8] of the four points, separated by the separator ","

The *OverlapPolyVar* interface: A variance-covariance matrix of two polygons is calculated using the simulation mode. The variance-covariance matrix of the two calculated polygons, including that of the perimeter, area, and gravity in a XML document, are returned to the client.

Name of Parameter	Data Type	Description of Function
URL	String	Input an XML document of the polygons to be performed with the overlay analysis
LayerName1	String	Input layer name of the first polygon
O_ID1	Long	Input the O_ID of the first polygon
LayerName2	String	Input the layer name of the second polygon
O_ID2	Long	Input the O_ID of the second polygon

Name of Parameter	Data Type	Description of Function
CovPoly1	String	Variance-covariance matrix of vertices of the first polygon with the separator, "," to separate the inputted values
CovPolyrow1	Int	Row number of the variance-covariance matrix of the vertices of the first polygon
CovPolycol1	Int	Column number of the variance-covariance matrix of the vertices of the first polygon
CovPoly2	String	Variance-covariance matrix of vertices of the second polygon with the separator "," to separate the inputted values
CovPolyrow2	Integer	Row number of the variance-covariance matrix of the vertices of the second polygon
CovPolycol2	Integer	Column number of the variance-covariance matrix of the second polygon
SimulNum	Long	Number of simulative calculation

17.4   SUMMARY AND COMMENT

The Web service, one of the latest developed Web technologies, has been adopted for disseminating the quality information on spatial data and spatial analyses. Based on the technology, the Web Services-Based Spatial Data Quality Information System, designed by the author, has been developed for delivering two types of quality information: (a) numerical information on the quality of spatial data (point, line, and area objects) and spatial analyses (buffer and overlay), and (b) visual representation of quality information. DQIS can be further extended in the following three areas: (a) covering attributes, logic consistency, completeness aspects of data quality; (b) handling complex geoanalysis based on geostatistics; and (c) being extended as a generic metadata information service system.

Section VII

Epilogue

18 Epilogue

The principles of modeling uncertainties in spatial data and analyses have been introduced, developed, and described in the previous chapters, and it is now appropriate to turn our attention to the future developmental trend of these principles in the interest of pushing forward the boundaries of knowledge in this field. Hence, an attempt is made to address this issue in the final chapter. The following four aspects are considered:

- Research issues for the future
- Additional mathematical theories for uncertainty modeling of spatial data
- Implementation issues
- Interaction with other disciplines

18.1 RESEARCH ISSUES FOR THE FUTURE

The framework for modeling uncertainty in spatial data and spatial analyses covers the following: understanding the complex and uncertain nature of the natural world; the quantification of the uncertainties in the cognition of the natural world; the modeling of errors when measuring recognized objects; the determination of uncertainties in spatial object representation; the modeling of uncertainty propagation and amplification in spatial analyses; quality control for spatial data; and GISci applications based on uncertain and incomplete geographic information.

The following research areas have been identified for further study, based on the above and the related theories so far developed.

18.1.1 Understanding More about the Complexity of the Natural World

The complexities of the natural world are still a mystery to mankind. Indeed, our understanding is probably limited to a certain range—say, less than 20% of the total. The exact and precise nature of reality in terms of whether it is determined or uncertain, is unknown. If the natural world is uncertain, the nature of those uncertainties need to be identified, a related question being, for instance, whether the nature of uncertainty, once defined, can be true for all aspects of reality. A further question concerns whether an accepted definition is true for the recognition and measurement of everything in the real world. Indeed, is it possible for everything in the real world to be recognized and measured? Such questions have not, as yet, been fully addressed in studies on modeling uncertainties in spatial data and analysis, and possibly the answers to some of the queries need the extra input of philosophical analysis.

18.1.2 INVESTIGATION OF UNCERTAINTIES IN THE COGNITION OF GEOGRAPHIC OBJECTS

As indicated above, cognitive uncertainties of geographic objects have not, as yet, been well addressed in modeling uncertainty in spatial data and analysis. Geographic uncertainty cognition can be achieved either by human reasoning or by computer-based analysis of selected data. Examples of the latter are object cognition by computer vision or the classification of remotely sensed satellite images. Cognition uncertainty may result from (a) limited knowledge of the natural world or insufficient domain knowledge necessary for the cognition, and (b) limited information about the object to be recognized in spatial, temporal, and attribute domains.

However, the major cause of uncertainty or even mistakes in cognition is limited human knowledge of the natural world. A particular toy is available for children to enable understanding of shapes. The child is asked to select a specific shape and match it to the space of the same shape. When Daniel (a boy at 1½ years old) first played with such a toy, he could not tell the difference between any of the shapes. After two months of play, he was able to match the shapes correctly and therefore able to recognize such as a triangle and hexagon. His ability regarding shape cognition had improved, shape recognition uncertainty reduced and therefore shape knowledge gained. This simple example illustrates the uncertainty of human cognition of the complexities of the real world.

Cognition of the natural world is normally based on limited spatial, temporal, and attribute information and knowledge at any one time, this limitation being responsible for cognition uncertainty. For example, when using multispectral satellite images for land cover classification (a form of automatic cognition of land cover classes by computer) from a Landsat satellite image with a spatial resolution of 15m by 15m, a specific pixel (an area of 15m by 15m), encompassing a mixture of 30% area of grass, 60% area of tree, and 10% area of water could finally be classified as a pixel of a tree. Cognition uncertainty is from the classification. A better classification would give the percentiles of the classification components, rather than that of the majority class. In addition, the spatial distribution pattern of the component classes within the pixel should also be indicated. Another example of classification uncertainty is that arising from different spectrum reflections appearing from any one type of land cover at different times of the day or in different locations. Such phenomena lead to attribute classification uncertainty. Similarly, temporal uncertainty results if the land cover changes, from such as dry vegetation to flooded vegetation and back within, say, five days, given that the time interval of the obtained satellite images, for this change detection, is 14 days. This land cover change cannot be well detected because of the limited temporal resolution of the obtained satellite images.

Another source causing spatial cognition uncertainty is the cognition difference of an object, evidenced by different groups of people. This form of uncertainty can be potentially resolved if formal and consistent definitions have been given to spatial objects prior to any cognition necessity. It is the unrecorded definitions of objects by different people groups that is likely to cause such cognition uncertainty. For example, a building will be defined and classified differently according to the context of the users' experience. Such users could be land survey groups, urban planning groups, building development groups, and tax evaluation groups. Cognition uncertainty in this case is generated because of the different background knowledge of each of the groups. This form of uncertainty causes problems in spatial data interoperability and information transfer between the data provider and user groups.

Essential scientific issues to be resolved are the improvement of the cognition level of the natural world based on the real or imposed limited knowledge or information gained from the available spatial, temporal, and attribute domains, and also the quantification of uncertainties in spatial object cognition. How to achieve this knowledge is the subject of further research.

18.1.3 Modeling Integrated Measurement Errors

In position error modeling of spatial objects in GIS, current main stream research involves random error modeling. In fact, three types of measurement errors exist: random error, systematic error, and gross error. Several solutions for determining gross measurement errors, such as those in photogrammetry, have been developed. What needs further exploration are the methods for modeling integrated random, systematic, and gross errors.

Another type of uncertainty to be further studied is uncertainty in the temporal domain. Research in this area, so far, is limited. Temporal measurement, in many cases, is associated with spatial measurement. Therefore, a solution could be to model integrated temporal and positional uncertainty using geographic data. To a certain extent, methods developed for measuring positional error for spatial data can be applied if the nature (such as the nature of random error) of temporal error is similar to that of positional error.

18.1.4 Uncertainty Modeling for Complex Spatial Analyses

The methods developed for modeling positional uncertainties in spatial analysis, such as for overlay, buffer, and line simplification in GIS, are domain independent. Modeling attribute, topologic consistency, and ascertaining the completeness aspects of uncertainty is an extension of this research to be tackled in the future. Another important issue is domain-dependent spatial analysis, such as uncertainty modeling for land use change detection, spatial analysis based on multisource data with uncertainties. A model integrating attribute and positional uncertainties needs to be developed and used for this area.

18.1.5 Data Quality Control for GIS with Complex Uncertainties

Research on spatial data quality control has recently begun. So far, the aspect of errors has been addressed, and it is likely that further studies will focus on spatial data with complex uncertainties, such as quality control of spatial data with multi-category uncertainties (both spatial and attribute) and uncertainties in multiscale spatial data. Quality control of complex GIS analysis also needs to be addressed. Examples include such as change detection or logic inconsistency checks. Currently, least square adjustment has been adopted as the main stream theory for spatial data quality control analysis. Hence, alternative solutions need to be studied, considering the limitations of this adjustment method.

18.1.6   THREE FUNDAMENTAL ISSUES

The following three issues are fundamental to further research in the area of error modeling for spatial data and spatial analysis.

- Uncertainty in spatial data may possibly be the result of a combination of several error sources. For example, the uncertainty of a polygon boundary may be the result of both positional and attribute errors. Therefore, the commonly used classification of data quality components may have limitations. In this event, a reformulation of the existing foundations of data quality component classification, such as positional, attribute, logical consistency, completeness, and linage, may be necessary.
- Most GIS data is collected from only one single time measurement and as such could have limitations that could lead to difficulties during the application of statistics theory for modeling error in spatial data.
- Spatial distribution and spatial error dependency in spatial data is still an open issue and ripe for further investigation.

18.2   MORE MATHEMATICAL THEORIES FOR ERROR MODELING

In current research, probability theories and statistics are used as the mathematical bases for uncertainty modeling of spatial data and analyses, mainly regarding random error. Errors of another nature, such as systematic error and gross error also need to be modeled. It has been indicated above that spatial dependence of the errors in spatial data is still an open issue, and that spatial data quality control, particularly, needs further investigation. In this respect, there is a need for new mathematical theories that would, at the same time, also enrich the theoretical foundations for modeling spatial data and analyses uncertainty. The following are brief examples of several mathematical theories that have potential in this area.

18.2.1   GEOSTATISTICS

Statistics and probability theories have been used in modeling such as positional error of points, lines, and polygons. Within the framework of object-based models, for example, line errors can be modeled based on component point errors. In this field, the error correlation between line segments of a polyline need to be further studied. The spatial correlation between the component units, which affects the nature of the error distribution in the resultant field-based data, has not, to date, been well addressed within the framework of field-based data models. Example of such spatial correlation is correlations of error between pixels of the classified satellite image from an image classification. These issues related to spatial correlation and spatial distributions of the errors can potentially be resolved, based on statistics for spatial data.

18.2.2 LEAST SQUARES ADJUSTMENT

Least squares adjustment is an optimization method, based on statistics used in surveying data processing. The aim of the least squares method is to estimate the optimized value of the unknown parameters, and to estimate the accuracy of the observations, including the parameters. The characteristics of the least squares method are: (a) the estimators are unbiased, (b) results have a minimum variance, and (c) the estimation is independent of the error distribution of the original measurements. For example, they are not required to follow a normal distribution.

This method can potentially be applied for spatial data quality control, and in particular, to solve data quality control problems, where multisources of spatial data are in multiple scales and with different accuracy, for example, such as when logic inconsistency checking between multiple spatial data is involved.

Error propagation law in statistics is often used to analyze error propagation in spatial data and analyses. If the error in the independent variables, and a function that defines the relationship between dependent variables and the independent variables, are known, the error propagation law can be applied to estimate error in a dependent variable. However, the error propagation law has a restriction in that the above function should be linear. A nonlinear function should be linearized before the error propagation law is applied. In addition, independent variable errors are normally assumed to be random in nature. However, the nature of many relationships in spatial analyses or spatial dynamics is not as simple as linear, and sometimes is too complex to model. Furthermore, the dependent variable errors may not be simply random, but rather a combination of random, systematic, and gross errors. In such situations, the error propagation law for the estimation of uncertainty in the dependent variables cannot be applied to solve the problems. Other mathematical solutions have to be found. One possible solution is the approximation theory, which enables error analysis problems in any nonlinear functions to be addressed. Additionally, the nature of the error may not be limited to random error. In a case where a functional relationship cannot easily be built between the independent and dependent variables, other solutions such as a simulation approach or sensitivity analysis can be considered as alternatives.

18.3 IMPLEMENTATION ISSUES

18.3.1 ERROR AWARENESS

GIS users should be aware that errors and uncertainties are present in the spatial data used. With a new user this is not necessarily the case. However, this awareness can be achieved through education. Such education is even more important when spatial data are widely accessible and used day to day. Examples include such as high resolution satellite images, three-dimensional city models accessed from Google Earth, and the acquisition of the shortest path or hotel location searched from an individual navigation system. A user should be aware that these data may not be error free. In fact, the geometric position of the satellite image can possess positional error; for instance, the hotel name can be missing from the database, or the short path indicated by the navigation system may not be the shortest one in reality due to spatial data errors or data updating problems

Knowledge of the above inconsistencies gives users an awareness of the necessity to be assured that data are not only reliable and of high accuracy but also totally meet user requirements. GIS providers' awareness of these demands is a driving force to make further improvements in the data quality currently available. Hence, it is necessary for GIS software vendors to provide data quality control and analysis functions as one of the core components of their GIS software. The component would be used for (1) quality checking when spatial data are input into GIS and (2) uncertainty propagation estimation when spatial analyses are conducted, based on spatial analysis functions in GIS, as discussed below.

18.3.2 SOFTWARE FOR DATA QUALITY CONTROL

The developed theoretical methods and models on uncertainty modeling in spatial data and analyses can be materialized through GIS software. Hence, the development of GIS functions that can implement the developed models and methods for measuring errors, accuracy, and uncertainty in spatial data and analysis is necessary.

With such GIS data quality modules, users can capture, manage, and visualize not only geographic data but also, at the same time, acquire data quality information for both object-based and field-based data in GIS. Preliminary progress has been made by the GIS research community in relevant software developments; however, comprehensive modules from main stream GIS software products are still not available to GIS users.

Furthermore, uncertainty modeling software for spatial analyses, especially for complex spatial analyses, has not yet been developed. When such software for uncertainty modeling in spatial analysis is available, GIS users will not only be able to achieve the results of a spatial analysis but also enable obtain information about uncertainty propagation and even amplification in such spatial analyses.

18.3.3 QUALITY MARKS ON GIS DATA PRODUCTS

Quality marks are often found on both sophisticated and everyday products. Such a practice is normal and a marketing tool to assure the customer of the product's high standard. For example, on the back side of the telephone in my hotel, Best Western South Coast Inn, I found the quality mark: "QC PASSED." GIS data products do not have quality assurance marks attached. Currently, this practice does not exist, and hence it is not unreasonable for the customer to be unsettled about the exact quality standard of the product in which he is interested. He may even feel that the provider is not mindful of customer need for evidence of GIS products' quality control and assurance. This state of affairs should be addressed with quality marks provided for all purchased GIS data products.

However, quality assurance provided by quality marks is not as simple as the provision of "QC PASSED," as an example. Specific quality classifications, such as a classification of "excellent," "very good," and "good" level of quality related to object or feature level, map sheet level, coverage level, or overall data set are required. To enable quality to be determined by the users prior to purchase, such classifications require development and implementation of quality measurement methods for GIS data products. Obviously, the classification of GIS data products should take the users' application requirements into consideration. The data quality mark should be attached to the most basic level of GIS products.

18.3.4 DATA QUALITY STANDARDS

Data quality standards are the guidelines by which GIS data providers can assess and report the quality of their spatial data products. Data quality related standards have been developed by both international organizations, such as ISO/TC211, regional or national organizations. However, two major limitations are evident.

Firstly, the published standards are not specific enough to guide GIS data providers regarding the implementation of these standards. For example, the demand of a data quality standard may be to measure positional accuracy of line objects in GIS. However, specific error measurement indicators and corresponding methods on how to measure the accuracy of line objects are not provided.

Secondly, classification data quality elements in the current standards are based mainly on an understanding of the spatial data quality. In the infancy of such understanding, classification was somewhat limited and primitive. For example, a forest land parcel contains both positional and attribute errors. However, no error indicator to measure such integrated (positional and attribute) errors existed. Therefore, in order for the user to receive comprehensive spatial data quality information, a need was evident to better categorize the present data quality elements in GIS data quality standards in terms of positional accuracy, attribute accuracy, completeness, logic consistency, and lineage. There could indeed be a need to propose new foundation for the quality elements for spatial data.

18.4 INTERACTION WITH OTHER DISCIPLINES

Spatial concepts are not only used in geography. Indeed, spatial thinking is used in a wide range of disciplines, such as engineering, environmental sciences, ecology, medicine, economics, and psychology, beside geography. Therefore, methods developed for modeling geographic space can potentially be applied to other spaces, such as the human body and brain, social space, and even the cosmos.

For example, regarding the human body, GPS techniques have been reported to help surgeons position operating tools, enabling greater cutting precision and accuracy. Three-dimensional images of the human brain can be visualized in virtual reality, enabling detailed views of the human brain space, from any chosen view point. Here, the method for building three-dimensional space models in GIS, such as the three-dimensional TIN model, can also be used to build three-dimensional human brain space, three-dimensional positioning technology of GPS can also, as indicated above, be used for determining knife location within human body.

There is also a need for accuracy and uncertainty analysis in other spaces, such as the human brain space. Therefore, the knowledge developed for modeling uncertainty in spatial data and spatial analysis in geographic space can potentially be used for uncertainty modeling in human brain space. The theories developed for uncertainty modeling in geographic space should, in further research, be transformed to fit other "spaces" such as the human body, brain, and the cosmos, thus preventing the origins of similar knowledge from having to be reinvented for other disciplines.

18.5 CONCLUDING REMARKS

Research effort over the past two decades has produced a better understanding of uncertainties in spatial data and analyses. However, there is still a wide range of research issues for further development. As indicated above, these issues fall into two categories:

1. Modeling uncertainties in spatial data and analyses. This area requires the input of new theories from other disciplines, such as modern mathematical theories. Selection of such theories obviously requires intensive broad-based research.
2. The transformation of the knowledge already developed on uncertainty modeling for geographic space to other "spaces," such as the human body, brain, and the cosmos, etc.

The objective of research on the uncertainty phenomena in GISci is firstly to provide spatial data and spatial analysis results with uncertainty indicators so that the results can at least be categorized and, secondly, to clearly identify, control, and solve spatial data quality problems. However, as is clearly stated at the beginning of this chapter, our knowledge of the natural world is still very limited, hence it is reasonable to assume that without perfect knowledge of the natural world, a state of data uncertainty is to be expected, and the difficulty in addressing the problem of data quality is an obvious consequence.

At the present time, although progress has been made to answer questions raised in the field of uncertainty in spatial data and analysis, the results have limitations. However, clearly further progress can be made by building on the already established research results, the aim being to possibly establish at least initial universal certainty boundaries or parameters to enable further progress.

The research objective of the author of this book concerning the modeling of uncertainties in spatial data and analyses is to describe, control, and even eventually eliminate such uncertainties.

ACKNOWLEDGMENT

The author would like to thank Mike Goodchild for his valuable input in discussions on the future of spatial data quality research.

Index

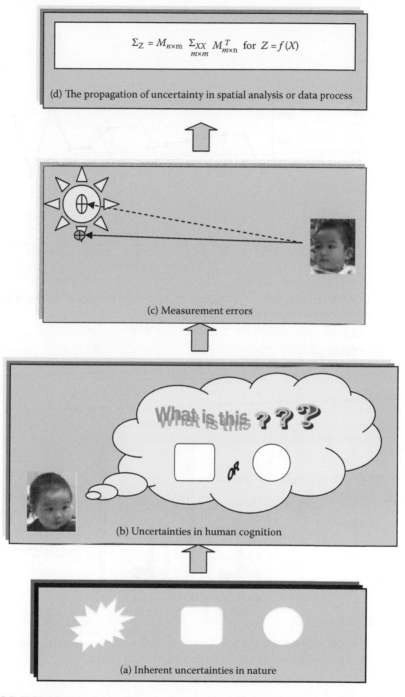

COLOR FIGURE 2.1 Main sources of uncertainty in spatial data and analyses (Adapted from Shi, W. Z., 2005. *Principle of Modeling Uncertainties in Spatial Data and Analysis.* *Science* Press, Beijing [in Chinese]).

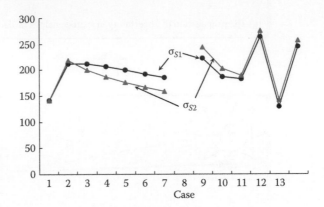

COLOR FIGURE 4.15 Comparison of the error in the area of a polygon and the area of the error band for the polygon from case 1 to 13.

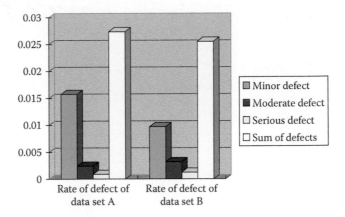

COLOR FIGURE 5.2 A bar chart of attribute quality of data sets *A* and *B* with three types of defect: minor moderate and serious defects.

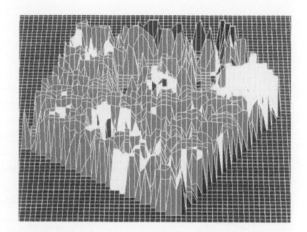

COLOR FIGURE 6.6 Attributes and uncertainty of a classified image. (From Shi, W. Z., 1994. *Modeling Positional and Thematic Uncertainties in Integration of Remote Sensing and Geographic Information Systems*. Enschede, The Netherlands.)

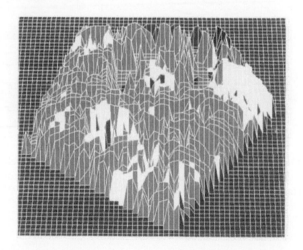

COLOR FIGURE 6.7 Integrated positional and attribute uncertainty.

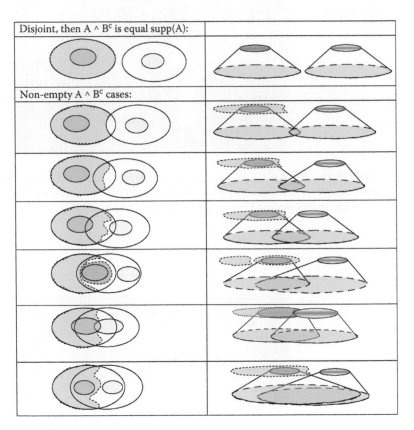

COLOR FIGURE 7.4(a) Quasi coincident fuzzy topological relations between two objects in GISci. (From Shi, W. Z. and Liu, K. F., 2004. Modeling fuzzy topological relations between uncertain objects in GIS, *Photogram. Eng. Remote Sens.* 70(8): 921–929. With permission.)

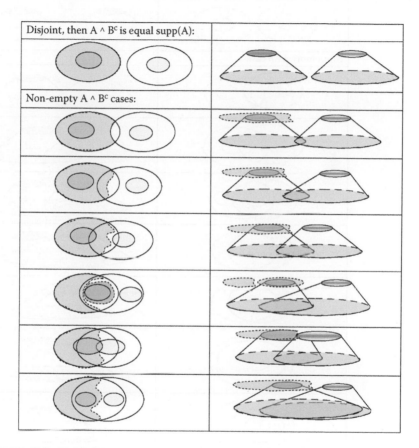

COLOR FIGURE 7.4(b)   Quasi coincident fuzzy topological relations between two objects in GISci. (From Shi, W. Z. and Liu, K. F., 2004. Modeling fuzzy topological relations between uncertain objects in GIS, *Photogram. Eng. Remote Sens.* 70(8): 921–929. With permission.)

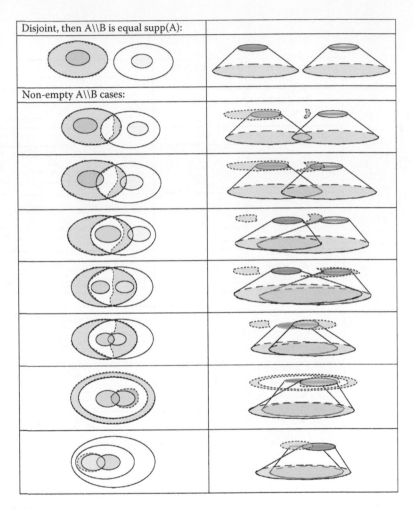

COLOR FIGURE 7.5 Different cases of quasi different fuzzy topological relations between two objects in GIS. (From Shi, W. Z. and Liu, K. F., 2004. Modeling fuzzy topological relations between uncertain objects in GIS, *Photogram. Eng. Remote Sens.* 70(8): 921–929. With permission.)

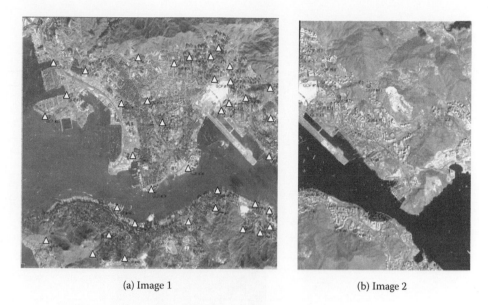

(a) Image 1   (b) Image 2

COLOR FIGURE 13.2   Test images with ground control points.

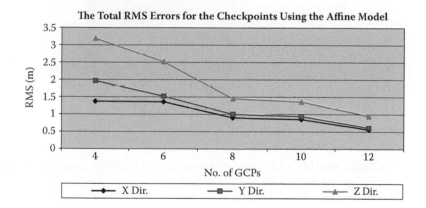

COLOR FIGURE 13.4   Shows the accuracy of the checkpoints from the three-dimensional affine transformation model. (From Shaker, A., and Shi, W. Z., 2003. *Proceedings of Asian Conference on Remote Sensing* [ACRS 2003] and International Symposium on Remote Sensing, Korea. On CD. With permission.)

COLOR FIGURE 13.3 The overlapped area of IKONOS image with GCPs for building stereo model.

COLOR FIGURE 13.5 Configurations of the ground control lines (GCLs) and checkpoints over the Hong Kong data set.

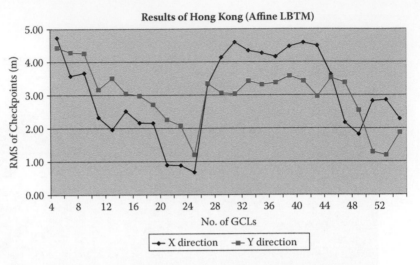

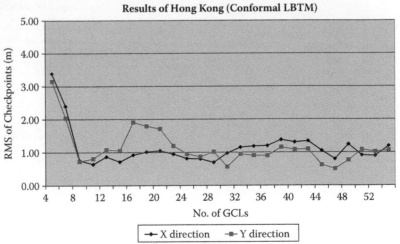

COLOR FIGURE 13.6 Results of image rectification of different data sets. (From Shi, W. Z. and Shaker, A., 2006. *Int. J. Rem. Sens.* 27(14): 3001–3012. With permission.)

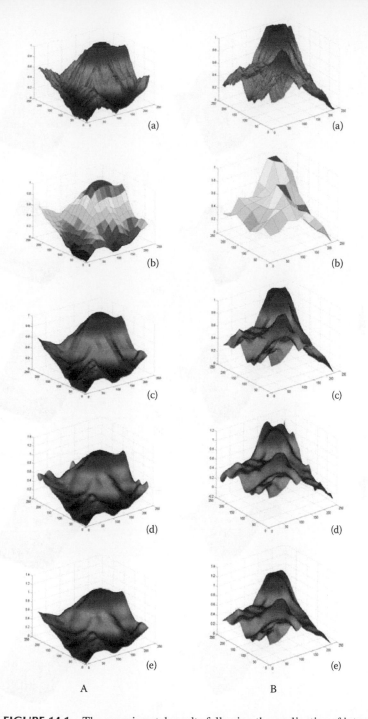

A   B

COLOR FIGURE 14.1 The experimental results following the application of interpolation methods for the data sets A and B. Where A-(a) is the original data set A, A-(b) is the resampled data set of data set A; A-(c) is the result of bilinear interpolation; A-(d) is the result of bicubic interpolation; and A-(e) is the result following the hybrid interpolation for data set A, with $\rho = 0.2$. B-(a) is the original data set B; B-(b) the resampled data set of B; B-(c) the result of bilinear interpolation; B-(d) is the result of bicubic interpolation; and B-(e) is the result following the hybrid interpolation for data set B.

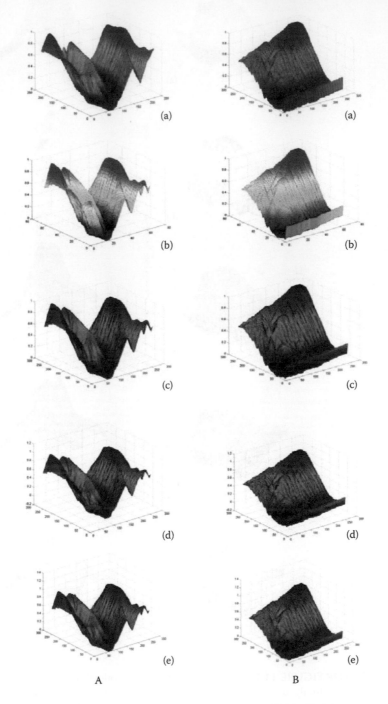

COLOR FIGURE 14.6 Experimental results for bidirectional interpolations based on data sets A and B.

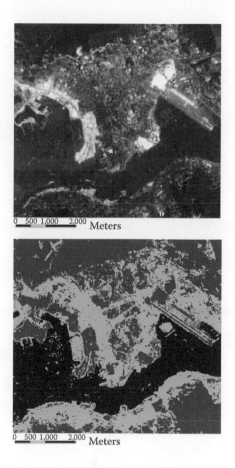

COLOR FIGURE 15.6 (a) At top, Landsat image of Hong Kong's Victoria Harbor taken in 1996, and (b) at bottom, the result of the maximum likelihood classification method. The classification of maximum membership functions is as follows: vegetation (blue), low-reflection urban area (green), high-reflection urban area (red), and water (black).

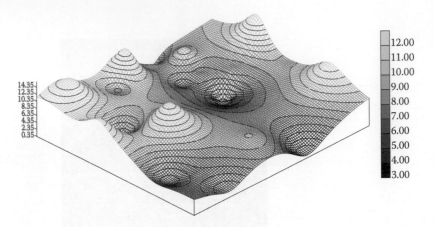

COLOR FIGURE 15.16 A three-dimensional uncertainty visualization in the rectified image.

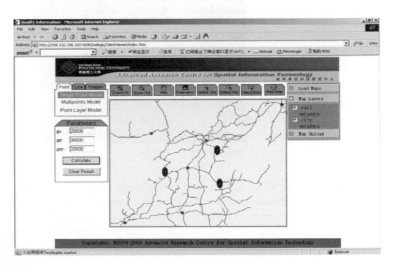

COLOR FIGURE 17.4 Visualization of quality of points with error ellipses model.

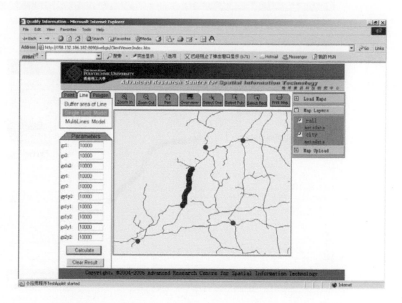

COLOR FIGURE 17.5 Visualization of line quality with an error band model.

COLOR FIGURE 17.6 Quality information of an area object.

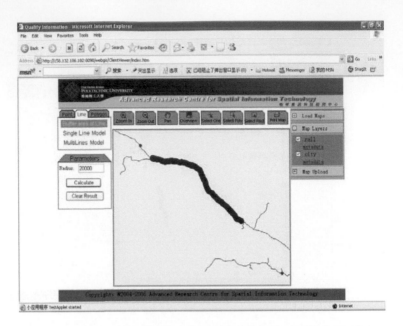

COLOR FIGURE 17.7   Visualization of quality information of a buffer spatial analysis.

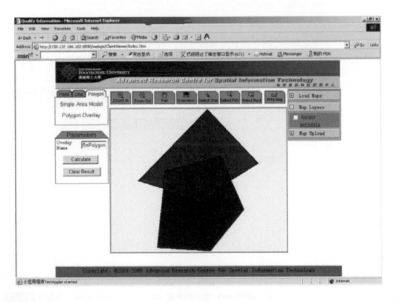

COLOR FIGURE 17.8   A visual description of overlay for providing quality information in the analysis.

Printed and bound by CPI Group (UK) Ltd, Croydon, CR0 4YY

24/10/2024

01778302-0013